Soft Ground Improvement
IN LOWLAND AND OTHER ENVIRONMENTS

Published by
ASCE PRESS
American Society of Civil Engineers
345 East 47th Street
New York, New York 10017-239

Allen County Public Library
900 Webster Street
PO Box 2270
Fort Wayne, IN 46801-2270

ABSTRACT:

The presence of thick deposits of soft clay combined with the effects of ground subsidence cause problems for engineering constructions for lowland areas such as the Central Plain (Chao Phraya) of Thailand. The ground subsidence is caused by the excessive extraction of ground water. To mitigate such natural geological hazards, different soil/ground improvement methods have been studied, namely: mechanically stabilized earth (MSE) embankments, granular or sand compaction piles, vertical drains, and the lime/cement deep mixing method. In view of their proven performance, durability, constructability, short time schedule, and low costs, mechanically stabilized earth (MSE) or earth reinforcement techniques seem to be very suitable and favorable aboveground soil improvement methods of embankment fills on subsiding ground. This method can be most appropriately combined with the various belowground improvement techniques such as vertical drains, sand compaction piles, and the lime/cement deep mixing method. Thus, the combined aboveground soil improvement with belowground subsoil improvement schemes can be a viable alternative to the existing method of supporting earth embankments and landfills with precast concrete piles, which may prove to be detrimental and more expensive in a subsiding environment.

Library of Congress Cataloging-in-Publication Data

Soft ground improvement in lowland and other environments / by D.T. Bergado ... [et al.].
  p.   cm.
Includes bibliographical references (p.).
ISBN 0-7844-0151-9
1. Soil stabilization. I. Bergado, D. T.
TA749.S65 1996   96-3270
624.1'51363'09144—dc20   CIP

The material presented in this publication has been prepared in accordance with generally recognized engineering principles and practices, and is for general information only. This information should not be used without first securing competent advice with respect to its suitability for any general or specific application.

The contents of this publication are not intended to be and should not be construed to be a standard of the American Society of Civil Engineers (ASCE) and are not intended for use as a reference in purchase specifications, contracts, regulations, statutes, or any other legal document.

No reference made in this publication to any specific method, product, process or service constitutes or implies an endorsement, recommendation, or warranty thereof by ASCE.

ASCE makes no representation or warranty of any kind, whether express or implied, concerning the accuracy, completeness, suitability, or utility of any information, apparatus, product, or process discussed in this publication, and assumes no liability therefore.

Anyone utilizing this information assumes all liability arising from such use, including but not limited to infringement of any patent or patents.

Photocopies. Authorization to photocopy material for internal or personal use under circumstances not falling within the fair use provisions of the Copyright Act is granted by ASCE to libraries and other users registered with the Copyright Clearance Center (CCC) Transactional Reporting Service, provided that the base fee of $4.00 per article plus $.25 per page is paid directly to CCC, 222 Rosewood Drive, Danvers, MA 01923. The identification for ASCE Books is 0-7844-0151-9/96 $4.00 + $.25 per page. Requests for special permission or bulk copying should be addressed to Permissions & Copyright Dept., ASCE.

Reprinted from: D.T. Bergado, J.C. Chai, M.C. Alfaro and A.S. Balasubramaniam, Improvement Techniques of Soft Ground in Subsiding and Lowland Environment, 1994, 232 pp., Hfl.165/US$95.00. Student edition: Hfl.95/US$55.00., A.A. Balkema, P.O. Box 1675, Rotterdam, Netherlands.

Copyright © 1996 by the American Society of Civil Engineers,
All Rights Reserved.
Library of Congress Catalog Card No: 96-3270
ISBN 0-7844-0151-9
Manufactured in the United States of America.

PREFACE

This text is written mainly from the Authors' extensive experience for more than two decades on soft ground improvement techniques mainly in lowland environment such as the coastal deposits in Southeast Asia, Japan, U.S.A., and elsewhere. However, soft ground improvement is also applicable in similar conditions of other environments. In particular, the text includes a large number of fully documented case histories in which full scale field tests are carried out and field observations are made on the in-situ behavior. The text is most appropriate to graduate students, researchers, and practitioners in the design and execution of airfield and highway embankments in soft clay deposits as well as in reclamation works. Also, selected sections of the text can be taught at undergraduate level, so that gradually, students can appreciate the well-defined classical soil mechanics on saturated soils as opposed to realistic situations in which geotechnical engineers and specialists are faced with challenging geotechnical problems which rely on numerical analyses in addition to well-defined limit state analysis. In most of the case histories presented here, attempts were made to predict field deformation using both limit equilibrium and finite element analyses. The latter includes an elasto-plastic analysis based on critical state soil mechanics models. For ease of reference, the text is presented in seven chapters.

Chapter 1 elaborates on the usefulness and need for ground improvement techniques for embankment on soft ground. Included in this chapter is a chart showing the procedures for modelling and selecting shallow and deep ground improvement techniques.

The surface compaction, which is the most traditional and the cheapest ground improvement technique, is then dealt with in Chapter 2. The material presented in this chapter, though dealt with in similar other texts, is written here with an emphasis on the fill material used in dams and embankments as well as in clay liners for waste containment structures.

The deep compaction is dealt with in Chapter 3. It includes such equipments and techniques as terra-probe, vibroflotation, resonance compaction, and dynamic compaction. These methods based on impact and vibration, while more applicable to cohesionless soils, are also used in cohesive soils using dynamic compaction in conjunction with dynamic replacement and mixing similar to sand compaction piles. In most instances, the text includes case histories which illustrate the success and the limitations of the methods. The resonance compaction method described here is seldom found in similar texts.

Chapter 4 deals exclusively with the use of prefabricated vertical drains (PVD) and its advantage over other types of drains in accelerating the consolidation settlement in most soft clay deposits under preloading in Southeast and Far East Asia. Besides the soft Bangkok clay, these applications include the Manila Bay Reclamation; Changi Airport and other reclamation works in Singapore; Kansai International Airport site in Japan; and several others in Hongkong, Taipei, Malaysia, Indonesia, etc. This chapter deals with the theories of consolidation with vertical drains, preloading techniques, PVD characteristics, soil parameters needed for PVD design, and the modelling of full-scale field test embankments and fills.

The granular piles and their merits are described in Chapter 5. The term "granular piles" used in this text refers to the components of compacted gravel and/or sand inserted into the soft ground. It also refers to the "stone column" and the popular "sand compaction piles". Various methods of granular pile construction are illustrated and the engineering behavior of composite ground is dealt with in a comprehensive manner.

Chapter 6 is devoted to lime/cement stabilization. Chemical admixture stabilization including lime or cement treatment has been used extensively for roadwork, especially in increasing the bearing capacity of soft subsoils and in the reduction of base course thickness in pavements. The material presented here concentrates more on deep stabilization of soft clays with lime or cement stabilized columns. The deep mixing method originally developed in Japan among other countries to improve the soft ground in port and harbor works now find wide applications in Southeast Asia to foundations of structures built on land such as embankments, buildings, and storage tanks. The fundamental concepts of soil-cement and soil-lime stabilization is described in the text and the consolidation and deformation characteristics of such stabilized soils are further described for a wider range of stress paths as applied under triaxial stress conditions. Design criteria for the use of lime or cement column in the settlement and stability analyses of embankments and other loaded areas are also included.

Finally, the material presented in Chapter 7 is on mechanically stabilized earth (MSE) or earth reinforcement. Under Chapter 7, the use of extensible and inextensible reinforcements in earth fills is presented. The basic principle of reinforced earth is first dealt with in greater detail and then laboratory and full-scale field test data are then presented extensively to clarify the roles of these basic principles as applied in actual design. The case studies illustrate the field behavior and modelling of reinforcing materials as well as the fill materials founded on the natural soft subsoils.

Thus, the text contained in this book is an updated material on ground improvement techniques including valuable and well-documented experiences applicable to lowland environment and other similar environments. A bridge between theory and practice is presented and further supplemented with comprehensive laboratory and field test data in the design and construction of earth structures on improved soft grounds.

The Authors

CONTENTS

Chapter 1	INTRODUCTION	1
Chapter 2	SURFACE COMPACTION	10
Chapter 3	DEEP COMPACTION	35
Chapter 4	PREFABRICATED VERTICAL DRAINS (PVD)	88
Chapter 5	GRANULAR PILES	186
Chapter 6	LIME/CEMENT STABILIZATION	234
Chapter 7	MECHANICALLY STABILIZED EARTH (MSE)	305

Chapter 1

INTRODUCTION

1.1 GENERAL

In general, the term soft ground includes soft clay soils, soils with large fractions of fine particles such as silts, clay soils which have high moisture content, peat foundations, and loose sand deposits near or under the water table (Kamon and Bergado 1991). For clayey soils, the softness of the ground can be assessed by its undrained strength, S_u, or by its unconfined compression strength, q_u. On the other hand, the SPT N-values are utilized to ascertain the consistency of the ground and its relative density. Table 1.1 outlines the identification of soft ground according to the types of the structures using the aforementioned assessment methods. Considering such factors as the significance of the structure, applied loading, site conditions, period of construction, etc., it becomes important to select appropriate method suitable for specific soil types as tabulated in Table 1.2. For soft and cohesive soils in subsiding environments, ground improvement by reinforcement (i.e. sand compaction piles), by admixtures (i.e. deep mixing method), and by dewatering (i.e. vertical drains) are applicable. For loose sand deposits, various in-situ compaction methods are applicable such as dynamic compaction, resonance compaction, and vibroflotation. Above the ground, such techniques as earth reinforcement or mechanically stabilized earth (MSE) and the utilization of lightweight synthetic materials are applicable.

1.2 EXISTING PROBLEMS

The Central Plain of Thailand is situated on a flat, deltaic-marine deposit with a north-south dimension of about 300 km and with a width of about 200 m. The stratigraphy consists of sedimentary deposits forming alternative layers of clay and sand with gravel down to 1000 m depth. The three uppermost layers consist of weathered clay, soft clay and stiff clay. The thicknesses of these layers at the location of the Asian Institute of Technology (AIT) located 42 km north of Bangkok, consist of 2 m thick weathered clay, about 6 m soft clay, and about 5 m thick first stiff clay layer. The thickness of the compressible and weak soft clay layer increases toward the southern direction to the Gulf of Thailand to about 15 m in Bangkok.

Most of the earth structure constructions in the coastal plains are road embankments, airport runways, flood control dikes, landfills, embankments along irrigation canals, etc. A slope failure along an irrigation canal is shown in Fig. 1.1. Due to the weak subsoil condition, only low embankments with gentle slopes can be constructed. Otherwise, slope stability failures can occur (Fig. 1.2). The main foundation problems for infrastructure construction is the presence of thick deposits of soft clay which will consolidate and cause large settlements when loaded (see Figs. 1.3 and 1.4). The load may be in the form of engineering structures causing compression in the clay or in the form of piezometric drawdown due to the withdrawal of groundwater for water supply causing land subsidence. As a matter of illustration, a case of Bangna-Bangpakong highway is described herein. Bangna-Bangpakong highway is a primary

Table 1.1 Outline for Identification of Soft Ground (Kamon and Bergado, 1991)

Structure	Soil Condition	N-value (SPT)	q_u (kPa)	q_c (kPa)	Water Content (%)
Road	A: Very soft B: Soft C: Moderate	Less than 2 2 to 4 4 to 8	Less than 25 25 to 50 50 to 100	Less than 125 125 to 250 250 to 500	
Express Highway	A: Peat soil B: Clayey soil C: Sandy soil	Less than 4 Less than 4 Less than 10	Less than 50 Less than 50 -		More than 100 More than 50 More than 30
Railway	(Thickness of layer) More than 2m More than 5m More than 10m	0 Less than 2 Less than 4			
Bullet train	A B	Less than 2 2 to 5		Less than 200 200 to 500	
River dike	A: Clayey soil B: Sandy soil	Less than 3 Less than 10	Less than 60		More than 40
Fill dam		Less than 20			

Table 1.2 Applicability of Ground Improvement for Different Soil Types (Kamon and Bergado, 1991)

Improvement Mechanism	Reinforcement	Admixtures or grouting	Compaction	Dewatering
Improving period	Depending on the life of the inclusion	Relatively short-term	Long-term	Long-term
Organic soil Volcanic clay soil Highly plastic soil Lowly plastic soil Silty soil Sandy soil Gravel soil	↑ ↓	↑ ↓	↑ ↓	↑ ↓
Improved state of soil	Interaction between soil and inclusion (No change in soil state)	Cementation (Change in soil state)	High density by decreasing void ratio (Change in soil state)	High density by decreasing void ratio (Change in soil state)

Fig. 1.1　　Embankment Failures along an Irrigation Canal

Fig. 1.2　　Typical Failure of Road Embankment on Soft Ground

Fig. 1.3   Road Embankment Failure on Soft Ground

Fig. 1.4   Pavement Failures due to the Sinking of Road Embankment

highway, No. 34, and 50 km long, built in 1960 linking Bangkok to the eastern part of Thailand (Fig. 1.5). The highway progresses from km 0 in Bangna to km 50 in Bangpakong. From 1979 to 1983, the land subsidence due to groundwater pumping contributed to the settlement of the highway embankment with a maximum of 35 cm. At that time, a maximum subsidence rate of 10 cm per year was observed in Bangna. Due to the combined effects of the consolidation of weak subsoil and land subsidence, settlement magnitudes in a 10 year period amounted to more than 2.0 meters in some sections of the highway (Fig. 1.6). Thus, after 10 years, some portions of the highway were already below maximum flood level and costly reconstruction was necessary to raise the embankment height.

The subsidence of Chao Phraya Plain in Thailand, where the city of Bangkok and the campus of the Asian Institute of Technology (AIT) are located, has been blamed on the drawdown of the underground piezometer levels due to excessive pumping of the groundwater. An average rate of 2 to 3 cm per year has been measured on the AIT campus, 75% of which occur in the uppermost 10 m year of the subsoil. The ground subsidence is mainly due to the compression of the soft Bangkok clay as a direct result of the piezometric drawdown. Due to the different compressibility characteristics and different densities of fine sand and silt lenses of this layer, the rate of subsidence is not uniform (Fig. 1.7). These differential settlements are the main cause of pavement cracks on subsiding ground. Moreover, the compression of the soft clay due to ground subsidence causes large differential movements between pile-supported and ground-supported structures which is the main disadvantage of using concrete pile foundation. The compression of the uppermost clay layer is illustrated in Fig. 1.8. This figure shows the suspended concrete cap surrounding the deep settlement marker which used to be on the ground surface. The deep settlement marker was installed to 8 m depth to the bottom of the soft clay layer and on top of the stiff clay layer. The differential elevation illustrates the amount of ground subsidence.

In the Eastern Seaboard of Thailand, two large reclamation projects for deep seaports in Laem Chabang and Map Ta Put have been constructed. The hydraulic fill deposits derived from dredging of seabed were utilized in these projects which require densification. Moreover, the infrastructure constructions along the adjacent sea coast also need in-situ densification and drainage for improving the weak ground foundation.

1.3　SOIL/GROUND IMPROVEMENT TECHNIQUES

Soil/ground improvement in geotechnical engineering means the increase on soil shear strength, the reduction of soil compressibility, and the reduction of soil permeability. As outlined in Table 1.2, the ground improvement techniques can be classified broadly into two categories, namely: a) those techniques involving the work on soil only such as dewatering and compaction and b) those methods that require foreign materials such as the use of the chemical admixtures and the utilization of various reinforcements. Various soil/ground improvement methods are presented that have been tested to provide soil strength improvement, mitigation of total and differential settlements, shorter construction time, reduced construction costs, and other characteristics which may impact on their utilization to specific projects on soft ground. A chart

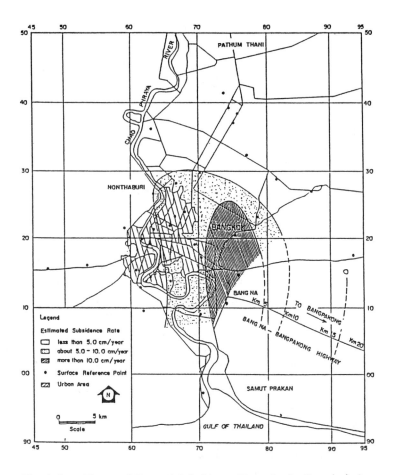

Fig. 1.5   Zones of Ground Subsidence Rates in the Bangkok Area

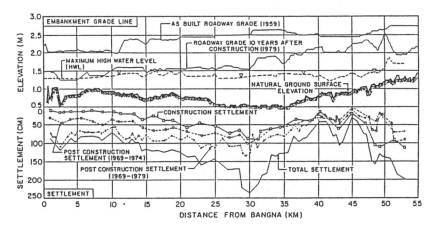

Fig. 1.6   Settlement Profiles along Bangna-Bangpakong Highway

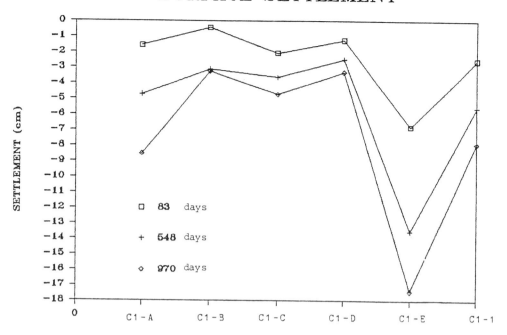

Fig. 1.7   Trend of Surface Settlements

Fig. 1.8   Deep Settlement Marker Showing Compression of Soft Clay Layer

showing the procedures for modelling and selecting shallow ground improvement techniques has been presented by Kamon and Bergado (1991) as given in Fig. 1.9. Figure 1.10 shows the selection procedures for deep ground improvement techniques. For infrastructures in embankment fill on soft ground, soil/ground improvement is not limited to portions below ground, but also includes improvement of fill soils above ground by reinforcing with grids and/or geotextiles as well as by the use of lightweight materials such as expanded polystyrene and other related products. Thus, the combined effects of ground improvement on both above and below ground locations yield optimum and ultimate advantages in the applications and utilization of ground improvement techniques.

1.4 REFERENCES

Kamon, M., and Bergado, D.T. (1992), Ground improvement techniques, Proc. 9th Asian Regional Conf. on Soil Mech. and Found. Eng'g., Bangkok, Vol. 2, pp. 526-546.

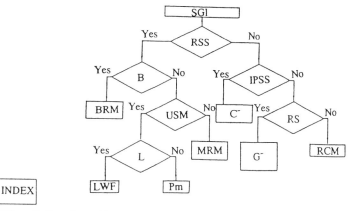

INDEX

SGI: Shallow Ground Improvement
B: Do you use Blast?
IPSS: Do you Improve Properties of Soft Soils?
BRM: Blasting Replacement Method
C¯: Cement and/or Lime Stabilization Method
RS: Do you Reinforce the Soils?
MRM: Mechanically Replacement Method
G¯: Geotextile – Sheet, Net and/or Grid Reinforcing Method
RCM: Roller Compacted Method
LWF: Light Weight Fill Method
Pm: Pre-mixed Soil Method

RSS: Do you Replace the Soft Soils?

USM: Do you Use Special Material?

L: Is it Light?

Fig. 1.9 Selection Flow of Shallow Ground Improvement Technique

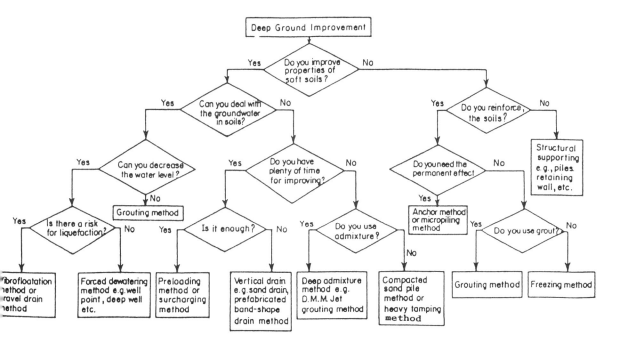

Fig. 1.10 Selection Flow of Deep Ground Improvement Technique
(Kamon and Bergado, 1991)

Chapter 2

SURFACE COMPACTION

2.1 INTRODUCTION

Surface compaction is the cheapest and the simplest method of ground improvement. It is the process of increasing the soil unit weight by forcing the soil particles into a tighter state and reducing air voids by the addition of either static or dynamic forces. Figure 2.1 shows the schematic diagram of the reduction of voids of the soil due to compaction. In the case of cohesionless soils, compaction leads to higher density and higher internal friction angle. For cohesive soils, the compaction process leads to closer particle arrangement and more cohesion. The reduction of voids also means less potential for deformation and less changes in moisture contents. Furthermore, the voids reduction will directly reduce soil permeability mainly because of restricted channels of flow, especially for cohesive-frictional soils.

2.2 TWO BASIC COMPACTION TESTS

Standard compaction test procedures were developed by Proctor (1933). Presently, there are two basic compaction tests, namely: standard Proctor compaction and modified Proctor compaction. Details of these tests are given in Table 2.1. Other details can be obtained from ASTM or AASHTO standards. The main difference between these two tests is the amount of applied energy. The tests are conducted by compacting moist soil into a standard mold by a specified hammer and drop height. The dry unit weight and water content of the soil in the mold are measured and plotted in Fig. 2.2.

2.3 COMPACTION CURVE

Starting at low moisture content, thin capillary films form around the soil particles that tend to develop tensile stresses. With the addition of water, the particles develop larger and thicker water films around them, which tends to lubricate and make it easier to move them around and reorient into denser configuration. Eventually, a water content is reached at which the density does not increase any further. At this point, water replaces the soil particles in the mold and since the density of water is less than that of the soil mineral grains, the dry density curve starts to fall. Figure 2.3 shows the water content versus density relationship curing compaction process. An example of the typical compaction test data sheet is illustrated in Table 2.2.

The plot of dry unit weight as a function of water content at 100% saturation (zero voids) can be useful, as shown in Fig. 2.2. This curve represents an upperbound for dry unit weight. The dry unit weight curve for any given saturation can be plotted using the following expression:

$$\gamma_d = \frac{G}{1+e}(\gamma_w) = \frac{G}{1 + \left[\frac{\omega G}{S_r}\right]} \gamma_w \qquad (2.1)$$

Table 2.3 shows the approximate range of moisture contents for some soil types.

Table 2.1 Details of Proctor Compaction Tests

Test	ASTM Number	AASHTO Number	Mold Volume		Hammer		Drop Height		Compactive Effort (Energy)
			(m³)	(ft³)	Mass (kg)	Weight (lb)	(m)	(ft)	
Standard Proctor	D698-70	T-90	945x10⁶	1/30	2.5	5.5	0.30	1.0	25 blows/layer 3 layers (590 kJ/m³ or 12,375 ft lb/ft³)
Modified Proctor	D1557-70	T-180	945x10⁶	1/30	4.5	10.0	0.46	1.5	25 blows/layer 5 layers (2,700 kJ/m³ or 56,250 ft lb/ft³)

Table 2.2 Compaction Test Data Sheet

TRIAL	1	2	3	4
Wet Density Determination				
Weight of mold and wet soil	13.84	14.30	14.00	13.89
Weight of mold	9.32	9.32	9.32	9.32
Weight of wet soil (W_T)	4.52	4.98	4.68	4.57
Wet density	135.60	149.40	140.40	137.90
Moisture Content Determination				
Cup identification	B-1	B-2	B-3	B-4
Weight of cup plus wet soil	39.10	55.30	66.60	75.48
Weight of cup plus dry soil	38.15	52.81	62.28	68.53
Weight of cup	15.10	14.21	14.43	14.33
Weight of dry soil	23.10	38.60	47.85	54.20
Weight of water	0.95	2.49	4.32	6.95
Water content (%)	4.00	6.50	9.00	12.80
Dry density (lb/ft³)	139.30	140.30	129.00	121.50

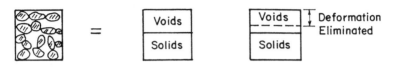

Fig. 2.1 Compaction Process

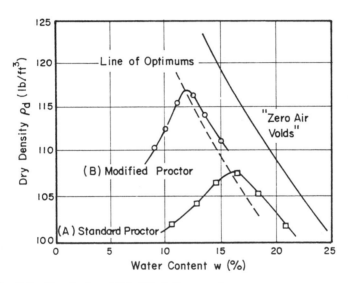

Fig. 2.2 Standard and Modified Compaction Curves for Crosby B Till (Holtz and Kovacs, 1981)

The laboratory compaction tests are designed to simulate the unit weight of soils compacted by field methods. Figure 2.4 shows the comparison of field and laboratory compaction showing good agreement with the standard Proctor compaction. For control of field compaction, construction specifications require that the dry unit weight of field compacted soils be equal to or greater than a given percentage of the maximum dry unit weight, which typically ranges from 90% to 100%. An illustration of 95% specified compaction is shown in Fig. 2.5. The standard Proctor is adequate compaction for most applications such as retaining wall backfill, highway embankments, and low earth dams. The modified Proctor, on the other hand, is utilized for heavier load applications such as airport runways, highway basecourses, and high earth dams.

2.5 FACTORS INFLUENCING THE COMPACTION OF SOIL

a) **Dry Density, γ_d**

The total soil unit weight, γ_t, is defined as the total soil mass divided by the specified volume. The dry density, γ_d, is defined as:

$$\gamma_d = \frac{\gamma_t}{1+\omega} \qquad (2.2)$$

where ω is the moisture content of the soil. For a given compactive effort, the dry unit weight of compacted soil will vary with the water content.

b) **Moisture Content, ω**

With a given amount of compaction, the water content at which the dry unit weight is maximum is termed as the optimum moisture content. At low moisture content, the soil is stiff and difficult to compact. With the addition of water, serving as "lubricant", the soil softens and becomes workable. At higher moisture content, the soil-air combination prevents any further decrease in air content. The total voids continue to increase with the increase in water content resulting in the reduction of dry density.

c) **Amount of Compaction**

With a given amount of moisture content, increasing the amount of compaction (compactive effort) results in a closer spacing and packing of soil particles and increased dry density, until the volume of air is reduced that further compaction produces no substantial change in volume. Compactive effort is measured in terms of energy per unit volume of compacted soil. For all types of soil, increasing the amount of compaction results in the increase of maximum dry density, γ_{max}, and decrease in optimum moisture content, ω_{opt}.

d) **Soil Type**

Figure 2.6 demonstrates the influence of the different soil types on compaction characteristics. Flat compaction curves refer to uniform (poorly-graded) sands. Being of similar

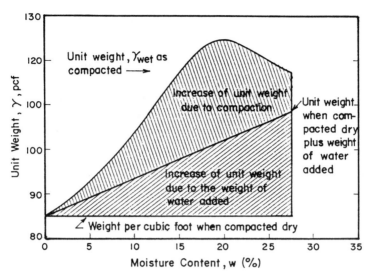

Fig. 2.3 Water Content-Density Relationship Indicating the Increased Density Resulting from the Addition of Water and Applied Compaction Effort. *Soil Is Silty Clay, LL = 37, PI = 14, Standard Proctor Compaction* (Johnson and Sallberg, 1960)

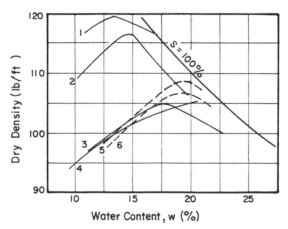

Fig. 2.4 Comparison of Field and Laboratory Compaction. *(1) Laboratory Static Compaction, 2000 psi; (2) Modified Proctor; (3) Standard Proctor; (4) Laboratory static Compaction, 200 psi; (5) Field compaction, rubber-tired load, six coverages; (6) Field compaction, sheepsfoot roller, six passes.* (Turnbull, 1950)

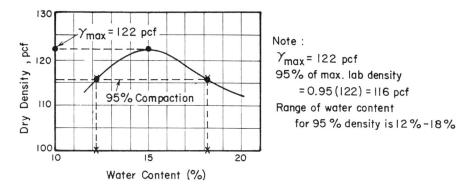

Fig. 2.5 Example to Compute Density for a Specified Percentage of Compaction, and Related Range of Water Content

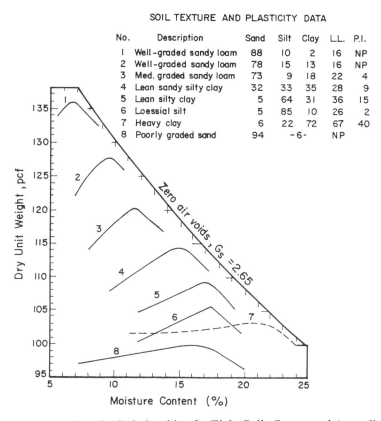

Fig. 2.6 Water Content-Dry Density Relationships for Eight Soils Compacted According to the Standard Proctor Method (Johnson and Sallberg, 1960)

particle sizes, different manners of packing produce not much difference in density (see Fig. 2.7). On the other hand, a pronounced peak is observed in the compaction curves of well-graded sands. In this case, ideal packing is achieved when the smaller particles lodged themselves in the voids between larger particles (see Fig. 2.8). Similarly, as shown in Fig. 2.6, the compaction curves for silty soil demonstrate peak values while the corresponding curve for clayey soil is flat. Silts are moisture sensitive, with a small change in moisture content affecting major changes in dry densities for a given compaction energy. Clays are energy sensitive, wherein a small change in compaction energy can produce large changes in dry densities.

2.5 EFFECT OF COMPACTION ON SOIL STRUCTURE

The compacted soil tends to have flocculated structure when compacted in the dry side of optimum. In this structure, the net force between particles is attractive and particles tend to be more attached to each other. Increasing the moisture content tends to increase the interparticle repulsions and, with the addition of force during compaction, results in a more orderly and parallel arrangement called dispersed structure. In this structure, the negative charge in the surface of the clay is balanced by the cations in the surrounding double layer. Figure 2.9 illustrates the effect of compaction on soil structure.

2.6 EFFECT OF COMPACTION ON ENGINEERING BEHAVIOR

a) Strength

Samples compacted in the dry side of optimum tend to be more rigid and stronger than samples compacted wet-of-optimum. Samples 1 and 2 in Fig. 2.10 show superior strength compared to samples 5 and 6.

b) Compressibility

During one-dimensional consolidation, and at low stresses, soil samples compacted at the wet side of optimum tend to be more compressible. In the same manner of consolidation and at high stresses, soil samples compacted at the dry side of optimum tend to be more compressible due to collapse of soil structure. Figure 2.11 illustrates the compressibility characteristics of compacted soil.

c) Permeability

Increasing water content results in a decrease in permeability in the dry side and a slight increase in the wet side (Fig. 2.12). Increasing the compactive effort decreases the permeability.

d) Water Absorption

Since there is water deficiency in the dry side of optimum compaction, there is more potential for water absorption. As shown in Fig. 2.13, the swelling potential due to water absorption is larger in the dry side of optimum compaction with flocculated structure.

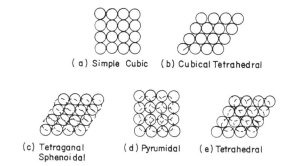

Fig. 2.7 Models for Regular Packing of Equal Sphere

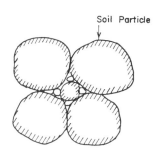

Fig. 2.8 Ideal Particle Size Distribution for Optimum Packing

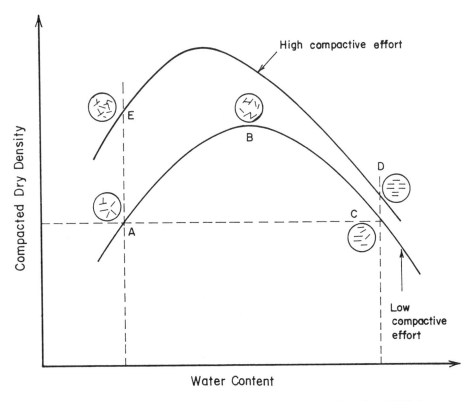

Fig. 2.9 Effect of Compaction on Soil Structure (Lambe, 1958a)

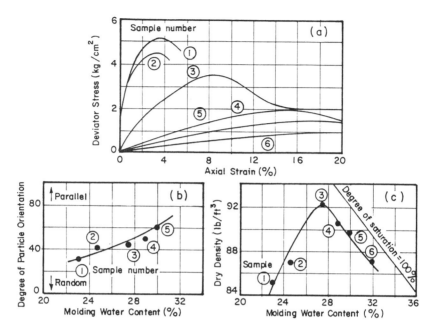

Fig. 2.10 Influence of Water Content on Structure and Stress-Strain Relationship for Compacted Samples of Kaolinite. *(a) Stress vs. strain relationships for compacted samples; (b) Degree of particle orientation vs. water content; (c) Dry density vs. water content* (Seed and Chan, 1959)

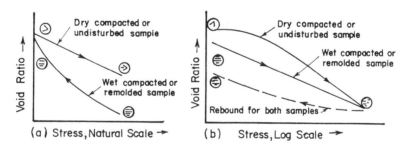

Fig. 2.11 Effect of One-Dimensional Compression Structure. *(a) Low-stress consolidation; (b) High-stress consolidation* (Lambe, 1958b)

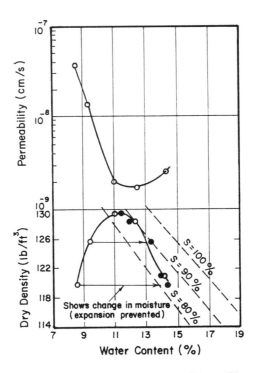

Fig. 2.12 Compaction-Permeability Tests on Sibura Clay (Lambe, 1958b)

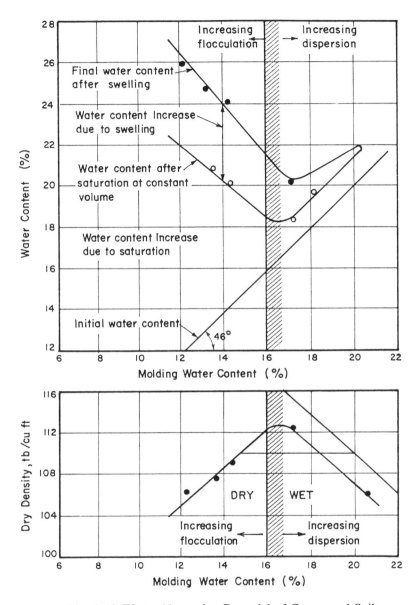

Fig. 2.13 Water Absorption Potential of Compacted Soils

2.7 TWO PHILOSOPHIES OF COMPACTION IN DAMS AND EMBANKMENTS

a) Dry Side of Optimum Compaction

For good and sound foundation conditions, dry side of optimum compaction is favored. In this compaction, not only is less construction pore pressure generated, but also higher strength and more rigid compacted mass is obtained. However, the resulting brittle compacted mass does not tolerate large differential settlements.

b) Wet Side of Optimum Compaction

In areas with weaker and compressible foundation conditions, the resulting flexible compacted mass from wet of optimum compaction is preferred. Although the compacted mass has lower strength with large construction pore pressure in this manner of compaction, however, the compacted mass is flexible and can tolerate large differential settlements.

For most earthwork construction, for water retaining structures, wet side compaction is favored to minimize soil permeability and to minimize strength reduction (collapse) when saturated with water. A summary of the comparison between dry side and wet side compaction is given in Table 2.4.

2.8 FIELD SURFACE COMPACTION

In the field, the soil is compacted by applying energy in three ways, namely: apply pressure by rolling and kneading, by ramming, and by vibrations. The types of field equipment include rollers, rammers and vibrators. The rollers consist of smooth wheel, pneumatic, and sheepsfoot. The rammers include dropping weights by internal combustion or pneumatic types. The vibrators comprise of out-of-balance type or pulsating hydraulic type mounted on plates or rollers. Table 2.5 summarizes the soil compaction characteristics and the corresponding compaction equipments.

a) Conventional Smooth Wheel Rollers

The conventional smooth drum rollers are not well-suited for compacting earthfill because of low pressures due to the large soil to drum contact area. This equipment is usually used to compact granular base for highways and airfields and to seal the surface of clayey fill.

b) Pneumatic Tire Rollers

Pneumatic tire rollers compact primarily by kneading, especially with the weaving (wobble) wheel path. Various numbers of wheels per axle can be used. Compaction load can be varied by adjusting the ballast. The contact pressure can be varied by adjusting the tire air pressure. This equipment is effective in compacting both cohesive and cohesionless soils.

Table 2.3   Approximate Range of Optimum Moisture Content vs. Soil Type

Soil Type	Probable Value of Optimum Moisture, %, Modified Proctor Test
Sand	6 to 10
Sand-silt mixture	8 to 12
Silt	11 to 15
Clay	13 to 21

Table 2.4   Comparison of Dry-of-Optimum and Wet-of-Optimum

Comparison	Comparison
Structure - particle arrangement - water deficiency - permanence	Dry side more random Dry side more deficient, more swell Dry side more sensitive to change
Strength - as molded (drained and undrained) - after saturation - pore pressure - stress-strain modulus - sensitivity	Dry side higher Dry side higher if swelling is prevented Wet side higher Dry side higher Dry side more sensitive
Compressibility - magnitude	Wet side more compressible at low stress Dry side more compressible at high stress
Permeability - magnitude - permanence	Dry side more permeable Dry side permeability reduces more by percolation

Table 2.5 Soil Compaction Characteristics and Recommended Compaction Equipment

General Soil Description	Unified Soil Classification	Compaction Characteristics	Recommended Compaction Equipment
Sand and sand-gravel mixtures (no silts or clay	SW, SP, GW, GP	Good	Vibratory drum roller, vibratory rubber tire, pneumatic tire equipment.
Sand or sand-gravel with silt	SM, GM	Good	Vibratory drum roller, vibratory rubber tire, pneumatic tire equipment.
Sand or sand-gravel with clay	SC, GC	Good to fair	Pneumatic tire, vibratory rubber tire, vibratory sheepsfoot.
Silt	ML	Good to poor	Pneumatic tire, vibratory rubber tire, vibratory sheepsfoot.
	MH	Fair to poor	Pneumatic tire, vibratory rubber tire, vibratory sheepsfoot, sheepsfoot type.
Clay	CL	Good to fair	Pneumatic tire, sheepsfoot, vibratory sheepsfoot, and rubber tire.
	CH	Fair to poor	Rubber tire.
Organic Soil	OL, OH, PT	Not recommended for structural earth fill	

c) **Sheepsfoot Rollers**

Sheepsfoot rollers consist of drums with projecting studs. It compacts the soil by a combination of tamping and kneading. The drum can be filled with water or sand to increase the weight. This equipment is quite suited for silty and clayey soil. Due to the presence of studs, this equipment yields good bonding between lifts of compaction. Thus, sheepsfoot rollers are advisable for compacting water retaining structures such as earth and earth-rock dams, irrigation and flood control dikes, etc.

d) **Vibratory Rollers and Rammers**

Vibratory rollers are available as vibrating drum, vibrating pneumatic tire, and vibrating plate. A separate motor drives an arrangement of eccentric weights so that high frequency, low amplitude, up and down oscillation of the drum occurs. The frequency range is preferable within the natural frequency of most soils in order to have resonance. The vibratory drum rollers are best suited for compacting granular soils with little or no fines and with lift thickness up to 1.0 m thick. The vibrating plates are suitable for compacting granular base course of pavements. For compaction with limited spaces or at areas close to existing structures, vibrating plates as well as vibratory rammers (Fig. 2.14) can be used.

2.9 MEASUREMENT OF COMPACTION IN THE FIELD

a) **Core Cutter Method**

The cutter is rammed into the compacted soil. The cutter containing the soil is then dug out of the ground and trimmed off. The density is computed from the volume and weight of the soil inside the cutter.

b) **Sand Replacement Method**

A hole of about 4 inches (101.6 mm) in diameter is excavated with suitable tools. The weight of the soil removed is determined and the moisture content taken. Sand of uniform grain size is run into the hole. The weight of the sand is determined and the density is computed.

c) **Rubber-Balloon Method**

This is similar to the sand replacement method except that the volume of the hole is determined by means of balloon filled with water.

d) **Nuclear Moisture-Density Method**

This method is becoming popular because of rapid results that can be obtained. The principal elements are, namely: nuclear source that emits gamma rays, detector to pick up gamma rays, and counter or scaler to determine the rate at which gamma rays reach the detector.

The gamma rays penetrate the soil where some are absorbed and some reach the detector by direct transmission. The amount of gamma rays reaching the detector is inversely proportional to the soil density. Densities are determined by obtaining nuclear count rate received at the detector and relating such reading to calibration readings made on materials of known densities. Test procedures may utilize direct transmission, backscatter method or airgap methods. An example of a nuclear moisture density gauge is shown in Fig. 2.15.

2.10 PROCEDURE FOR COMPACTING EARTH WORKS

a) Preliminary Investigations

Proctor or modified Proctor compaction tests serves as guides for optimum moisture content (ω_{opt}). Index tests such as Atterberg limits (LL and PL) and grain sizes (GS), are used to classify soils to select the most suitable compaction method.

b) Field Trials

The test area of 15 x 20 m is prepared. Then, fill materials is spread over the area, in 3 strips of 5 m wide. The depths are varied from 6 to 18 inches (150 to 450 mm) in its natural water content. The dry density, γ_d, are measured at 4, 8, and 16 passes (5 tests each strip). Repeat at two other moisture contents. For estimation of moisture contents, it is recommended to use the optimum moisture content from the laboratory tests, the natural moisture content, and one value of moisture content in between. If the moisture contents are similar, use $\pm 3\%$ moisture content difference.

2.11 QUALITY CONTROL - PURPOSEFUL VERSUS RANDOM SAMPLING

a) Purposeful Sampling

This is a single acceptance to represent uniform compacted volume of 2,000 m^3. For 1,000,000 m^3 compacted volume, only 50 tests are necessary. A basic prerequisite for its proper application is to have experienced field inspectors continually observe all phases of laying, spreading, homogenizing and compacting material. The selection of test location is based on engineering judgment of the field inspectors reflecting the worst compaction achieved in the volume.

b) Random Sampling

The fundamental prerequisite of random sampling is that each part of the volume of soil to be accepted must have an equal chance of being sampled for testing. This scheme requires very extensive sampling or testing. If 12 tests were necessary for each lot of 2,000 m^3, then for 1,000,000 m^3 of volume, 6,000 tests are necessary for 10% probability of accepting substandard lot depending on the operating characteristic (OC) curve. The OC curve is a curve giving the probability of acceptable lot as a function of some estimated parameter of the lot (for instance, % compaction).

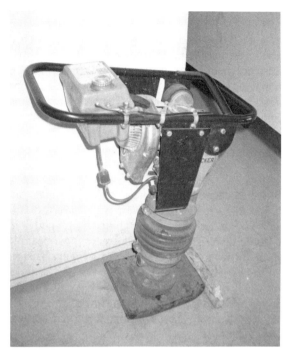

Fig. 2.14 Wacker BS 60Y Vibratory Rammer (Distributed by Wacker Machinery Co. Ltd., U.S.A.)

Fig. 2.15 Troxler 3400 Series Nuclear Moisture-Density Gauge (Distributed by Troxler Electronic Laboratories, Inc., U.S.A.)

2.12 APPLICATION OF COMPACTION TO WASTE CONTAINMENT STRUCTURES

Surface impoundments and landfills are typical structures in which either compacted clay liners (Fig. 2.16) or synthetic membrane liners or combination of both (Fig. 2.17) are used to contain waste. These man-made liner systems are typically composed of hydraulic barriers and drains. It is not uncommon to employ a mix of materials, including compacted clayey soils and geomembranes for barriers, and sand, gravel or geosynthetics for drains. Because the main purpose of the liner is to minimize leakage of liquid through the liner, the hydraulic conductivity or permeability (k) of the liner is its most important characteristic. Thus, special attention is given to compaction details and quality assurance of compaction. Often, the plasticity index and clay content of the soil are regarded as key index properties.

Daniel (1987) has collected data on 15 soils compacted in the laboratory following ASTM Standard D698 (Standard Proctor). All materials were compacted 0 to 2% wet of optimum and then permeated with water. The values of the permeability are plotted as a function of plasticity index (PI) in Fig. 2.18. While there is a trend for decreasing permeability, k, with increasing PI, the trend is weak, especially for low plastic clay over a highly plastic clay. Soils with low plasticity are often easier to mix, hydrate, and homogenize in the field and tend to be less susceptible to desiccation. Clay content can be important, but only for soils with less than 20% clay content. The sensitivity of k to clay content is illustrated in Fig. 2.19, which shows that one needs only to add a small amount of bentonite (montmorillonite clay) to silty sand to achieve low k.

2.13 CONSTRUCTION OF COMPACTED CLAY LINERS

Mitchell et al. (1965) showed in the laboratory that compaction water content, method of compaction, and compactive effort have major influences in the permeability (k) of compacted clay (Fig. 2.20). Numerous studies have since verified that in order to achieve a low k, the soil should be compacted wet-of-optimum water content using a high level of kneading-type compactive effort. There are theories explaining why compaction wet-of-optimum produces low k values. Olsen (1962) suggested that compacted soils consist of clusters called peds or clods. After the soil is compacted, a few interclod voids or macropores will exist such that most of the permeating liquid flows rather than through the remaining clods. Wet-side compaction produces low k values, because the soft (wet) clods are remolded during compaction minimizing the continuity of interclod pores. Thus, the ideal situation is small, soft, weak clods of clay that are easily remolded, and compaction with heavy footed rollers with fully penetrating feet that effectively remolds and molds the clods together (Fig. 2.21). Ideally, a heavy roller should be passed over tolerably thin lifts of soil with sufficient number of passes. Tests indicated that lighter sheepsfoot rollers would not produce low k values. A number of cases in which the actual k of compacted clay liner has been greater than 1×10^{-7} cm/sec have been reported (Daniel, 1987).

Soil liners are traditionally compacted to a minimum dry unit weight over a specified range in water content. This approach evolved from concern of adequate strength and permissible compressibility. With soil liners, permeability (hydraulic conductivity) is more important. The water content-density requirements over a range of compactive energy can be related to soil

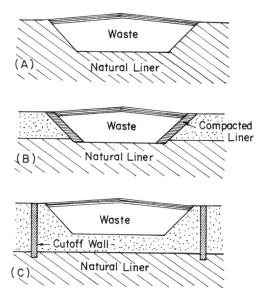

Fig. 2.16 Examples of Natural Liners

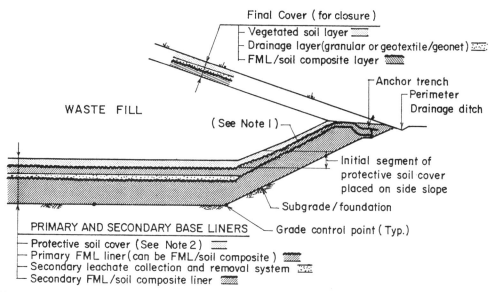

Notes: 1. For surface impoundments, the primary liner on side slopes may be a FML/soil composite.
2. For landfills, a primary leachate collection and removal system is placed between the protective soil cover and the primary liner.

Fig. 2.17 Double-Lined Waste Containment Unit

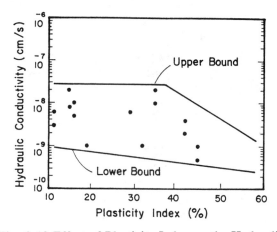

Fig. 2.18 Effect of Plasticity Index on the Hydraulic Conductivity of Laboratory-Compacted Soils

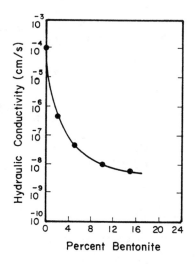

Fig. 2.19 Effect of Bentonite on Hydraulic Conductivity of Compacted Silty Sand

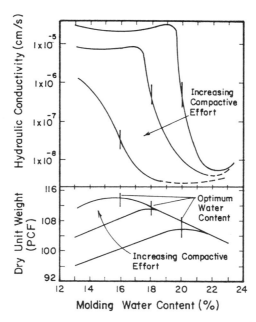

Fig. 2.20 Influence of Molding Water Content and Compacted Silty Clay (Mitchell, et al., 1965).

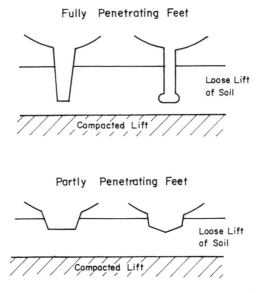

Fig. 2.21 Examples of Fully Penetrating Feet on Footed Rollers

permeability. Figure 2.22 shows the overlap of acceptable zones based on the permeability and strength requirements (Daniel and Benson, 1990).

2.14 ATTACK OF COMPACTED SOIL BY WASTE

Waste liquids may attack and effectively destroy earthen liners. Strong acids and bases can dissolve solid materials in the soil, form channels, and increase k. Hydrofluoric and phosphoric acids are particularly aggressive and dissolve soil readily. When concentrated acid is passed through clayey soils, the results depicted in Fig. 2.23 are commonly observed. Initially, there is a drop in k that is caused by precipitation of solid matter from the permeating liquid as the acid is neutralized by the dissolved soil. The precipitation plugs the pores in the soil specimen. With continued permeation, fresh acid enters the soil, redissolves the precipitates, and eventually causes an increase in k values.

The effect of neutral, inorganic liquids may be evaluated with the Gouy-Chapman theory (Mitchell, 1976), which states that the thickness (T) of the diffuse double layer varies with the dielectric constant of the pore fluid (D), the electrolyte concentration (n_o), and the cation valence (V) as follows:

$$T \propto \sqrt{\frac{D}{n_o V^2}} \qquad (2.3)$$

As the diffuse double layer of absorbed water and cations expands, k decreases because flow channels become constricted. Distilled water with few electrolytes tends to expand the double layer and produce low k. In similar manner, solutions with monovalent cations, e.g. Na^+, tend to produce lower k than solutions with polyvalent cations, e.g. Ca^{++}. Most organic chemicals have lower dielectric constants than water. Low D tends to cause low T, and thus high k value.

2.15 REFERENCES

Daniel, D.E. (1987), Earthen liners for land disposal facilities, Geotechnical Practice for Waste Disposal '87, ASCE Geotech Publ. No. 13, R.D. Woods, ed., New York, pp. 21-39.

Daniel, D.E., and Benson, C.H. (1990), Water content-density criteria for compacted soil liners, J. of Geotech. Eng'g., ASCE, Vol. 116, No. 12, pp. 1811-1830.

Holtz, R.D., and Kovacs, W.D (1981), An introduction to geotechnical engineering, Prentice Hall, Inc., Englewood Cliffs, N.J., 733 pp.

Johnson, A.W., and Sallberg, J.R. (1960), Factors that influence field compaction of soils, Bulletin 272, HRB, National Research Council, Washington D.C., 206 pp.

Lambe, T.W. (1958a), The structure of compacted clay, J. of Soil Mech. and Found. Div., ASCE, Vol. 84, No. SM2, pp. 1654-1 to 1654-34.

Lambe, T.W. (1958b), The engineering behavior of compacted clay, J. of Soil Mech. and Found. Div., ASCE, Vol. 84, No. SM2, pp. 1655-1 to 1655-35.

Mitchell, J.K., Hooper, D.R., and Campanella, R.G. (1965), Permeability of compacted clay, J. of Soil Mech. and Found. Division, ASCE, Vol. 91, No. SM4, pp. 41-65.

Mitchell, J.K. (1976), Fundamentals of soil behavior, John Wiley and Sons, Inc., New York, pp. 422.

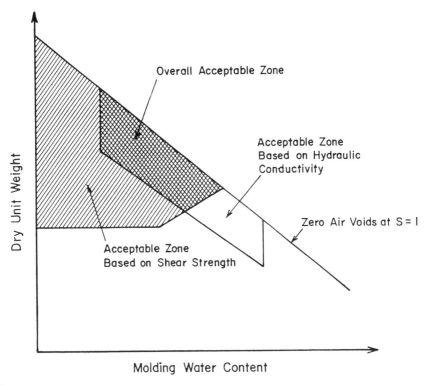

Fig. 2.22 Use of Hydraulic Conductivity and Shear Strength to Defined Single Acceptable Zone

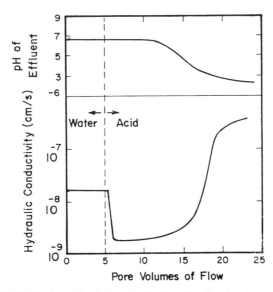

Fig. 2.23 Typical Variation in Hydraulic Conductivity for Sample of Compacted Clay Permeated with Concentrated Acid

Olsen, H.W. (1962), Hydraulic flow through saturated clays, Clays and Clay Minerals, Vol. 11, pp. 131-161.

Proctor, R.R. (1933), Four articles on the design and construction of rolled-earth dams, Engineering News Record, Vol. 3, pp. 245-248, 286-289, 348-351, 372-376.

Seed, H.B., and Chan, C.K. (1959), Structure and strength characteristics of compacted clays, J. of Soil Mech. and Found. Div., ASCE, Vol. 85, No. SM5, pp. 87-128.

Turnbull, W.J. (1950), Compaction and strength tests on soil, Presented at Annual Meeting, ASCE (cited by Lambe and Whitman, 1969).

Chapter 3

DEEP COMPACTION

3.1 INTRODUCTION

Deep compaction is the reduction of compressibility and increase of strength by packing soil particles together with high energy vibrations as the probe is progressively inserted to and withdrawn from thick soil deposits. No cohesion, however, is added. Deep compaction is needed for poor sites not previously considered for economic development. The site soil deposits can be categorized as natural soils and fill materials. Natural soils could be of cohesionless and cohesive type of materials. Granular (cohesionless) soils consist of sand and gravel while cohesive soils refer to silty clayey sands and silty clays. Fill materials comprise ash, slag, demolition rubble, quarry waste, and hydraulic landfill. Cohesive fill materials could be mixture of rubber and ash with clay, clayey sand and gravel, silty and clayey sands. In relation to deep compaction techniques, several factors affect the choice of deep ground compaction such as the following:

a) Ground conditions: type of subsoil, depth of treatment, and groundwater level.

b) Requirements: foundations, temporary or permanent additional strength, and temporary or permanent permeability reduction.

c) Size of area: dynamic consolidation, for instance, requires 8,000 to 10,000 m^2 to be economical.

d) Required increase in bearing capacity of treated ground.

e) Acceptable settlement of treated ground.

f) Proximity to existing structures and buried services due to possible effects of transmitted vibrations.

g) Availability of contracting entities and construction equipments.

3.2 TERRA-PROBE

Terra-probe consists of a vertical vibrator energizing a heavy cylindrical probe up to 14 m long (see Fig. 3.1a). The vibratory pile driver which makes up the body of the vibrator usually operates at 15 Hertz. The probe has an outside diameter of 760 mm (30") and a wall thickness of 9.5 mm. The vibrator and the probe unit are suspended about the ground surface and then vibrated to penetrate the sand to the desired depth (Fig. 3.1a). Compaction is effected by the continuous vibration during penetration and steady withdrawal. Compared with the vibroflotation technique, the production rate is much more rapid. However, the extent of compaction achieved per probe is significantly less than the vibroflot due to some inefficiency in the transfer of vibration energies. Consequently, this technique is now replaced by the more efficient resonance compaction technique as discussed later in this chapter. A square pattern is best for loose and saturated sand deposits. Table 3.1 gives details of the Terra-Probe as compared with the vibroflotation technique which is discussed in the next section.

3.3 VIBROFLOTATION TECHNIQUE

The vibroflotation technique employs mechanical vibrations together with simultaneous saturation with water to rearrange loose sand and gravel particles into a denser state. Vibration and shock waves in loose saturated deposits can cause liquefaction followed by densification and settlement accompanying dissipation of pore water pressures. In this technique, a cylindrical probe, as shown in Fig. 3.1b, which is approximately 400 mm diameter is lowered into the soil layer by a combination of vibration and jetting high pressure water through the orifices at the base of the probe (Brown and Ralph, 1977). When the required depth is achieved, the water flow is reduced and diverted to a set of jets at the top of the probe. The resulting upward flow of water maintains a channel around the probe allowing coarse fill fed from the surface to descend to the tip level. The probe is progressively raised to the surface (typically at 300 mm lifts) as the filling continued. When the feeding channel collapses, the probe is raised and lowered until the system is restored. Table 3.1 presents some significant information for both Vibroflot and Terra-Probe. Vibroflotation process can be done using either wet or dry process. The wet method uses high pressure water jets and the dry method utilizes compressed air.

The sequence of operation varies according to the type of soil deposits at the site. Vibroflotation is best suited for cohesionless deposits with less than 20% gravel content and with less than 18% silts and clay (0.074 mm particle diameter) as shown in Fig. 3.2. The process gradually transfers to vibro-replacement resulting in the formation of stone columns or granular piles for cohesionless soils with more than 18% clays and silts (see Fig. 3.2).

The vibro-replacement method is used to improve cohesive soils with more than 18% passing no. 200 U.S. standard sieve. The equipment used is similar to that for vibro-compaction. The vibroflot sinks into the ground under its own weight assisted by water or air jets as a flushing medium until it reaches the predetermined depth (Baumann and Bauer, 1974; Engelhardt and Kirsch, 1977). The process is illustrated in Fig. 3.3. The method can be carried out either with the wet or dry process. In the wet process, a hole is formed in the ground by jetting a vibroflot down to the desired depth with water. When the vibroflot is withdrawn, it leaves a borehole of greater diameter than the vibrator. The uncased hole is flushed out and filled in stages with 12 mm to 75 mm size imported gravel. The densification is provided by an electrically or hydraulically actuated vibrator near the bottom of the vibroflot. The wet process is generally suited for unstable borehole and a high groundwater table. In the dry process, the borehole must be able to stand open upon extraction of the vibroflot, which requires the soil under treatment to have an undrained shear strength of more than 40 kN/m^2 and a relatively low groundwater level. The main difference between the dry process and the wet process is the absence of jetting water during the initial formation of the hole in the former.

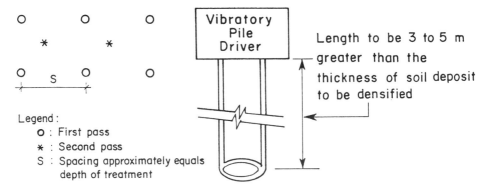

Fig. 3.1a The Terra-Probe Technique (Glover, 1982)

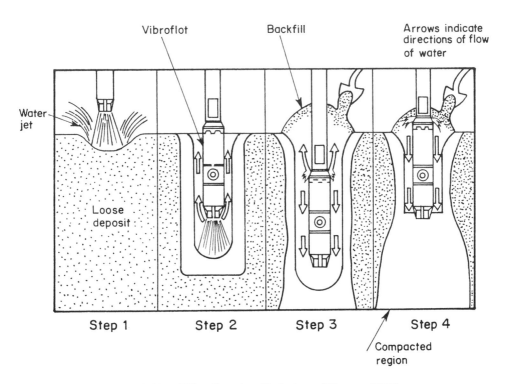

Fig. 3.1b Vibroflotation Technique (Glover, 1982)

Table 3.1 Details of Vibroflot and Terra-Probe (Brown and Ralph, 1977)

Characteristics	Vibroflot	Terra-Probe
Diameter (mm)	410	760
Length (m)	2.1	13.7
Probe mass (kg)	1800	-
Total mass (kg)	5400	9000
Frequency	1800	700-1000
Motor (kW)	75[+]	-
Amplitude[*] (mm)	30	9.5-25
Penetration rate (m/min)	5	-
Extraction rate (m/min)	0.4	-
Production rate[#] (m/min)	0.32	1.4
Production time[#] (min)	11	-

* - Restrained vibration
\# - Total time for insertion and withdrawal but not including setting up and traveling times.
+- Introduced to U.S.A. in 1971.

3.3.1 Ground Suitable for Vibroflotation

The range of soil types suitable for vibroflotation is illustrated in Fig. 3.4 designated as Zones A, B, and C.

Zone A: Particle size distribution with zone A are excellent soils for densification by vibroflotation. When the gravel content, however, exceeds 20%, the penetration to depths greater than 10 m may be difficult.

Zone B: Vibroflotation is best suited for densifying loose sands in Zone B.

Zone C: The presence of silt and clay layers hinders densification by vibroflotation and require a modification of the method. The modified method is called vibro-replacement method.

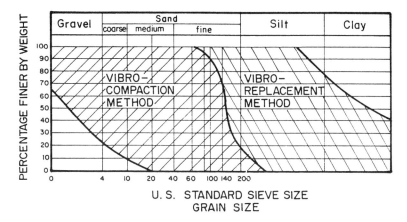

Fig. 3.2 Range of Soils Suitable for Vibro-Compaction and Vibro-Replacement Method (Bauman & Bauer, 1974)

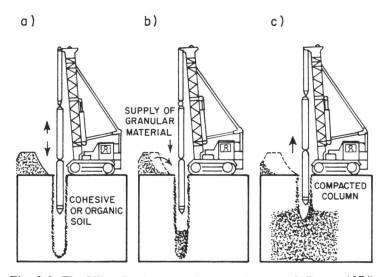

Fig. 3.3 The Vibro-Replacement Process (Bauman & Bauer, 1974)

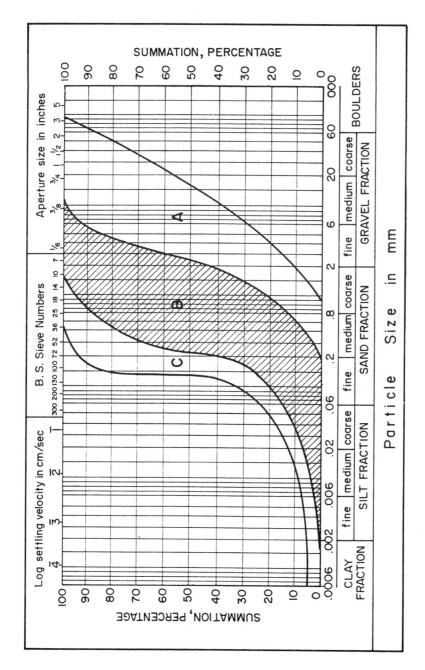

Fig. 3.4 Range of Soils Suitable for Improvement by Vibroflotation (Glover, 1982)

3.3.2 Design of Vibroflotation Technique

The design of D'Appolonia (1953) still remains probably the best method for estimating spacing and patterns. Figure 3.5 shows the relative densities activated at various radial distances from the vibroflot. Coarse material with little or no backfill provides the best backfill materials evaluated according to Brown's suitability number given in Eq. 3.1, where BSN is Brown's Suitability Number and D_{50}, D_{20}, D_{10} are at 50, 20 and 10% finer by weight (Brown and Ralph, 1977).

$$BSN = 1.7 \sqrt{\frac{3}{(D_{50})^2} + \frac{1}{(D_{20})^2} + \frac{1}{(D_{10})^2}} \qquad (3.1)$$

The rating of BSN is given as follows:

BSN	RATING
0-10	Excellent
10-20	Good
20-30	Fair
30-50	Poor
>50	Unsuitable

A single probe of 75 kW increases the density of an initially loose sand to a distance of the order of 2.7 m. There is some variation of the relative density with depth, but if one considers the average relative density over the depth of the compaction for a range of probe spacings, the results plotted in Fig. 3.6 can be obtained. The depth of compaction was 11 m and the relative density was averaged within the hexagonal area and over the depth range of 3 to 11 m. The in-situ sands marked H and T were both poorly graded. Figure 3.6 also shows the effect of grid spacing on relative density for the Terra-Probe based on probing to approximately the same as the vibroflot. The data suggests that the compaction is not as effective as the vibroflot. However, the production of the Terra-Probe is likely to be about four times that of the vibroflot.

3.3.3 Economics and Advantages of Vibroflotation over Conventional Piling

a) Stress concentration

   Conventional - loads are concentrated at the pile caps and transferred through the rigid pile shaft to firm strata by bypassing the uppermost weaker layer.

   Vibroflotation - strengthens the soil locally beneath spread foundations near ground surface. This allows a less concentrated stress to be dispersed more uniformly through the upper layers of ground, thereby, avoiding sharp stress concentrations at the top and bottom of treated ground.

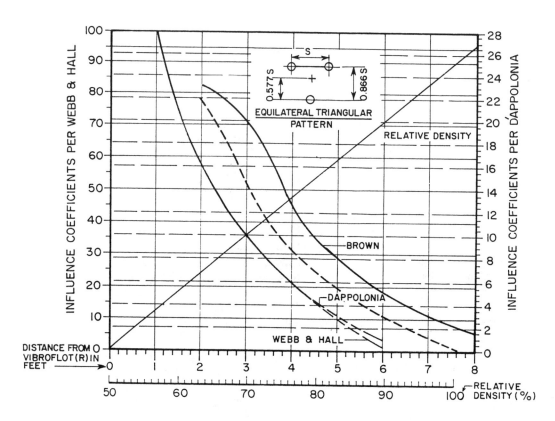

Fig. 3.5 Area Pattern Design Chart (D'Appolonia, 1953)

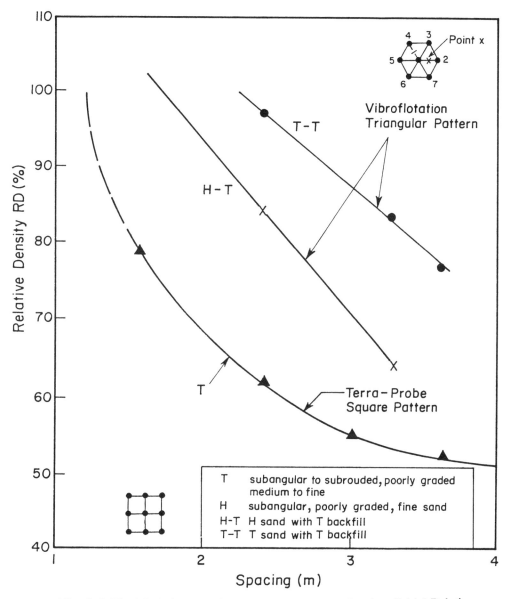

Fig. 3.6 Final Relative Density for Range of Probe Spacing, Initial Relative Density of 45% (Brown and Glenn, 1976).

b) Lesser depth

> Where alternating hard and soft layers are found, vibroflotation treatment can be curtailed in hard layers located at a much lesser depth than conventional piling, which brings considerable economy.

c) Avoid heavily reinforced foundations

> Conventional shallow footings can be adopted avoiding the need for heavily reinforced foundations required for piling.

d) Increased stability and reduced settlement

> The installation of stone columns beneath earth and rockfill dams and embankments has the advantage of reducing consolidation settlement to 1/3 or 1/4 of what would occur without treatment. The coarser material within the columns increases the shearing resistance along any potential failure surface.

3.3.4 Application of Vibroflotation

<u>LNG Plan, Sumatra, Indonesia</u> Three LPG storage tanks, designated as tanks F6401 to F6403, with outside diameters of 70 m, were constructed from 1986 to 1988 (Gouw, 1989). These tanks are used as temporary storage of LPG before transferring them to LPG tankers.

The site was originally tidal flatland adjacent to the beach which was reclaimed by hydraulic filling with dredged sandy materials from adjacent areas. The ground surface was raised from near mean sea level (MSL) to 4 m above MSL. The soil profile is shown in Fig. 3.7 together with standard penetration test (SPT) blowcounts which is dominated by loose deposits of poorly-graded fine to medium sands with maximum of 18% fines. The water table is near the ground surface.

Vibroflotation was applied to tank F6401 site to densify the soil down to 16 m depth. Vibroflots with 30 Hp and 1800 rpm were utilized (Gouw, 1989). It was intended to increase the overall SPT n-values to a level above the boundary of liquefaction potential. The treated area was extended 3 m outside the tank dimensions covering an area of 77.2 m in diameter. The layout of the treated area including the SPT location before and after treatment is shown in Fig. 3.8. The intersections of the diagonal lines indicated the point vibroflot locations with spacing of 2.4 m in triangular pattern.

As shown in Fig. 3.9, the vibroflotation was very effective in increasing the SPT N-values above the liquefaction boundary values. The overall post-treatment SPT N-values were averaged at 35 blows per foot which is higher than the required value of 15 blows per foot (Gouw, 1989). From the design chart of D'Appolonia (1953) as given in Fig. 3.5, the relative density of 75% was obtained. Using the post-treatment average SPT N-values, the relative

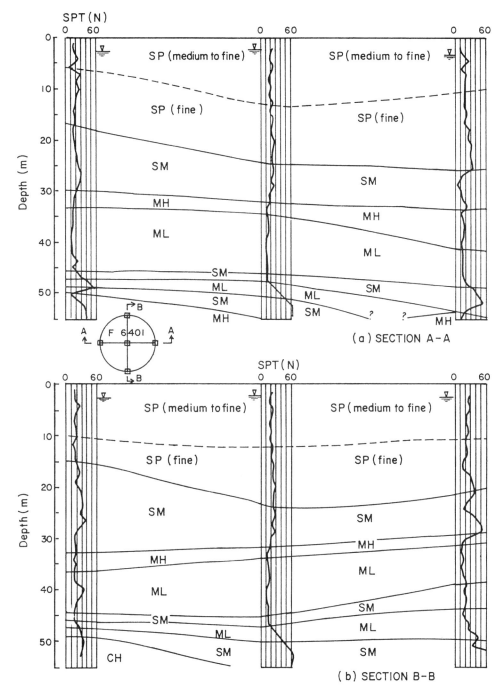

Fig. 3.7 Soil Profile at F6401 Site (Vibroflotation (VBF) Area)

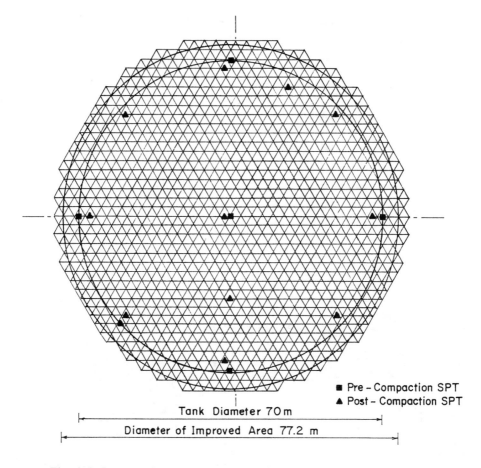

Fig. 3.8 Layout of Area Improved by Vibroflotation (F6401 Site) (VBF Points at Grid Intersection, with 2.4 m Spacing)

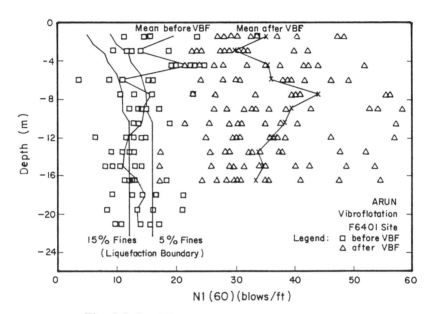

Fig. 3.9 Pre-VBF and Post-VBF SPT at F6401 Site

density of 77% was obtained using the equation of Skempton (1986). Thus, the post-compaction relative density is quite close to the design value obtained from the chart of D'Appolonia (1953).

3.4 MUELLER RESONANCE COMPACTION (MRC)

Vibratory waves, which propagate through soil layers, can be amplified by resonance vibration. Resonance occurs when the dominating frequency of the propagating wave coincides with the natural frequency of one or several soil layers. At resonance, the probe achieves an optimal transfer of vibration energy to the surrounding soil. Resonance is achieved by adjusting the vibrator frequency to one of the resonant frequencies of the soil-probe system.

The dynamic characteristics, consisting of frequency and amplitude, of modern vibrators can be varied and monitored continuously (Massarsch and Heppel, 1991). The soil response can be measured by vibration sensors which are placed on the ground surface. The frequency and amplitude of the vibrator can be adjusted to attain resonance with different soil layers. Figure 3.10 shows a fully automated electronic vibrator monitoring system called Mueller Process Control (MPC) System (Massarsch, 1992) which records and analyzes simultaneously, the performance of the vibrator and the dynamic response of the ground. The vibratory probe consists of the patented MRC Flexi Probe which is especially designed to achieve optimal transfer of compaction energy from the vibrator to soil. As shown in Fig. 3.11, flexible, low impedance probe can achieve dramatically increased ground improvement results compared with conventional, rigid compaction probes. Figure 3.12 shows an MPC computer print out. The left section of this figure displays as a function of time (vertical direction) the penetration depth of the vibratory probe, the hydraulic pressure and the operating frequency of the vibrator (thick line). On the right side of this figure, the penetration speed (dotted line) and the vibration velocity of the ground (thick line) are recorded.

Figure 3.13 shows the result of a resonance test, indicating a strong vibration amplification or resonance at 15.5 hertz. The vibration measurements were done at a distance of 6 m from the compaction probe. As shown in the figure, the vertical vibration velocity at resonance using the Flexi Probe is about 40 mm/s. This value is about 5 times higher compared to compaction with stiff (high impedance) compaction probe. Figure 3.14 shows the influence of compaction probe stiffness (impedance) on ground amplification. A reduction of probe stiffness by 50% results in an increase of ground vibrations by a factor of between 3 and 10.

It is possible to calculate theoretically the resonance frequency, f, of an elastic horizontal layer resting on an infinitely rigid base as follows:

$$f = \frac{C}{4H} \qquad (3.2)$$

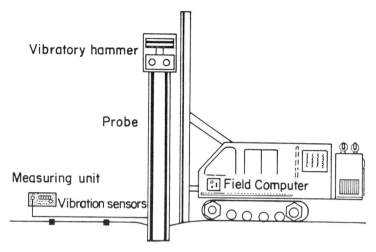

Fig. 3.10 Electronic Vibrator and Ground Response Control System (MPC System) (Massarsch and Heppel, 1991)

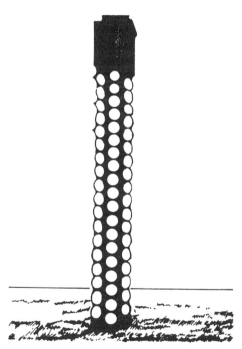

Fig. 3.11 The "Low-Impedance" MRC FLEXI PROBE for Deep Soil Compaction (Massarsch and Heppel, 1991)

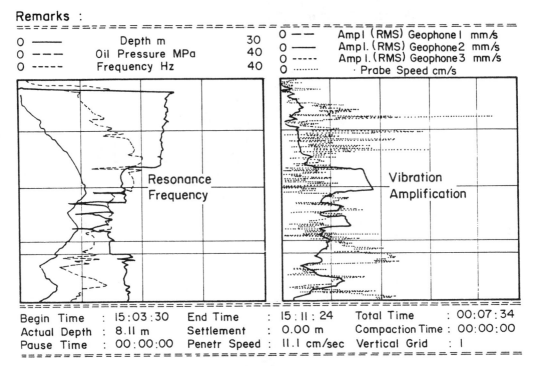

Fig. 3.12 Muller Process Control (MPC) Record of Vibrator Performance (left side) and Ground Response (right side) During Vibratory Driving of a Steel Pile in Dense Sand with an MS 50 Variable Frequency Vibrator (Massarsch and Heppel, 1991)

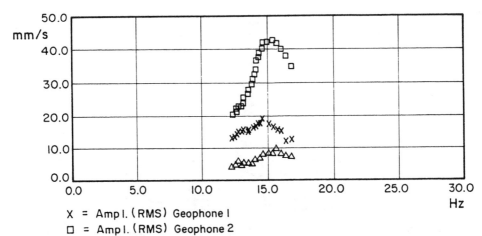

Fig. 3.13 Result of Resonance Test Performed During Deep Soil Compaction (Massarsch and Heppel, 1991)

where C is the wave velocity of the soil and H is the layer thickness. However, in practice, it is more convenient and reliable to measure the resonant frequency directly on the site during various phases of compaction.

Field vibration measurements from several compaction projects in sand are the basis for the empirical relationship required to induce liquefaction (Fig. 3.15). The ground acceleration required to cause liquefaction can be estimated, if the initial cone penetration resistance and the soil layer thickness are known.

3.4.1 Vibratory Compaction Procedure

Four important parameters that must be determined before the start of the compaction project are, namely: the spacing between the compaction points, the duration of compaction in each point, the mode of probe movement (insertion and extraction), and the sequence of compaction in the grid. The spacing between compaction points (grid spacing) ranges typically between 1.5 and 4.5 meters and is affected by the shape and size of the compaction probe. In general, it is preferable to choose a closer grid spacing, as soil densification will then be more homogeneous.

The duration of vibration in each compaction point is another important factor and can vary typically between 5 to 25 minutes. It depends on the soil layer thickness, the grid spacing, and the required degree of soil improvement. The duration of compaction should be determined whenever possible by field test.

Also the mode of probe insertion and extraction in the soil layer to be compacted plays a significant role, but is the most difficult parameter to determine. The compaction efficiency and the optimal compaction procedure can be monitored with the aid of vibration sensors placed on the ground surface, and by monitoring the variation of ground response during probe movement. Figure 3.16 shows the vertical ground vibration velocity (RMS-values) during penetration, suspension, and extraction of the probe in loose saturated sand. Initially, when the probe penetrates into the soil, the vibration amplitudes increase. Suddenly, however, the sand liquefies and the vibration amplitude drops sharply. When the soil consolidates and gradually regains its strength, the vibration amplitude increases again. During the stepwise extraction of the probe, the vibration amplitude shows peak, but is generally lower than during insertion.

Figure 3.17 shows vibration measurements at the same site during probe insertion in an already compacted zone. In this case, the soil does not liquefy. It is apparent that even relatively simple vibration measurements can provide valuable information about the compaction process and the behavior of the soil during vibrocompaction. It might be possible in the future to use the information gained during vibrocompaction to assess the liquefaction potential of soil deposits in situ.

Another important parameter of the compaction process is the sequence in which compaction is performed within an area. Based on extensive experience from projects in

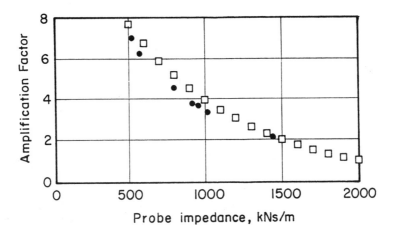

Fig. 3.14 Vibration Amplification in the Ground as a Function of Residual Compaction Probe Stiffness (Low Impedance Probe) (Massarsch, 1992)

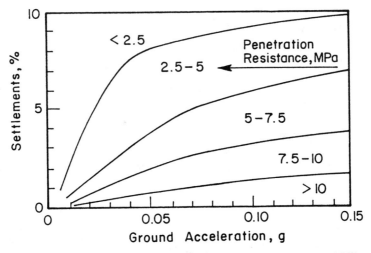

Fig. 3.15 Settlements Caused by Vibratory Pile Driving and Vibratory Soil Compaction (Massarsch and Heppel, 1991)

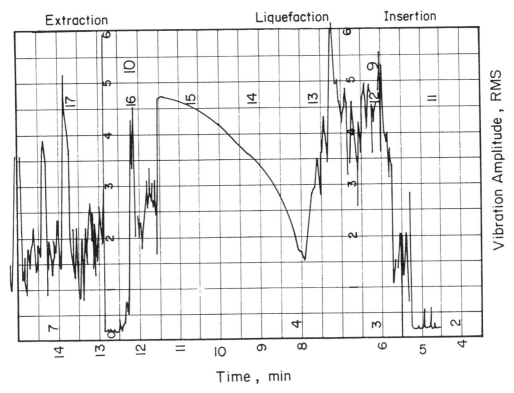

Fig. 3.16 Variation of Ground Vibration Velocity during Insertion, Suspension and Extraction of a Compaction Probe in a Loose Sand which Liquefies during Compaction (note that chart starts from the right) (Massarsch and Heppel, 1991)

different soils, it is recommended to carry out the compaction project in two phases. During the first phase, a coarse grid spacing used, i.e. by compacting every second point in a grid. After the first pass, the soil deposit should be given sufficient time for consolidation before the second pass is started. During the second phase, compaction is then carried out in the intermediate points. In most cases, during the second pass the required compaction time is significantly shorter than during the first pass. This stepwise procedure allows also a comparison of ground settlements during the first and second compaction pass, which can help to optimize the compaction time during the second pass.

During vibratory probe penetration, the vibratory energy should be transferred efficiently from the probe to the surrounding soil. This can be achieved by adjusting the operating speed of the vibrator to correspond to one of the resonant frequencies of the probe-soil system. At resonance, probe penetration is markedly reduced and ground vibration response increases (Massarsch, 1990).

The monitoring equipment required for resonant compaction consists of vibration sensor, an amplitude recording device, and a frequency analyzer. In field practice, velocity transducers (geophones) are commonly used. The optimal compaction frequency can be readily determined on site by varying the vibrator speed and measuring its effects on ground response. The optimal compaction frequency can change during the compaction process depending on the soil type, layer thickness, and soil density.

The number of insertion cycles, their duration and respective penetration depth are influenced by several factors such as the geotechnical conditions, soil strata, and the compaction effect to be achieved in the respective layers. Ground vibration measurements from field trials provide valuable guide for most effective compaction procedure. Figure 3.18 shows the vibration of vertical ground vibration velocity during penetration, suspension, and extraction of the probe.

3.4.2 Ground Vibration Velocity

The availability of accurate seismic field measuring equipment in the recent past has made it possible to monitor and control soil compaction accurately. The monitoring equipment required for resonant compaction consists of vibration sensor, amplifiers, an amplitude recording device (analog and/or digital) and a frequency analyzer. The ground vibration velocity can be used to establish the required grid spacing and the optimal compaction procedure, as discussed above. However, this information is also useful to evaluate the effect of ground vibrations on slopes, installations in the ground or on adjacent buildings. The vibration attenuation can be approximately estimated from the following theoretical relationship:

$$A_2 = A_1 \left[\frac{R_1}{R_2}\right]^{1/2} e^{-\alpha (R_2 - R_1)} \qquad (3.3)$$

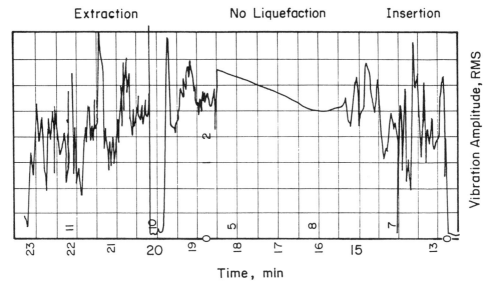

Fig. 3.17 Variation of Ground Vibration Velocity during Insertion, Suspension and Extraction of a Compaction Probe in a Densified Sand (note that chart starts from the right) (Massarsch and Heppel, 1991)

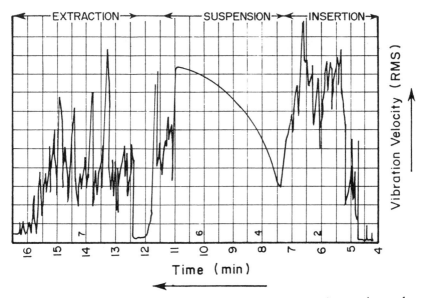

Fig. 3.18 Variation of Vibration Velocity during Penetration, Suspension and Extraction (Massarsch and Heppel, 1991)

where A_1 and A_2 are the vibration amplitudes at distances R_1 and R_2 from the probe, respectively. The coefficient of wave attenuation α is a measure of soil damping and depends on the vibration frequency (wave length) and the dynamic properties (wave propagation velocity) of the soil. The attenuation of vibration energy in the soil can be difficult to predict. Figure 3.19 shows typical results from vibration measurements obtained from several sites, indicating an approximate relationship between the vertical vibration velocity (and acceleration) as a function of distance from the compaction point. The attenuation coefficient α for vibratory compaction in saturated sands typically varies between 0.05 and 0.10 m^{-1}.

3.4.3 Wave Velocity Measurements

The compaction effect can also be monitored by cross-hole tests before and after compaction (Massarsch and Lindberg, 1984). The shear wave velocity can be used to estimate increase in soil stiffness. If the shear wave (C_s) is known, the shear modulus (G_s) at small strain can be obtained as follows:

$$G_s = C_s^2 \rho \qquad (3.4)$$

where ρ is the bulk density of the soil. Figure 3.20 shows the results from cross-hole measurements with shear wave velocities before and after compaction (Massarsch and Broms, 1983).

3.4.4 Vibration Amplitude

Based on settlement measurements, a direct correlation can be established between observed settlements and maximum ground acceleration. Settlements after compaction vary typically from 4% to 10% of layer thickness. Figure 3.21 represents an empirical relationship between relative settlements, initial cone penetration resistance, and the required ground acceleration.

3.4.5 Recent Application in Thailand

Recently, the resonance compaction technique was utilized in a reclaimed area within the Map Ta Put Deep Sea Port in Eastern Seaboard in Thailand. As shown in Fig. 3.22, the reclaimed area was filled up of hydraulic fill dredged from nearby seabed up to 12 m thick of relatively clean sands and corals. The resonance compaction equipment is shown in Fig. 3.23. The compacted area is used for foundation of oil storage tanks as shown in Fig. 3.24. The site is mostly underlain by sandy deposit as illustrated in the soil profile in Fig. 3.25. Figure 3.22 also shows the SPT N-values before and after compaction. Cone Penetration Tests (CPT) were also conducted to evaluate the magnitude of improvement as presented in Fig. 3.26. The improvement is quite significant, as shown in the aforementioned results.

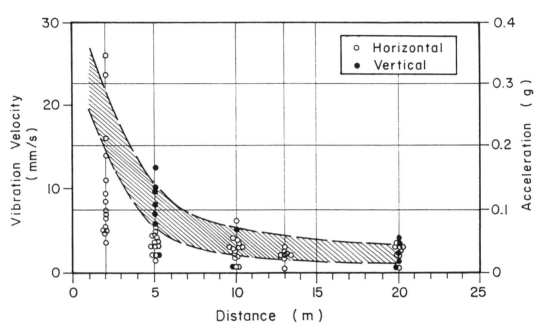

Fig. 3.19 Variation of Vertical and Horizontal Vibration Amplitude as a Function of Distance from the Compaction Point (Massarsch and Heppel, 1991)

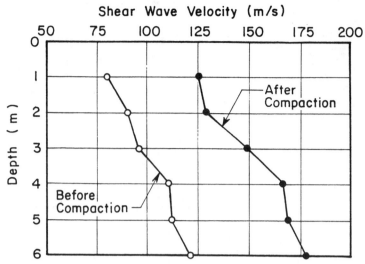

Fig. 3.20 Variation of Shear Wave Velocity with Depth Determined from Cross-Hole Tests (Massarsch and Heppel, 1991)

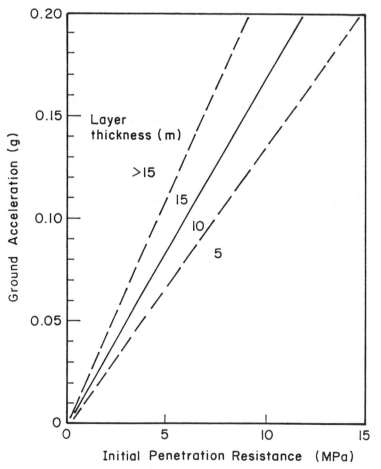

Fig. 3.21 Ground Acceleration Required to Induce Liquefaction in Sand during Vibratory Compaction as a Function of Initial Cone Penetration Resistance and Compaction Depth (Massarsch and Heppel, 1991)

Fig. 3.22 Reclaimed Area in Map Ta Put Deep Port, Thailand

Fig. 3.23 The Resonance Compaction Equipment

Fig. 3.24 Oil Storage Tanks Founded on Compacted Foundation

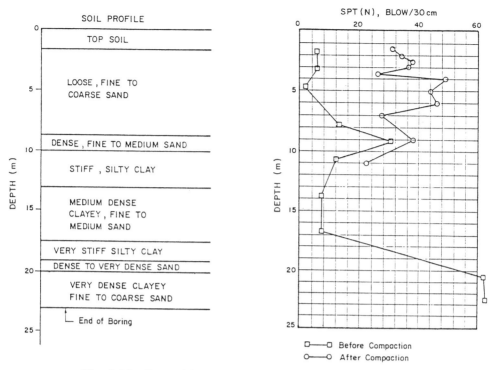

Fig. 3.25 General Soil Profile at Map Ta Put Deep Sea Port

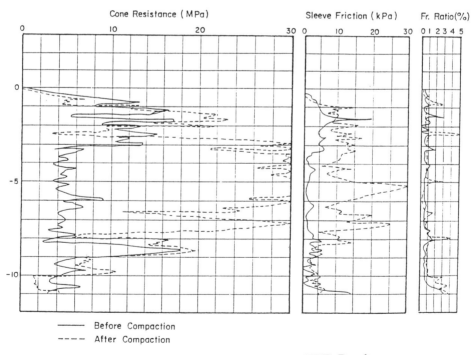

Fig. 3.26 Cone Penetration Tests (CPT) Results

3.5 DYNAMIC COMPACTION

The objective of the Dynamic Compaction (DC) method is to stabilize and densify granular soils deposited both above and below the groundwater table. It is aimed to improve the soil bearing capacity, decrease the amount of settlement, and decrease the potential of liquefaction. Dynamic compaction was first popularized by Menard in the early 70's (Menard and Broise, 1975). The process consists essentially of dropping large weight, 10 to 20 tons, up to 150 tons, into the ground to be compacted. The height of drop generally ranges from 10 to 20 m, up to 40 m (Fig. 3.27).

High technical knowledge and experience is required to densify the ground of various characteristics below the water table uniformly to a desired density. Unless properly guided with technical management, simple tamping with a heavy hammer may result in non-uniform ground densified only in a few meters below the ground surface or other defects in the ground treated by tamping.

Properly planned tamping work based on technical management is therefore important for the success of the DC method. The flow of work consisting of tamping work and technical management, as shown in Fig. 3.28, is summarized in the flow chart shown in Fig. 3.29.

3.5.1 Condition Applicable to Dynamic Consolidation

The technique of dynamic consolidation is suitable for fill materials with low moisture and below water, rockfill, end-dump in water, loose sand and gravel alluvium, non-cohesive fill including industrial and city refuse, silt and silty sandy clay with and without preloading and/or vertical band drains. This method is also applicable to granular soils containing a limited amount of fine grained soil, soil containing stones and rock fragments, and waste products. More details as to the factors that influence the effectiveness of dynamic compaction are discussed below:

Coarse-grained pervious deposits: Dynamic compaction works well in these deposits both above and below the water table. Above the water table the soil particles move into a denser state of packing similar to that which occurs from Proctor compaction. Below the water table, densification occurs because the excess pore pressures generated during tamping dissipate almost immediately following impact. Thus, good interparticle contact and interlocking is achieved as a result of densification.

Semi-impervious deposits: Excess pore pressures are generated in both saturated or nearly saturated deposits. Some period of time, frequently on the order of minutes or days, is required for excess pore water pressures to dissipate. Sand or silt boils can form in the ground surface. Because of this, the energy that is applied is only partially effective in densifying the soil. In order to achieve a significant improvement, the field operation must be properly controlled to allow excess pore pressure dissipation between tamping. Usually, this requires the installation of pore pressure monitoring devices, multiple passes over the area, and wide spacing

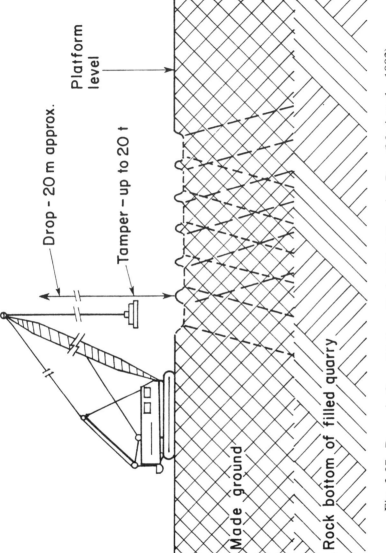

Fig. 3.27 Dynamic Consolidation: Typical First Tamping Pass (Mori et al., 1992)

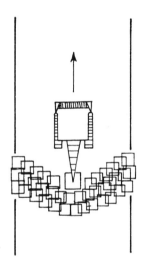

Fig. 3.28 Flow of Tamping Work (Mori et al., 1992)

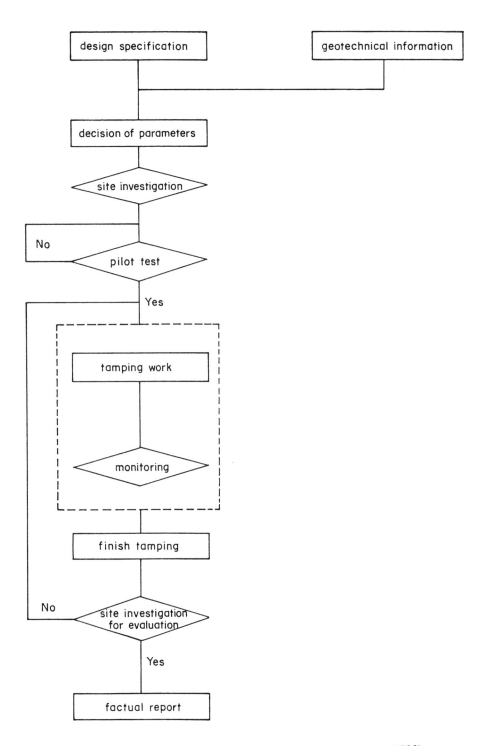

Fig. 3.29 Flow of Dynamic Compaction Work (Mori et al., 1992)

between impact points.

<u>Saturated impervious deposits</u>: Dynamic compaction is generally not effective for this deposit. Excess pore pressures which develop under impact cannot be rapidly relieved because of the low permeability and long drainage paths.

<u>Partially saturated impervious deposits</u>: Modest improvements in shear strength and a reduction in compressibility can be attained. Clay fills above the water table have been improved by dynamic compaction when water content of the clay is near or below the plastic limit.

<u>Soft and weak soil</u>: Dynamic compaction can be applied to the ground covered with soft and compressible soil, but the ground surface should be covered with sand fill of a thickness more than one meter in order to enable a crawler crane to maneuver and to recover the hammer driven into the soft soil easily. Sand fill is also required at sites where the groundwater table is high to keep the ground surface 1.5 to 2.0 m higher than the groundwater table. Locally distributed spots of soft soil near the ground surface should be replaced with crushed stones.

<u>Soil properties</u>: Dynamic compaction can be applied to gravel, sandy soil, rock fragments, and waste products with good results.

<u>Environmental conditions</u>: Tamping produces vibration and noise. Noise can be considerably reduced by surrounding the tamping machine with soundproof sheets. To reduce vibration, damping from a distance is required. The relationships between vibration/noise and distance are shown in Fig. 3.30 (Narumi, 1987).

To meet the regulation of vibration, that is, 75 dB at the border of a construction site, a distance of 50 to 100 m from the tamping machine is required depending on the subground conditions. Special care to minimize vibration effects on residents around the site is needed. The influence of vibration against existing structures should be controlled so that the vibration level due to tamping may not exceed the allowable acceleration in the seismic design of the structures.

3.5.2 Suitability of Deposits for Dynamic Compaction

The important parameters for the soil that affect the suitability of dynamic compaction include the soil classification, degree of saturation, and permeability plus length of drainage paths. Table 3.2 indicates the suitability of various deposits for dynamic compaction based upon these factors. The four first categories, described in Table 3.2, fall into the broad category of deposits that can be classified by conventional index tests. This is followed by a discussion of recent landfill and organic soil deposits where conventional index tests are not appropriate for identification.

Table 3.2 Suitability of Deposits for Dynamic Compaction (Lukas, 1986)

General Soil Type	Most Likely Fill Classification	Most Likely AASHTO Soil Type	Degree of Saturation	Suitability for D.C.
Pervious deposits in the grain size range of boulders to sand with 0% passing the #200 sieve Coarse portion of Zone 1*	Building Rubble Boulders Broken Concrete	A-1-a A-1-b A-3	High or Low	Excellent
Pervious deposits containing not more than 35% silt Fine portion of Zone 1*	Decomposed Landfills	A-1-6	High	Good
		A-2-4 A-2-5	Low	Excellent
Semi-pervious soil deposits, generally silty soils containing some sand but less than 25% clay with PI<8 Zone 2*	Flyash	A-5	High	Fair
	Mine Spoil		Low	Good
Impervious soil deposits, generally clayey soils where PI>8 Zone 3*	Clay Fill Mine Spoil	A-6 A-7-5 A-7-6 A-2-6	High	Not recommended
			Low	Fair - minor improvements - water content should be less than plastic limit.
Miscellaneous fill including paper, organic deposits, metal and wood	Recent Municipal Landfill	None	Low	Fair - long-term settlement anticipated due to decomposition. Limit use to embankments
Highly organic deposits, peat-organic silts		None	High	Not recommended unless sufficient granular fill added and energy applied to mix granular with organic deposits.

*These zones are identified on Fig. 3.30

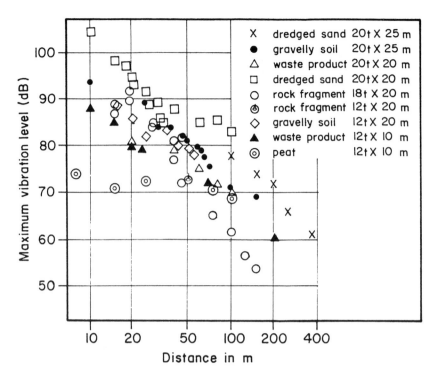

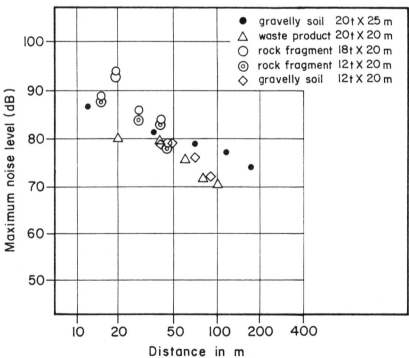

Fig. 3.30 Damping of Vibration and Noise with Respect to Distance (Narumi, 1987)

3.5.3 Deposits Classified by Index Tests

Most natural soil or fill deposits can be characterized by index tests including grain size gradation and/or Atterberg Limits. On a typical gradation chart, three zones are shown in Fig. 3.31 as supplement to Table 3.2 (Lukas, 1986). Zone 1 represents the gradation range where dynamic compaction is the most appropriate. Zone 3 is the gradation of fully or nearly saturated. Zone 2 is the transition range where dynamic compaction will work but multiple passes are required to allow excess pore pressures to dissipate before more energy is applied.

3.5.4 Deposits not Characterized by Index Tests

Certain deposits could not be classified by conventional index tests, namely: sanitary landfills, highly organic peat deposits, and waste fill deposits. The use of dynamic compaction on these deposits is a relatively new development and will require additional experimentation and observations before actual performance can be evaluated properly. Usually, secondary compression in these deposits could be significant.

3.5.5 Design of Dynamic Consolidation

In order to compact the ground uniformly with varying characteristics above and below the water table to a desired density, stiffness, and strength, tamping work has to be guided by well programmed engineering management. The design of dynamic compaction is defined to assess various parameters of the method and specify the points of engineering management during construction. The parameters and the management points are presented as follows:

<u>Tamping energy per blow</u>: The weight, W, of a hammer and the drop height, H, are selected depending on the depth, D, to which the ground is to be improved. Menard and Broise (1975) proposed to select an energy WH per blow larger than H^2. Summarizing a number of field experiences, Mitchell and Solymar (1984) concluded that the use of the equation proposed by Leonards et al. (1980) as shown below would provide a conservative estimate of depth D.

$$D = n\sqrt{WH} \qquad (3.5)$$

where D is the depth of improvement in meters, W is the weight in tons, H is the drop height in meters. The coefficient, n, accounts for factors that can affect the depth of improvement other than the weight of tamper and drop height. Some investigators have suggested using $n=0.5$ for all soil deposits as shown in Fig. 3.32 (Mayne et al. 1984). There is evidence that the factor n is affected by many variables including: the types of soil deposit, the applied energy, the contact pressure, the influence of cable drag, and the presence of energy absorbing layers. The suggested values for the empirical factor, n, to be used in predicting the depth of ground improvement are listed in Table 3.3 (Lukas, 1986).

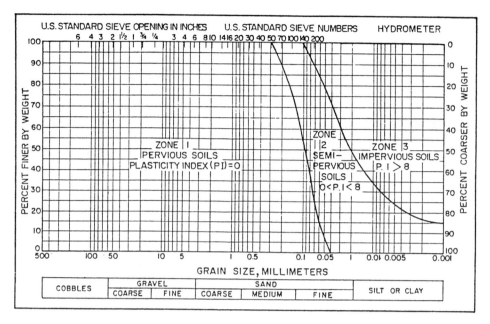

Fig. 3.31 Grouping of Soils for Dynamic Compaction (Lukas, 1986)

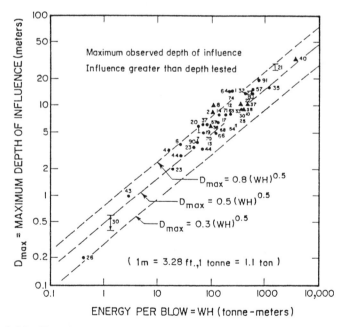

Fig. 3.32 Trend Between Apparent Maximum Depth of Influence and Energy Per Blow (Mayne et al., 1984)

Table 3.3 Recommended n Value for Different Soil Types (Lukas, 1986)

Soil Type	Degree of Saturation	Recommended n value*
Pervious soil deposit-granular soils	High Low	0.5 0.5 to 0.6
Semi-pervious soil deposits, primarily silts with PI < 8	High Low	0.35 to 0.4 0.4 to 0.5
Impervious deposits, primarily clayey soils with PI > 8	High Low	Not recommended 0.35 to 0.4 Soils should be at a water content less than the plastic limit.

* For an applied energy of 34 to 100 ton-ft/ft^2 (100 to 300 txm/m^2) and for a weight dropped using a single cable with a free spool drum.

3.5.6 Depth of Improvement

If dynamic compaction (DC) is to be undertaken with readily available cranes, there is limitation on the depth of improvement that can be achieved. The maximum capacity cranes normally available are rated as 150 to 175 tons (136 to 158 metric tons) and these cranes can lift weights up to 20 to 22 tons (18 to 20 metric tons) with a maximum height of 66 to 98 feet (20 to 30 m). Using an average value of n of 0.5, a maximum depth of improvement in the range of 33 to 39 ft (10 to 12 m) can be achieved.

Grid dimensions: In DC work, a site is marked off in a grid pattern, and one tamping point is located in each grid. The grid dimensions have to be selected to densify the ground uniformly. Tamping at too short a distance internally densifies only a shallow layer near the ground surface in the early stage of tamping work, and the compacted shallow layer will interrupt the propagation of the tamping energy generated by subsequent passes. As a common practice, a larger dimension is selected in the first pass, and in the next pass tamping points are selected between the imprints in the first pass. As initial estimate the grid spacing for the first pass can be taken equal to the desired depth of treatment.

Time interval between passes: Excess pore water pressure develops in soil below the groundwater table when the ground is subjected to tamping. Part of the shock pore pressure developed by the impact of a hammer remains. This residual pore pressure is accumulated and increases with the progress of tamping work.

The excess pore water pressure developed by tamping varies widely, depending on density and permeability of the ground. The developed pore pressure is large in loose soil and small in dense soil. In the ground of low permeability, accumulated excess pore water pressure remains and increases during tamping. In very pervious ground, however, pore pressure caused by a blow of a hammer dissipates entirely before the next blow, and sometimes no residual pore pressure is found.

Figure 3.33 shows an example of residual excess pore water pressure in sandy soil reported by Kumagai et al. (1981). In this case, the ground consisted of loose fine sand of relatively poor permeability. To improve the soil to a depth of 5 m, a tamping energy of 12 tons x 20 m x 24 blows was proposed for each grid of 6 x 6 m^2. A considerably large accumulated pore water pressure was developed, and finally the subsoil was liquefied with boiling sand observed on the ground surface.

3.5.7 Limiting on Improvement and Total Energy Requirement

Leonards et al. (1980) indicated that there may be an upper limit to the densification that can be achieved by dynamic compaction. The cone penetration resistance increased up to a limiting value as the amount of energy increased. The anticipated relative improvements for different soil types is shown in Table 3.4. Based upon a review of a number of projects by Lukas (1986), the typical upper bound test values for SPT, CPT, and PMT tests following dynamic compaction are summarized in Table 3.5. The applied energy requirement is also presented in Table 3.6.

The applied energy is generally reported in terms of unit energy applied over the ground surface in ton-meter/square meter. The determination of the precise amount of energy is difficult because of many factors such as: the type of soil deposit; the initial relative density; the thickness of soil deposit; and the required degree of improvement.

Table 3.4 Anticipated Relative Improvements for Different Soil Types (Lukas, 1986)

Soil Type	Anticipated Amount of Improvement*
Pervious Coarse Grained Soils - Sand and Gravel	300 to 400%
Semi-pervious Soils A. Silty sands B. Silts and partially saturated clayey silts	100 to 400% 100 to 250%
Partially Saturated Impervious Soils - Clay fills and mine spoil	200 to 400%
Landfills	200 to 400%
Building Rubble	200 to 300%

* - For applied energies of 34 to 100 ton-ft/ft^2 (100 to 300 txm/m^2)

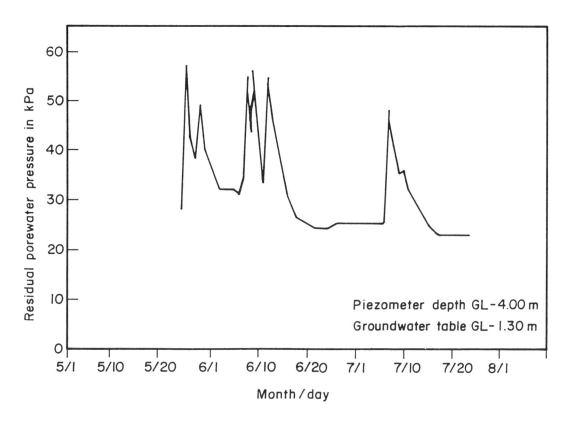

Fig. 3.33 Measurement of Residual Porewater Pressure in Sandy Soil (Kumagai et al., 1981)

Table 3.5 Upper Bound Test Values After Dynamic Compaction (Lukas, 1986)

Soil Type	SPT (blows/ft)	CPT (tsf)**	Limit Pressure (tsf)**
Pervious Coarse Grained Soil - Sand and gravel	40-50	200-300	20-25
Semi-pervious: - Sandy silts - Silts and Clayey Silts	34-45 25-35	140-180 100-140	15-20 10-15
Partially Saturated Impervious Deposits: - Clay fill, Mine spoil	30-40*	N/A	15-20
Landfills	20-40*	N/A	5-10

* - Higher test values will occur when sampling on large particles present in the soil mass.
**- 1 ton/ft^2 = 95.8 kN/m^2

Table 3.6 Applied Energy Requirements (Lukas, 1986)

Type of Deposit	Applied Energy Normally Used
Pervious Coarse Grained Soil: - Zone 1 of Figure 3.31	20 - 25 txm/m^3
Semi-pervious Fine Grained Soils: - Zone 2 and clay fills - Above the water table - Zone 3 of Figure 3.31	25 - 35 txm/m^3
Landfills	60 - 110 txm/m^3

NOTE: Standard Proctor energy equals 60.5 txm/m^3
1 txm/m^3 = 0.10225 ton-ft/ft^3

<u>Influence of tamping energy</u>: An example of the SPT N-values plotted against depth in Fig. 3.34 shows an increasing trend with the progress of tamping passes applied to a sandy gravel layer of about 10 m deep, reclaimed on the sea bed (Kawamura and Ikuta, 1983). In this case, the ground having N-value of about 5 was increased up to 10 to 25. As shown in this

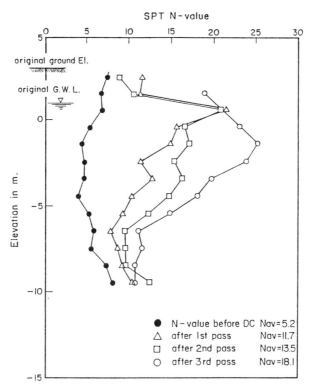

Fig. 3.34 Example of Ground Improvement Effect by Dynamic Compaction (Kawamura & Ikuta, 1983)

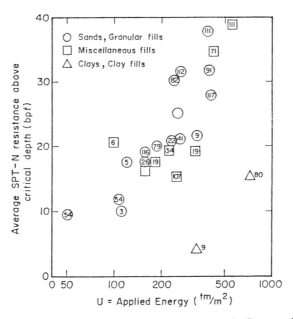

Fig. 3.35 Ground Response to Dynamic Compaction (Mayne et al., 1984)

figure, N-values increase with depth to a certain point, called the critical depth. Beyond this depth, however, it tends to decrease gradually with depth.

Summarizing the SPT results for the subground improved by the DC method, Mayne et al. (1984) presented a case in which the average N-values above critical depths are plotted against the applied energy in tm/m^2, as shown in Fig. 3.35. Although plotted data are widely scattered, the range of the average N-values which depend upon the applied energy may be estimated from the envelopes of these plotted points aside from the data for clay. It is to be noted, however, that the plotted N-values are those for the soil above the critical points (see Fig. 3.35).

The ratio of average settlement to the average energy per unit volume (tm/m^3) is defined as the tamping efficiency. Summarizing the data of the SPT N-values and the modulus of deformation, E_{av}, obtained from the pressuremeter for the ground improved by the DC method, Yuasa et al. (1987) found the correlation between the tamping efficiency and N-value or the modulus, E_{av}, as shown in Fig. 3.36.

3.5.8 Compression of Waste Products Due to Dynamic Consolidation

The experience in three improvement projects in Japan (Yamazaki et al. 1991) showed an increase in modulus E obtained from the pressuremeter test from 1500 to 4000 kPa before and after DC, respectively. The limit pressure P was improved, in the order from 200 to 500 kPa. The improved waste could be used as the subgrade of roads and runways or the bearing layer for footing foundations.

In Fig. 3.37, N-values are plotted against the content of fine grained soil having grain sizes less than 74μ. It is identified from this figure that N-values decrease remarkably with an increase of the fine content. Although the effectiveness of the DC method for sandy soil, in general, is reduced as its fine content increases, its resistance against liquefaction increases.

3.5.9 Applications of Dynamic Compaction

<u>Thap Salao Dam, Thailand</u> In 1986, a 40 m high, 4 km long earth dam was constructed at Thap Salao in the Uthai Thani Province of Central Western Thailand. The site investigations at the proposed dam site included the drilling of 20 test holes (to bedrock). The boring logs showed that the ancient river channel was filled up to 12 m of loose sand. Standard Penetration Tests (SPT) revealed results of N values ranging from 3 to 4 at over 10 m depth (Fig. 3.38). Loose sand was found over about 12 hectares beneath the highest section of the proposed 40 m high dam embankment. The geotechnical consultants recommended the use dynamic compaction technique for the densification of the sand deposit. This method was adopted due to the following merits:

(a)   Water is not necessary for this method.
(b)   The method is relatively simple.

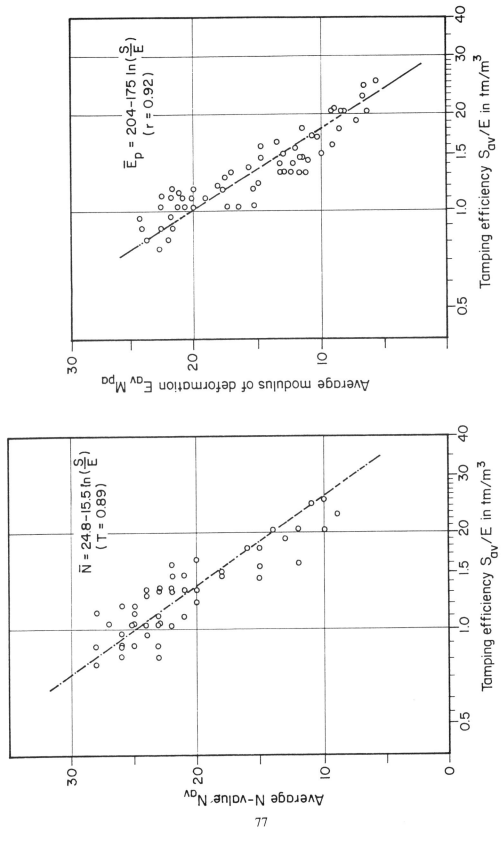

Fig. 3.36 Tamping Efficiency Versus N-Value/Modulus of Deformation (Mori et al., 1992)

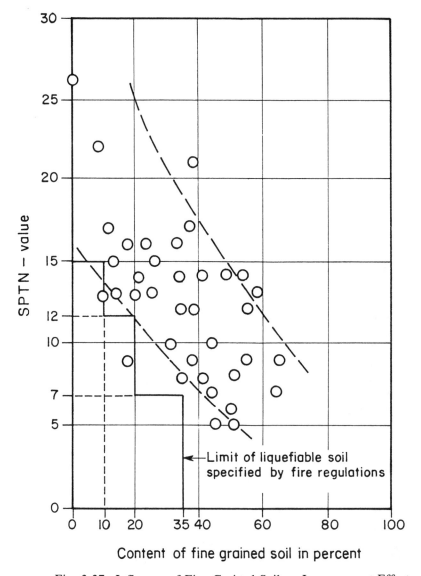

Fig. 3.37 Influence of Fine Grained Soil on Improvement Effect of Dynamic Compaction (Mori et al., 1992)

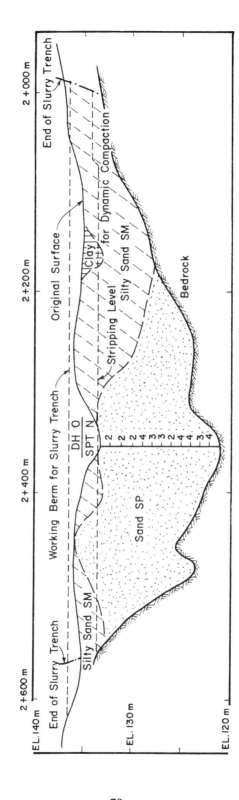

Fig. 3.38 Cross Section of Dam Foundation, Thap Salao Dam (Fitzhardinge, 1990)

(c) The method is relatively flexible. The pounder weight, height of drop, number of impacts per site, and spacing can all be varied to optimize the operation.
(d) The method has the better chance of successfully treating areas of silty sand.

The dropping of weight for the dynamic compaction trial was carried out on the pattern shown in Fig. 3.39. During the initial pass, the dropping sites were spaced at 12 m, the weight being dropped up to six times in each location. Measurements were made after each drop of dimensions of the crater and of the dilation of the surrounding soil. The effective dimensions of the crater formed by each impact followed a trend similar to Fig. 3.40. After four impacts, the crater left by the repeated pounding penetrated to the water table, and the increase with additional drops began to decrease; it was considered that energy was being lost in displacing water. Piezometer readings showed that the maximum pore pressure increase was 0.5 m of water and it dissipated quickly.

On the basis of the results of the dynamic compaction trial, the program shown in Table 3.7 was adopted. The needed total energy input was approximately 275 ton-meter per square meter. The dynamic compaction successfully increased the Standard Penetration Tests results (Fig. 3.41) to meet or considerably exceed the specifications in order to prevent liquefaction during the occurrence of the design earthquake magnitude. The original ground surface of the compacted area was depressed by an average of 2 m which demonstrates the effectiveness of the dynamic compaction method and indicates an increase in the density of the material by over 16%. Similar to the reports by Mitchell and Solymar (1984), the strength of the soil as measured by SPT was found to continue to increase significantly with time after compaction.

Table 3.7 Dynamic Compaction Program (Fitzhardinge, 1990)

Pass	Blows	Spacing (m)	Cumulative Energy (t.m/m^2)
1	4	12	17-18
2	4	12	34-37
3	6	8.5	85-91
4	7	6	205-220
5	1	3	273-292

<u>Bhumibol Dam, Thailand</u> The Bhumibol Hydroelectric Project, as reported by Buttling (1994), was constructed in the 1950s with the only arch concrete gravity dam in Thailand. The water level is at about 135 meters above mean sea level (mMSL) and the dam crest rises to 260 mMSL. The lack of rainfall, partly prompted by environmental concerns, has led to a serious drop in the water level within the reservoir. In order to mitigate this problem, a scheme was devised which involved a significant amount of civil work. Part of the scheme was the

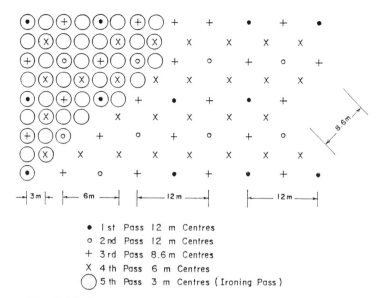

Fig. 3.39 Dropping Pattern for Dynamic Compaction, Thap Salao Dam (Fitzhardinge, 1990)

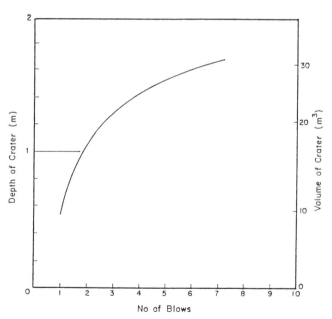

Fig. 3.40 Crater Formation in Dynamic Compaction, Thap Salao Dam (Fitzhardinge, 1990)

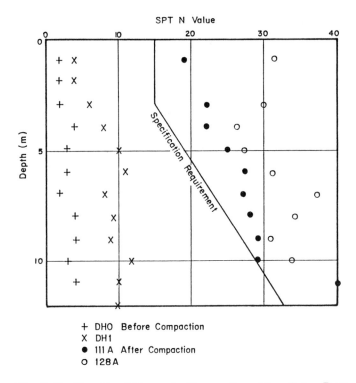

Fig. 3.41 Results of Dynamic Compaction, Thap Salao Dam (Fitzhardinge, 1990)

Fig. 3.42 Actual Application of Dynamic Compaction for Dam Rehabilitation in Thailand

construction of an After Bay Dam in order to retain some of the water downstream to be pumped back upstream to the main dam. The construction of the After Bay Dam, about 5 km downstream of the main dam, required that the subsoil be compacted.

The existing soil conditions were generally medium to coarse sand with layers of gravel and cobbles. The grading analyses for samples from boreholes showed that the amount of fine material, i.e. passing No. 200 sieve, was no more than 1% at all depths. The logs descriptions only showed sand and gravel, with no differentiation between layers.

It was decided to carry out the dynamic compaction using a weight which is lifted on a single fall of rope (Fig. 3.42). For the type of sand and gravel present at the site, an energy of about 20 ton.m/m^3 was required and therefore 240 ton.m/m^2 was needed for 12 m depth of treatment. The compaction was initially carried out in four passes, each on a 10 m grid offset by 5 m, as shown in Fig. 3.43. This could be achieved with two passes of 22 drops each of a 15 ton weight from 25 m height and 2 passes of 24 drops of a 12 ton weight from a 14 m height, with a total energy input of 246 ton.m/m^2. Pre-treatment and post-treatment tests were carried out using Cone Penetration Tests to evaluate the increase in strength (Fig. 3.44). The post-treatment showed the following observations when compared with the pre-treatment results:

a) Low densities near the surface as the points were within the perimeter of a crater produced during the second pass.

b) General improvement in the middle range of depths.

c) An extremely low value at 10 m depth in one location.

The very low results at 10 m depth were due to the presence of fines content in the sand layer, which caused a rise in pore pressure, thus reducing the effective stress.

3.6 SUMMARY

Deep compaction is achieved by applying vibration energy into the subsoil. The objectives are to reduce total and differential settlements, increase shear strength, increase resistance to soil liquefaction during earthquakes, and reduce the cost of foundation systems. The increase in density of the foundation subsoil will allow the utilization of a shallow foundation rather than the more expensive deep foundation.

The factors influencing the deep compaction include the type of in-situ soil, requirements for foundation design, depth of compaction, backfill materials, availability of construction equipments, site location, extent of site, time constraints, and compaction procedure. The application of deep compaction techniques varies from simple vibrating poker for clean sands such as the Terraprobe, Resonance Compaction, and Vibroflotation, to dropping of heavy weight such as Dynamic Compaction for more complicated sites with silty and clayey sands, old mine quarries, mine spoils, and landfills.

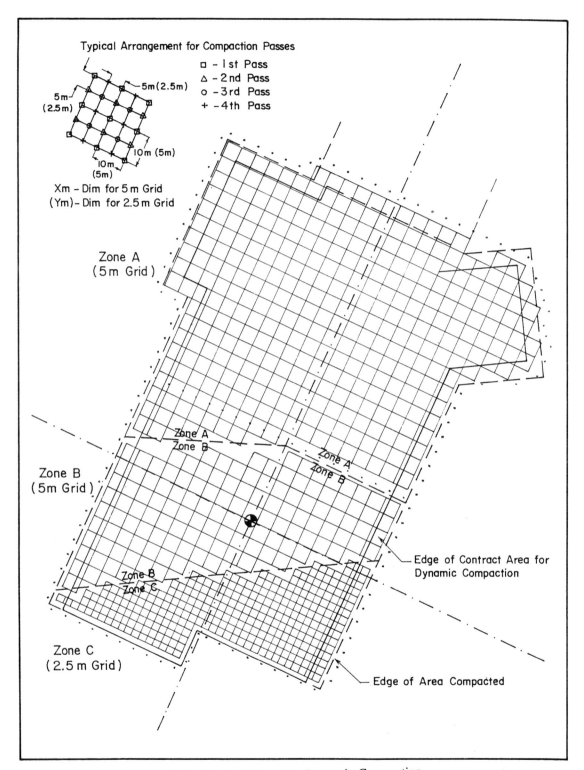

Fig. 3.43 Design Pattern for Dynamic Compaction, Bhumibol Dam (Buttling, 1994)

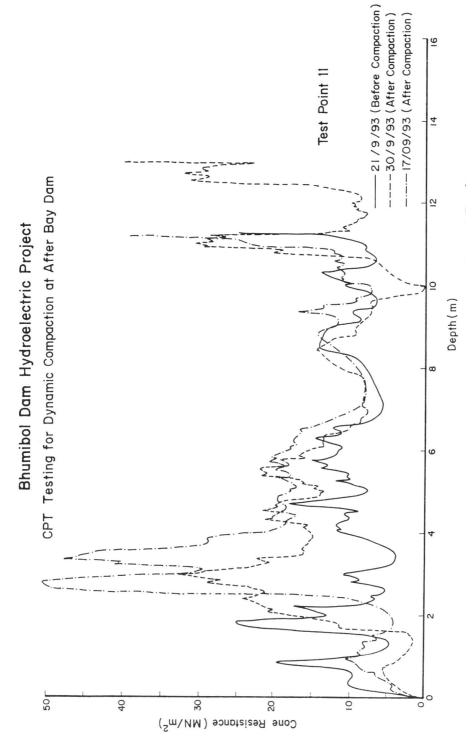

Fig. 3.44 Pre-treatment and Post Treatment Strength Results, Bhumibol Dam (Buttling, 1994)

3.7 REFERENCES

Baumann, R.D., and Bauer, G.E.A. (1974), The performance of foundations on various soils stabilized by the vibro-compaction method, Canadian Geotech., Vol. 11, No. 4, pp. 509-530.

Brown, R.E., and Glenn, A.J. (1976), Vibroflotation and terraproble comparison, J. of Geotech. Eng'g. Div., ASCE, Vol. 102, pp. 1059-1072.

Brown, R.E., and Ralph, E. (1977), Vibroflotation compaction of cohesionless soils, J. of the Geotech. Eng'g. Div., ASCE, Vol. 103, No. GT12, pp. 1327-1451.

Buttling, S. (1994), Ground improvement works at Bhumibol Dam, Proc. Conf. Deep Found. and Ground Improvement Schemes, Asian Institute of Technology, Bangkok, Thailand.

D'Appolonia, E. (1953), Loose sands - Their compaction by vibroflotation, Proc. Symp. on Dynamic Testing of Soils, STP 156, ASTM, Philadelphia, Pennsylvania.

Engelhardt, K., and Kirsch, K. (1977), Soil improvement by deep vibration technique, Proc. 5th Southeast Asian Conf. on Soil Eng'g., Bangkok, Thailand.

Fitzhardinge, C.F.R. (1990), Dynamic compaction and slurry trench cut-off at Thap Salao Dam, Thailand, Proc. 10th Southeast Asian Geotech. Conf., Taipei, ROC, Vol. 1, pp. 59-64.

Glover, J.C. (1982), Sand compaction and stone columns by the vibroflotation process, Vibroflotation International (H.K.) Ltd., Hong Kong.

Gouw, T. L. (1989), Deep compaction for ground improvement and for reduction of liquefaction potential in Sumatra, Indonesia, M. Eng'g. Thesis No. GT-88-11, AIT, Bangkok, Thailand.

Kawamura, Y., and Ikuta, R. (1983), Ground improvement under a sewerage in the South Osaka Bay by means of dynamic consolidation (In Japanese), J. of Sewerage Association, Vol. 20, No. 231, pp. 1-12.

Kumagai, S., Takiya, M., and Harada, S. (1981), Improvement of shallow sandy ground by DC method (In Japanese), Monthly of Found. Eng'g., pp. 76-83.

Leonards, G. A., Cutter, W. A., and Holtz, R. D. (1980), Dynamic compaction of granular soils, J. of Geotech. Eng'g., ASCE, Vol. 106, No. GT1, pp. 35-44.

Lukas, R.G. (1986), Dynamic compaction for highway construction, Vol. 1, Design and Construction Guidelines, Federal Highway Administration, Report No. FHWA/RD-86/133, National Technical Information Service, Washington, D.C.

Massarsch, K. R. (1990), Deep soil compaction using vibratory probes, Deep Foundation Improvements: Design, Construction, and Testing, STP 1089, ASTM, Philadelphia, Pennsylvania.

Massarsch, K. R. (1992), Static and dynamic soil displacements caused by pile driving, Proc. 4th Intl. Conf. on the Application of Stress-Wave Theory to Piles, The Hague, The Netherlands.

Massarsch, K. R., and Broms, B. B. (1983), Soil compaction by vibro wing method, Proc. 8th Europ. Conf. Soil Mech. and Found. Eng'g., Helsinki.

Massarsch, K. R., and Heppel G. (1991), Deep vibratory compaction of landfill using soil resonance, Proc. Infrastructure '91, Intl. Workshop on Technology for Hongkong's Infrastructure Development, pp. 677-697.

Massarsch, K. R., and Lindberg, B. (1984), Deep compaction by vibro wing method, Proc. 8th World Conf. on Earthquake Eng'g., San Francisco.

Mayne, P.W., Jones, J.S., and Dumas, J.C. (1984), Ground response to dynamic compaction, J. of Geotech. Eng'g., ASCE, Vol. 110, No. 6, pp. 757-774.

Menard, L., and Broise, Y. (1975), Theoretical and practical aspects of dynamic consolidation, Geotechnique, Vol. 4, Session 12, pp. 509-521.

Mitchell, J.K., and Solymar, Z.V. (1984), Time-dependent strength gain in freshly deposited or densified sand, J. Geotech. Eng'g., ASCE, Vol. 110, No. 11, pp. 1559-1576.

Mori, H., Tsuchiya, H., and Ohkura, T. (1992), Dynamic consolidation practice, Proc. Geotech '92 Workshop on Applied Ground Improvement Technique, AIT, Bangkok, Thailand.

Narumi, N. (1987), Dynamic consolidation method (In Japanese), Construction Monthly, pp.69-78.

Skempton, A.W. (1986), Standard penetration test procedures and the effects in sands of overburden pressure, relative density, particle size, aging, and overconsolidation, Geotechnique, Vol. 36, pp. 425-447.

Welsh, W.P., Anderson, R.D., Barksdale, R.P., Satyapriya, C.K., Tumay M.T., and Wahls, H.E. (1987), Soil improvement - A ten year update, Proc. Symp. on Placement and Improvement of Soils, ASCE Geotech. Special Publ. No. 12, pp. 67-73.

Yamazaki, Y., Ogawa, Y., Narumi, N., and Katsumata, T. (1991), Improvement of waste products by dynamic consolidation (In Japanese), Monthly of Found. Eng'g., pp. 109-115.

Yuasa, K., Sakamoto, Y., Yamazaki, H., and Okura, T. (1987), Dynamic consolidation for blasted rockfill improvement (In Japanese), J. of Japanese Soc. of Soil Mech. and Found. Eng'g., No. 1697, pp. 39-44.

Chapter 4

PREFABRICATED VERTICAL DRAINS (PVD)

4.1 GENERAL

The consolidation settlement of soft clay subsoil creates a lot of problems in foundation and infrastructure engineering. Because of the very low clay permeability, the primary consolidation takes a long time to complete. To shorten this consolidation time, vertical drains are installed together with preloading by surcharge embankment or vacuum pressure. Vertical drains are artificially-created drainage paths which can be installed by one of several methods and which can have a variety of physical characteristics. Figure 4.1 illustrates a typical vertical drain installation for highway embankments. In this method, pore water squeezed out during the consolidation of the clay due to the hydraulic gradients created by the preloading, can flow a lot faster in the horizontal direction toward the drain and then flow freely along the drains vertically towards the permeable drainage layers. Thus, the installation of the vertical drains in the clay reduces the length of the drainage paths and, thereby, reducing the time to complete the consolidation process. Consequently, the higher horizontal permeability of the clay is also taken advantage. Therefore, the purpose of vertical drain installation is twofold. Firstly, to accelerate the consolidation process of the clay subsoil, and, secondly, to gain rapid strength increase to improve the stability of structures on weak clay foundation. Vertical drains can be classified into 3 general types, namely: sand drains, fabric encased sand drains, and prefabricated sand drains. Table 4.1 shows the general types and subtypes of vertical drains.

Applications of sand drains for improvement of soft ground in the Southeast Asian region have been reported by Tominaga et al. (1979) in Manila Bay Reclamation Area, Philippines; by Choa et al. (1979) in Changi Airport, Singapore; by Chou et al. (1980) in Taiwan; by Akagi (1981), Balasubramaniam et al. (1980), Brenner and Prebaharan (1983), Moh and Woo (1987), and Woo et al. (1989) in Bangkok, Thailand. Recent sand drain applications in Japan were reported by Takai et al. (1989) and Suzuki and Yamada (1990) in the Kansai International Airport Project and by Tanimoto et al. (1979) in Kobe, Japan.

In Southeast Asia, various applications have been recently reported with regards to prefabricated vertical drains by Choa et al. (1981), Lee et al. (1989), and Woo et al. (1988) in Singapore; by Nicholls (1981) in Indonesia; by Volders (1984) and Rahman et al. (1990) in Malaysia; and by Belloni et al. (1979) in the Philippines. In the soft Bangkok clay in Thailand, prefabricated vertical band drains have been successfully applied and tested by full scale test embankments by Bergado et al. (1988, 1990a,b, 1991).

4.2 PRELOADING

Preloading refers to the process of compressing foundation soils under applied vertical stress prior to placement of the final permanent construction load. If the temporary applied load exceeds the final loading, the amount in excess is referred to as surcharge load. When a

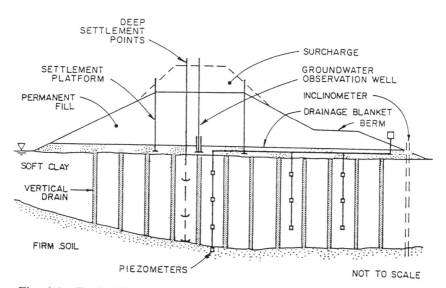

Fig. 4.1 Typical Vertical Drain Installation for a Highway Embankment (Rixner et al. 1986)

preload is rapidly applied to a saturated, soft clay deposit, the resulting settlement can be divided into three idealized components, namely: immediate, primary consolidation, and secondary consolidation. In actual condition, the settlement behavior is more complex. Figure 4.2 illustrates a general relationship of the three idealized components. The relative importance and magnitude of each type of settlement depends on many factors such as: the soil type and compressibility characteristics, the stress history, the magnitude and rate of loading, and the relationship between the area of loading and the thickness of the compressible soil. Generally, the primary consolidation settlement predominates and, for many preloading projects, is the only one considered in the preload design. Preloading techniques have been discussed in detail elsewhere (Jamiolkowski et al. 1983; Pilot, 1981). One very important key point is that the amount of preloading should provide surcharge stresses that exceed the maximum past pressure in the clay subsoil. Figure 4.3 shows the initial (σ_{vo}) and final (σ_{vf}) effective stresses under the centerline of the test embankment compared with the maximum past pressure obtained by Casagrande method (Bergado et al. 1991). The Poulos (1976) method assuming finite elastic layer with rigid base was found to be approximately 35% higher than the predictions of Janbu et al. (1956) assuming semi-infinite elastic layer of the soil mass.

4.3 SAND DRAINS

Early applications of vertical drains to accelerate consolidation of soft clay subsoils utilized sand drains. These are formed by infilling sand into a hole in the soft ground. There are two categories of installation methods, namely: displacement and non-displacement types. In the displacement type, a closed end mandrel is driven or pushed into the soft ground with resulting displacements in both vertical and lateral directions. The non-displacement type installation requires drilling the hole by means of power auger or water jets and is considered to have less disturbing effects on soft clay. Casagrande and Poulos (1969) concluded that driven sand drains are harmful in soft and sensitive clays due to the disturbance in driving the drains causing the reduction of shear strength and horizontal permeability. However, Akagi (1979) asserted that the mere installation of the sand drains alone results in the consolidation of the soft clay because of the large stresses induced during the installation. Thus, high excess pore pressure is generated (Brenner et al. 1979) and, after its subsequent dissipation, a gain in strength is achieved (Akagi, 1977a). Non-displacement sand drains by water jetting as shown in Fig. 4.4 have been tested during the 1984 SBIA Site Study (Dept. of Aviation, 1984).

4.4 CHARACTERISTICS OF PREFABRICATED DRAINS

A prefabricated vertical drain can be defined as any prefabricated material or product consisting of synthetic filter jacket surrounding a plastic core having the following characteristics: a) ability to permit porewater in the soil to seep into the drain; b) a means by which the collected porewater can be transmitted along the length of the drain.

The jacket material consists of non-woven polyester or polypropylene geotextiles or synthetic paper that function as physical barrier separating the flow channel from the surrounding soft clay soils and a filter to limit the passage of fine particles into the core to

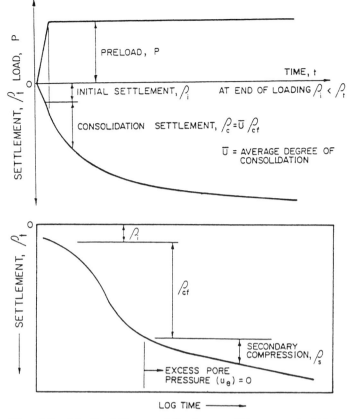

Fig. 4.2 Idealized Types of Settlement (Rixner et al. 1986)

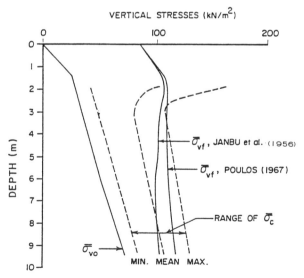

Fig. 4.3 Initial and Final Stresses Under the Center of the Test Embankment with Range of Preconsolidation Pressure (Bergado et al. 1991)

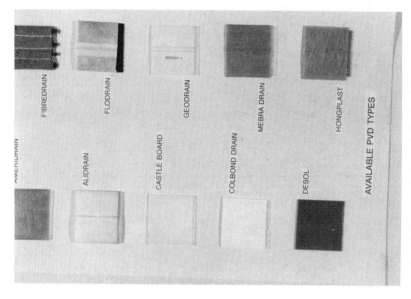

Fig. 4.5 Prefabricated Vertical Band Drains

Fig. 4.4 Non-Displacement Sand Drains Using Water Jets

Table 4.1 Types of Vertical Drains (Rixner, 1986)

General Types	Sub-Types	Remarks
Sand Drains	Closed end mandrel	Maximum displacement
	Screw type auger	Limited experience
	Continuous flight hollow stem auger	Limited displacement
	Internal jetting	Difficult to control
	Rotary jet	Can be non-displacement
	Dutch jet-bailer	Can be non-displacement
Fabric Encased Sand Drain	Sandwick, Pack Drain Fabridrain	Full displacement of relative small volume
Prefabricated Vertical Drain	Cardboard drain	Full displacement of small volume
	Fabric covered	Full displacement of small volume
	Plastic drain without jacket	Full displacement of small volume

prevent clogging. The plastic core serves two vital functions, namely: to support the filter jacket and to provide longitudinal flow paths along the drain even at large lateral pressures. Typical prefabricated drain products are shown in Fig. 4.5. Some details of various drain cores are also shown in Fig. 4.6. The configuration of different types of prefabricated vertical drains (PVD) are illustrated in Fig. 4.7. The PVD core can be classified into 3 main categories, namely: grooved core, studded core, and filament core.

4.5 CONSOLIDATION WITH VERTICAL DRAINS

Barron (1948) presented the first exhaustive solution to the problem of consolidation of a soil cylinder containing a central sand drain. His theory was based on the simplifying assumptions of one-dimensional consolidation theory (Terzaghi, 1943). Barron's theory enable one to solve the problem of consolidation under two conditions, namely: (i) free vertical strain assuming that the vertical surface stress remains constant and the surface displacements are non-uniform during the consolidation process; (ii) equal vertical strain assuming that the vertical

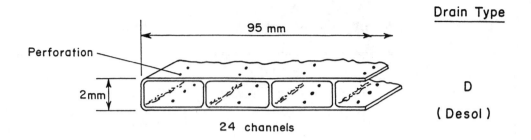
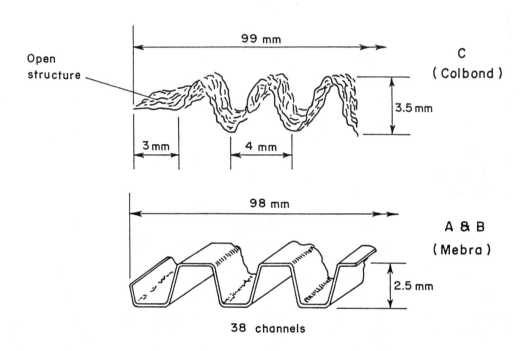
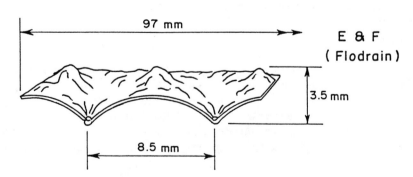

Fig. 4.6　Geometrical Shape of Various Drain Cores

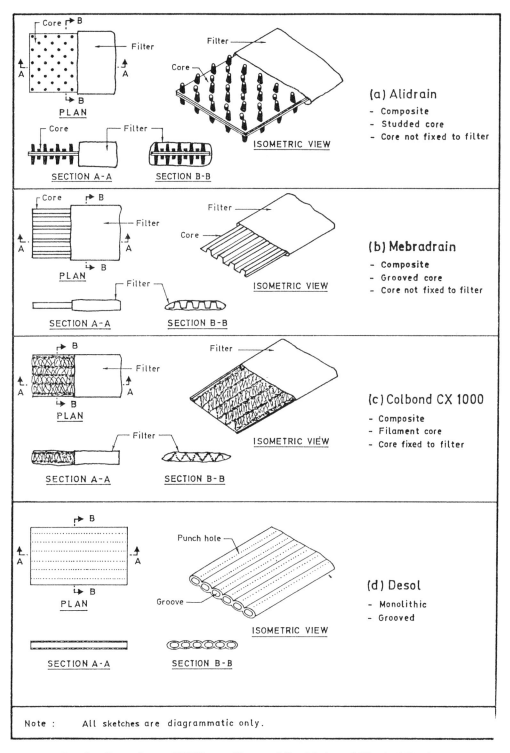

Fig. 4.7 Configurations of Different Types of Prefabricated Vertical Drain (PVD)

surface stress is non-uniform.

In the case of equal strain, the differential equation governing the consolidation process is given as:

$$\frac{\partial U}{\partial t} = C_h \left[\left(\frac{\partial^2 U}{\partial r^2} \right) + \frac{1}{r} \left(\frac{\partial u}{\partial r} \right) \right] \tag{4.1}$$

where u is the average excess pore pressure at any point and at any given time; r is the radial distance of the considered point from the center of the drained soil cylinder; t is the time after an instantaneous increase of the total vertical stress, and C_h is the horizontal coefficient of consolidation. For the case of radial drainage only, the solution of Barron (1948) under ideal conditions (no smear and no well resistance) is as follows:

$$U_h = 1 - \exp\left[\frac{-8T_h}{F(n)}\right] \tag{4.2}$$

where:

$$T_h = \frac{C_h t}{D_e} \tag{4.3}$$

$$F(\ ) = \left[\frac{n^2}{(1-n)}\right]\left[n(n) - \frac{3}{4} + \frac{1}{n^2}\right] \tag{4.4}$$

and D_e is the diameter of the equivalent soil cylinder, d_w is the equivalent diameter of the drain, and n ($n = D_e/d_w$) is the spacing ratio.

Hansbo (1979) modified the equations developed by Barron (1948) for prefabricated drain applications. The modifications dealt mainly with simplifying assumptions due to the physical dimensions, characteristics of the prefabricated drains, and effect of PVD installation. The modified general expression for average degree of consolidation is given as:

$$U_h = 1 - \exp\left(\frac{-8T_h}{F}\right) \tag{4.5}$$

and

$$F = F(n) + F_s + F_r \qquad (4.6)$$

where F is the factor which expresses the additive effect due to the spacing of the drains, F(n); smear effect, F_s; and well-resistance, F_r. For typical values of the spacing ratio, n, of 20 or more, the spacing factor simplifies to:

$$F(n) = \ln\left[\frac{D_e}{d_w}\right] - \frac{3}{4} \qquad (4.7)$$

To account for the effects of soil disturbance during installation, a zone of disturbance with a reduced permeability is assumed around the vicinity of the drain, as shown in Fig. 4.8. The smear effect factor is given as:

$$F_s = \left[\left[\frac{k_h}{k_s}\right] - 1\right] \ln\left[\frac{d_s}{d_w}\right] \qquad (4.8)$$

where d_s is the diameter of the disturbed zone around the drain; and k_h is the coefficient of permeability in the horizontal direction in the disturbed zone.

Since the prefabricated vertical drains have limited discharge capacities, Hansbo (1979) developed a drain resistance factor, F_r, assuming that Darcy's law can be applied for flow along the vertical axis of the drain. The well-resistance factor is given as:

$$F_r = \pi z (L-z) \frac{k_h}{q_w} \qquad (4.9)$$

where z is the distance from the drainage end of the drain; L is twice the length of the drain when drainage occurs at one end only; L is equal to the length of the drain when drainage occurs at both ends; k_h is the coefficient of permeability in the horizontal direction in the undisturbed soil; and q_w is the discharge capacity of the drain at hydraulic gradient of 1.

The schematic diagram of PVD with drain resistance and soil disturbance is shown in Fig. 4.8. Incorporating the effects of smear and well-resistance, the time, t, to obtain a given degree of consolidation at an assumed spacing of PVD, is given as follows:

$$t = \left[\frac{D_e^2}{8C_h}\right](F(n) + F_s + F_r) \ln\left[\frac{1}{1-U_h}\right] \qquad (4.10)$$

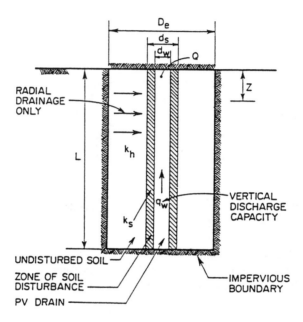

Fig. 4.8 Schematic of PVD with Drain Resistance and Soil Disturbance (Rixner et al. 1986)

For convenience on the part of users in designing vertical drain scheme, a design graph devised by Bergado et al. (1993a) is given in Fig. 4.9. This is the first design graph that incorporates both the effects of smear and well-resistance.

4.6 DRAIN PROPERTIES

The theory of consolidation with radial drainage assumes that the soil is drained by vertical drain with circular cross section. The equivalent diameter of a band-shaped drain is defined as the diameter of a circular drain which has the same theoretical radial drainage performance as the band-shaped drain. Subsequent finite element studies performed by Rixner et al. (1986) and supported by Hansbo (1987) suggested that the equivalent diameter preferable for use in practice can be obtained as:

$$d_w = \frac{(a+b)}{2} \quad (4.11)$$

The relative sizes of these equivalent diameters are compared to the band shaped cross-section of the prefabricated drain by Rixner et al. (1986).

The discharge capacity of prefabricated drains is required to analyze the drain resistance factor and is usually obtained from published results reported by manufacturers. Rixner et al. (1986) reported results of vertical discharge capacity tests and those obtained by others, as shown in Fig. 4.10. The results demonstrate the major influence of confining pressure.

4.7 DRAIN INFLUENCE ZONE

The time to achieve a given percent consolidation is a function of the square of the equivalent diameter of soil cylinder, D_e. This variable is controllable since it is a function of drain spacing and pattern. Vertical drains are usually installed in square or triangular patterns as shown in Fig. 4.11. The spacing between drains establishes D_e through the following relationships:

$D_e = 1.13S$ (square pattern)   **(4.12)**
$D_e = 1.05S$ (triangular pattern)   **(4.13)**

The square pattern has the advantage for easier layout and control. A square pattern is usually preferred. However, the triangular pattern provides more uniform consolidation between drains.

4.8 WELL-RESISTANCE

Hansbo (1979, 1981) presented, for equal-strain conditions, a closed-form solution which allows for ready computation of the effects of well-resistance on drain performance. The finite drain permeability (well-resistance) was considered by imposing on the continuity equation of

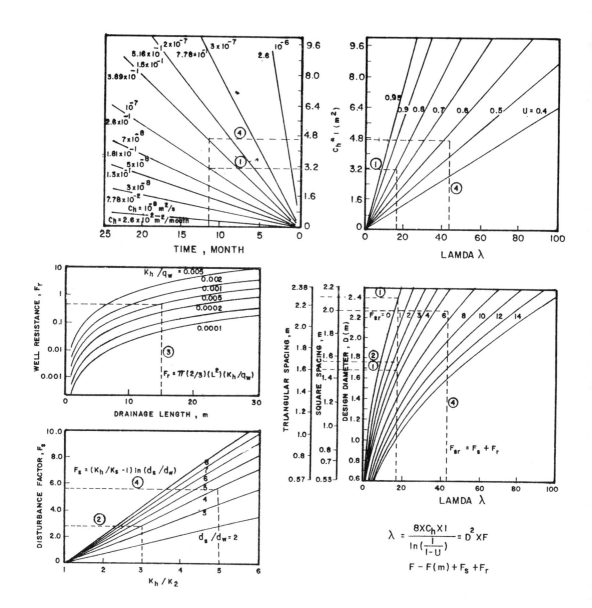

Fig. 4.9 Configuration Chart for Prefabricated Vertical Drain (PVD)

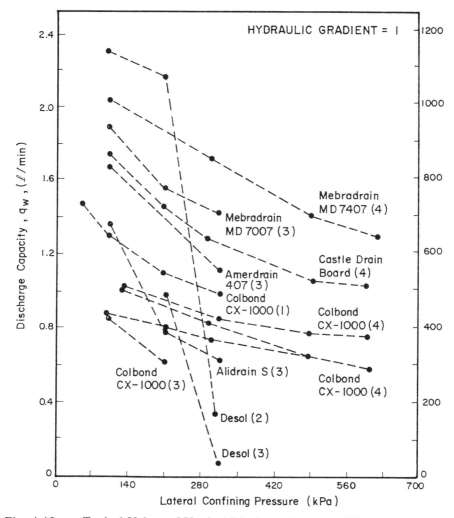

Fig. 4.10   Typical Values of Vertical Discharge Capacity (Rixner et al. 1986)

flow toward the drain. In this assumption, the flow rate in the considered section of the drain is equal to the maximum flow rate which can be discharged through the drain.

The discharge capacity of sand drains depends on the permeability of the sand. The sand used for the drains must be clean with good drainage and filtration characteristics. The discharge capacity of band drains varies considerably depending on the make of the drain and decreases with increasing lateral pressure (Fig. 4.10). This is caused by either the squeezing in of the filter sleeve into the core channels reducing the cross-sectional area of the channels, or, for drains without a filter sleeve, the channels themselves are squeezed together. Another important factor is the folding of the drain when subjected to large vertical strains. In this case the channels of flow would be reduced, thus reducing the discharge capacity. The sedimentation of small particles in the flow channels may also decrease the discharge capacity.

The introduction of the well-resistance concept affects the value of the degree of consolidation, U_h, which is no longer constant with depth (Fig. 4.12). Taking well-resistance into consideration, the rate of radial consolidation is controlled not only by C_h and D_e but also by the ratio q_w/k_h (Fig. 4.13). This factor may play a very important role when prefabricated band drains of great lengths are used with typical values of q_w/k_h less than 500 m^2 where the time necessary to achieve a specific degree of consolidation is increased (Jamiolkowski et al. 1983). The influence of well-resistance on the consolidation rate increases as the drain length increases. This is illustrated in Fig. 4.14 for a typical band-shaped drain (q_w/k_h = 400 m^2).

4.9 SMEAR EFFECTS AND DISTURBANCES

Although there are numerous variations in installation equipment for vertical drains, most of the equipment has fairly common features, some of which can directly influence the drain performance. The installation rigs are usually track-mounted boom cranes. The mandrel protects the drain during installation and creates the space for the drain by displacing the soil during the penetration. The mandrel is penetrated into the subsoil using either static or vibratory force. The drain installation results in shear strains and displacement of the soil surrounding the drain. An example of soil movements produced in Bangkok clay as a result of the installation of displacement sand drains is given in Fig. 4.15. The shearing is accompanied by increases in total stress and pore pressure.

The installation results in disturbance to the soil around the drain. The disturbance is most dependent on the mandrel size and shape, soil macrofabric, and installation procedure. The mandrel cross-section should be minimized, while at the same time, adequate stiffness of the mandrel is required. Bergado et al. (1991), from a full scale test embankment performance, obtained faster settlement rate in the small mandrel area than in the large mandrel area indicating lesser smeared zone in the former than the latter. For design purposes, it has been evaluated by Jamiolkowski et al. (1981) that the diameter of disturbed zone, d_s, can be related to the cross-sectional dimension of the mandrel as follows:

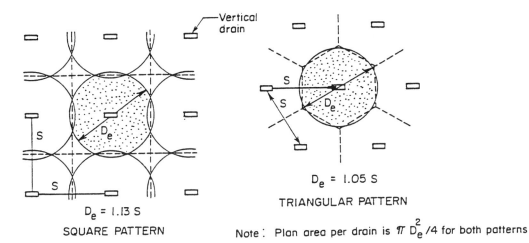

Fig. 4.11 Relationship of Drain Spacing (S) to Drain Influence Zone (D) (Rixner et al. 1986)

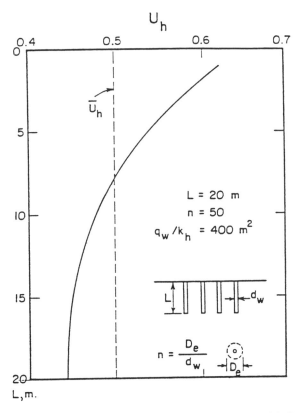

Fig. 4.12 Example of Variation of Degree of Consolidation with Depth for Drains with Well Resistance (Jamiolkowski et al. 1983)

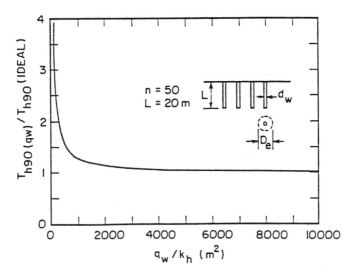

Fig. 4.13 Influence of Finite Drain Permeability on Consolidation Rate (Jamiolkowski et al. 1983)

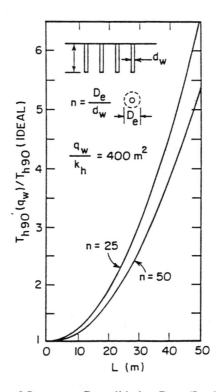

Fig. 4.14 Influence of Smear on Consolidation Rate (Jamiolkowski et al. 1983)

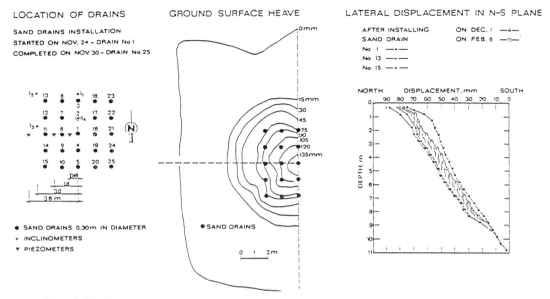

Fig. 4.15 Ground Movements During and After Installation of Driven Sand Drains in Soft Bangkok Clay (Akagi, 1981)

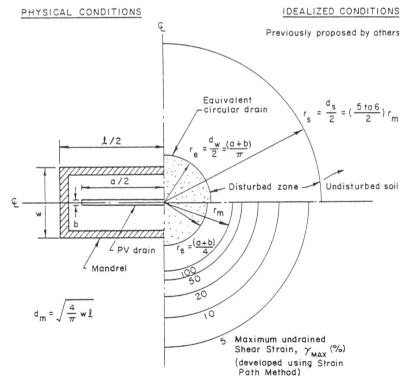

Fig. 4.16 Approximation of Disturbed Zone Around the Mandrel (Rixner et al. 1986)

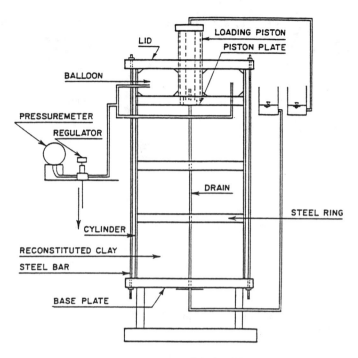

Fig. 4.17   Schematic of Large-Scale Consolidation Test Apparatus (Bergado et al. 1991)

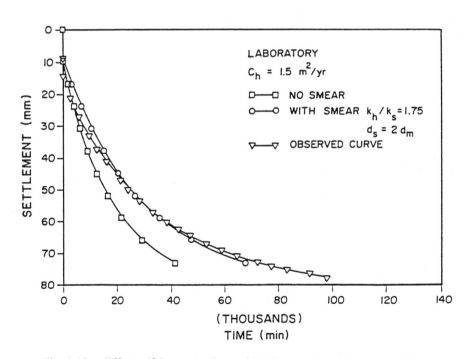

Fig. 4.18   Effects of Smear on Rate of Settlement (Bergado et al. 1991)

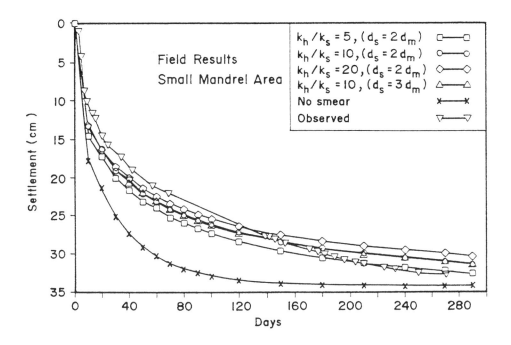

Fig. 4.19 Observed and Predicted Time-Settlement Relationship from Full Scale Field Test (Bergado et al. 1993)

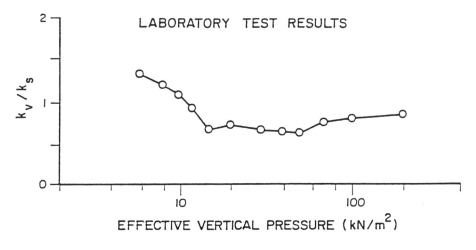

Fig. 4.20 K_v/k_s Values with Effective Pressure (Bergado et al. 1991)

$$d_s = \frac{(5-6)d_m}{2} \qquad (4.14)$$

where d_m is the diameter of a circle with an area equal to the cross-sectional area of the mandrel. At this diameter, the theoretical shear strain is approximately 5% as shown in Fig. 4.16. Hansbo (1987) recommended the following expression based on the results of Holtz and Holms (1973) and Akagi (1979):

$$d_s = 2d_m \qquad (4.15)$$

This relationship has been verified in the reconstituted soft Bangkok clay by Bergado et al. (1991) using a specially designed laboratory drain testing apparatus (Fig. 4.17) as plotted in Fig. 4.18. Thus, the influence of smear increases with increasing drain diameter for sand drain or mandrel diameter for prefabricated drains (Hansbo, 1981). The time-settlement relationships obtained from full scale field test embankment (Bergado et al. 1991) is shown in Fig. 4.19 for small mandrel area together with the settlement prediction. The performance of PVD is well predicted with smear effect taken into consideration using $k_h/k_v = 10$ and $d_s = 2d_m$ (Bergado et al. 1993b).

4.10   RATIO OF HORIZONTAL TO VERTICAL PERMEABILITY

For soils with pronounced macrofabric, the ratio of k_h/k_v can be very high, possibly up to 10. This beneficial effect of soil stratification (greater horizontal permeability) can be reduced or completely eliminated in the smeared zone. Thus, the permeability of this zone, k_s, can be put equal to k_v. Bergado et al. (1991) performed oedometer tests on samples taken shortly after drain installation on reconstituted soft Bangkok clay at different distances from the drain. The k_h/k_v values determined by square root time method is plotted against effective pressure in Fig. 4.20. The results closely agreed with the proposal of Hansbo (1987) wherein k_s is equal to k_v in the smeared zone.

Recently, Onoue et al. (1991) proposed 2 zones of disturbances (Zones II and III) as shown in Fig. 4.21 based on the measured void ratio data. Zone II is partially-remolded wherein the drain installation caused a decrease in void ratio, and hence permeability. Zone III is a fully remolded zone.

4.11   COEFFICIENT OF HORIZONTAL CONSOLIDATION

The coefficient of radial consolidation can be evaluated from C_v values by means of the approximate relationship:

$$C_h = \left(\frac{k_h}{k_v}\right) C_v \qquad (4.16)$$

Table 4.2 Range of Possible Field Values of the Ratio k_h/k_v for Soft Clays (Rixner et al. 1986)

Nature of Clay	k_h/k_v
No or slightly developed macrofabric, essentially homogeneous deposits	1 to 1.5
From fairly well to well developed macrofabric, e.g. sedimentary clays with discontinuous lenses and layers of more permeable material	2 to 4
Varved clays and other deposits containing embedded and more or less continuous permeable layers	3 to 15

where C_v is the coefficient of vertical consolidation. The ratio of k_h/k_v can be evaluated by back-analysis. Bergado et al. (1992) obtained k_h/k_v = 4 to 10 and $C_h(field)/C_h(lab)$ = 4 for the soft Bangkok clay. An average ratio of $C_v(field)/C_v(lab)$ = 26 was also found (Bergado et al. 1990a). A rough estimate of the in-situ anisotropy of the permeability of clays (k_h/k_v) can be made on the basis of the data given by Jamiolkowski et al. (1983), as tabulated in Table 4.2. In-situ piezometer probes and analysis of pore pressure dissipation curves can also be used to evaluate C_h and k_h (Jamiolkowski et al. 1985). In-situ determination of k_h by self-boring permeameters can also be used with laboratory m_v values to calculate C_h using the relationship (Jamiolkowski et al. 1983):

$$C_h = \frac{k_h}{(m_w \gamma_w)} \quad (4.17)$$

where γ_w is the unit weight of water and m_v is the coefficient of volume change. Albakri et al. (1990) obtained c_h values in the field by piezocone tests and compared them with the corresponding laboratory values in Fig. 4.22.

4.12 PARAMETER EFFECTS ON CONSOLIDATION TIME

Taking into consideration the smear effects, the time, t, corresponding to a given degree of consolidation is given as:

$$t = \frac{De}{8C_h}\left[\ln\left(\frac{D_e}{d_w}\right) - \frac{3}{4}\right] + \left(\frac{K_h}{K_s}-1\right)\ln\left(\frac{d_s}{d_w}\right)\ln\left(\frac{1}{1-U_h}\right) \quad (4.18)$$

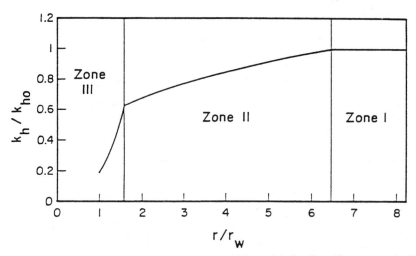

Fig. 4.21 Suggested Variation of Permeability with Radius (Onoue et al. 1991)

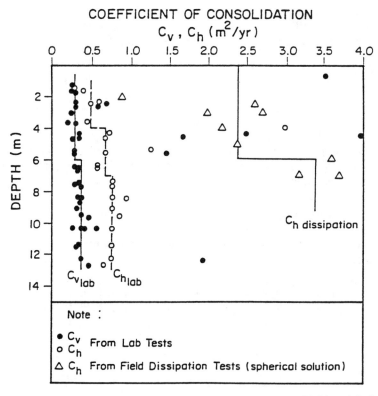

Fig. 4.22 Variation of Coefficient of Consolidation from Field and Laboratory Tests (Albakri et al. 1990)

The relative effects of key parameters of this equation is shown in Fig. 4.23 for a given case (Rixner et al. 1986). As shown in this figure, the greatest potential effect on consolidation time, t, is due to the variations of C_h and D_e. The value of C_h which can easily vary by a factor of 10, has the most dominant factor on t. D_e can vary by a factor of about 2 to 3. It reflects the influence of drain spacing. The effects of the properties of the disturbed zone (k_s and d_s) can also be significant. The equivalent diameter of the drain, d_w, has only a minimal influence.

4.13 RATE OF CONSOLIDATION

The principal objective of soil precompression with vertical drains is to achieve the desired degree of consolidation within a specified period of time. With vertical drains, the overall degree of consolidation, U, is the result of the combined effects of horizontal (radial) and vertical drainage. The combined effects is given by Carillo (1942) as:

$$U = 1 - (1-U_h)(1-U_v) \tag{4.19}$$

where U_h is the average degree of consolidation due to horizontal drainage, and U_v is the corresponding value due to vertical drainage. A comparison of one-dimensional consolidation due to vertical drainage, radial drainage, and combined drainage is shown in Fig. 4.24.

4.14 SELECTION OF DRAIN TYPE

The primary concerns in the selection of the type of prefabricated drain for a particular project include: equivalent diameter; discharge capacity; filter jacket characteristics and permeability; and material strength, flexibility, and durability.

For common prefabricated drains, d_w ranges from 50 to 75 mm. It is generally inappropriate to use a drain with an equivalent diameter of less than 50 mm (Rixner et al. 1986). Typical values of the discharge capacity, q_w, are given in Fig. 4.10 as a function of lateral confining pressures. Generally, the selected drain should have a vertical discharge capacity of at least 100 m^3/yr measured under gradient of one while confined to maximum in-situ effective horizontal stress.

Large permeability of the filter is desired but at the same time small particles should be prevented to pass through the filter. The basic requirement of the permeability criteria is that the geotextile filter must be and must remain more permeable than the adjacent soil (Holtz, 1987). For critical applications and severe conditions:

$$k_{geotextile} \geq 10\ k_{soil} \tag{4.20}$$

For less critical and less severe situations:

$$k_{geotextile} = k_{soil} \tag{4.21}$$

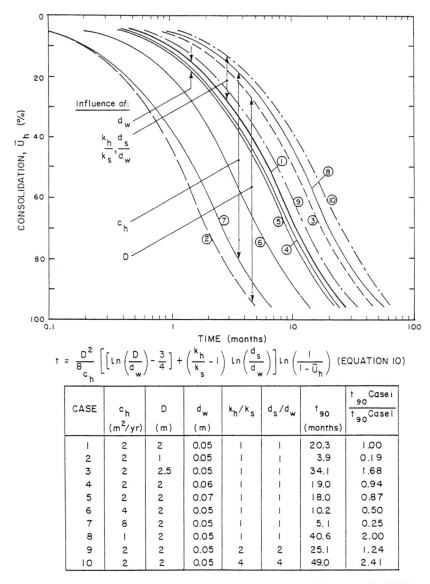

Fig. 4.23 Example of Parameter Effects on t_{90} (Rixner et al. 1986)

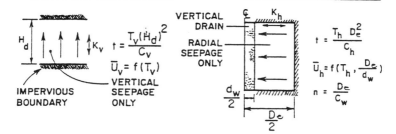
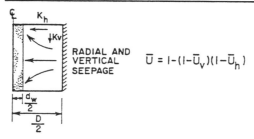
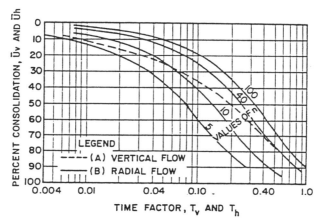

AVERAGE CONSOLIDATION RATES
(A) FOR VERTICAL FLOW IN A CLAY STRATUM OF THICKNESS H DRAINED ON BOTH UPPER AND LOWER SURFACES
(B) FOR RADIAL FLOW TO AXIAL DRAIN WELLS IN CLAY CYLINDERS HAVING VARIOUS VALUES OF n
(AFTER BARRON, 1948)

Fig. 4.24 Consolidation Due to Vertical and Radial Drainage (Rixner et al. 1986)

Almost all geotextile filters have sufficient permeability.

The retaining capacity and the resistance of a filter are mainly dependent on the particle size distribution of the soil and the pore pressure distribution of the filter. As illustrated in Fig. 4.25, three principal filter mechanisms can be distinguished, namely: cake filtration, blocking filtration, and deep filtration (Vreeken et al. 1983). Cake filtration takes place if the particles are larger than the filter pores. Blocking filtration occurs if the particles have the same diameter as the filter pores. Deep filtration happens if the particles are smaller than the filter pores and the particles adhere on the filter. Figure 4.26 shows typical pressure development for the different filtration mechanisms. The filter resistance is strongly influenced by small soil particles that may be concentrated in or against the filter by the porewater flow. Small particles should pass the filter freely. However, not too many particles should pass, because of risk of sedimentation and subsequent decrease of vertical discharge capacity.

The configurations of different types of prefabricated vertical drains (PVD) have been presented earlier in Fig. 4.7. The Asian Institute of Technology, in connection with the Second Bangkok International Airport (SBIA) Project, has conducted discharge capacity tests on various types of drain based on ASTM procedure. Figure 4.27 shows the drain discharge capacity plotted with lateral pressure at hydraulic gradient equal to one. The results in Fig. 4.27 implied that PVD with grooved cores yields higher discharge capacity.

Various laboratory tests were performed to determine the PVD performance at a given condition. Figure 4.28 exhibits the deformed shape of PVD for longitudinal discharge capacity tests. The comparison of the four types of drains at a given condition is illustrated in Table 4.3. The types of drains used in the tests are as follows (refer to Fig. 4.7):

1) Drains with separate core and filter

    - Grooved core (A)
    - Studded core (B)
    - Filament core (C)

2) Drain with filter fixed to the core

    - Grooved core (D)

4.15 FILTER CRITERIA

The first rational approach to deal with filter problems was the work of Terzaghi (1929). While there are slightly differences in opinion concerning quantitative criteria for satisfactory filters, the following rules are widely used:

1) D_{15} size of filter should be at least 5 times larger than the D_{15} of the soil being protected.

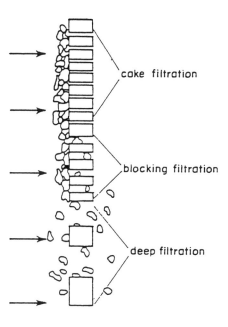

Fig. 4.25 The Filtration Mechanism (Vreeken et al. 1983)

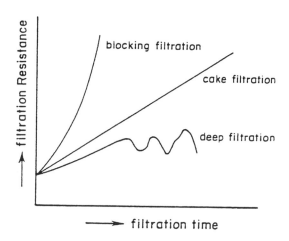

Fig. 4.26 Filtration Resistances During Filtration (Vreeken et al. 1983)

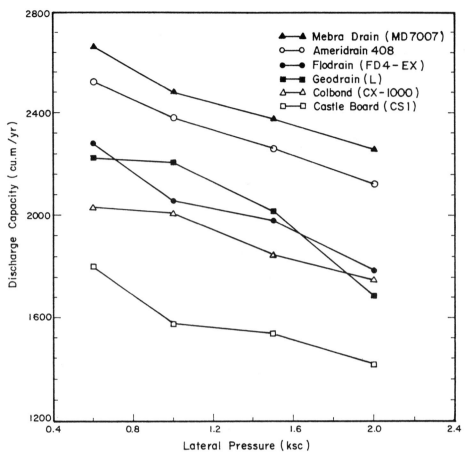

Fig. 4.27 Variation of Discharge Capacity with Lateral Pressure Based on ASTM Procedure (Hydraulic Gradient = 1.0)

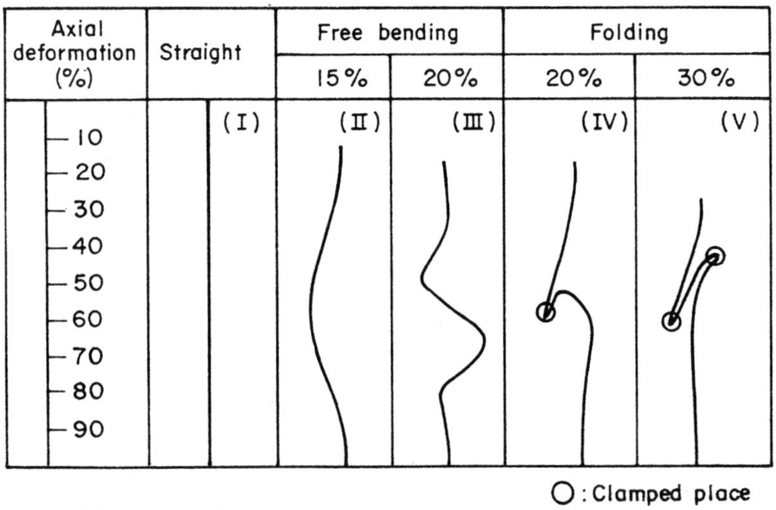

Fig. 4.28 Deformed Shapes of PVD for Longitudinal Permeability Test

Table 4.3 Summary of Laboratory Tests on Prefabricated Vertical Drains

Type of Test	Comparison of Results
A. Discharge Capacity	
- Straight condition	A > D > B and C
- 20% compression and free bending	A and D > B and C
- 20% compression and one clamped	D > A > B and C
- Twisted at 45°	D > A > B and C
B. Suitability for Safe Installation	
- Grab tensile strength	D > A and C > B
- Trapezoidal tear strength	A and D > B and C
- Puncture resistance	D > A and C > B
- Burst strength	A, C and D > B

2) The D_{15} size of filter should not be larger than 5 times the D_{85} of the protected soil.

These rules are too conservative for clays which have inherent resistance to piping because of their cohesion. As suggested by Cedergreen (1972), the piping ratio (D_{15}/D_{85}) can reach as high as 10 or even higher. Recently, Sherard and Dunnigan (1989) suggested a limiting value of 9 for the piping ratio.

4.16 GEOTEXTILE FILTER CRITERION

The filtration criterion depends on the critical nature of the project and on the severity of hydraulic loading conditions. For less critical applications, the geotextile filter with the largest opening size based on soil retention criterion should be specified. To evaluate the retention ability of the drains accurately, several factors have to be considered such as: electro-chemical forces of the geotextile, chemical properties of fibrous structure compound, and soil composition (Kellner et al. 1983). However, the retaining ability of the geotextile is too complicated to be determined. Hence, the rules of thumb are as follows:

$$\frac{O_{90}}{O_{50}} < 1.7 \text{ to } 3 \qquad \text{(Schober and Teindl, 1979)} \qquad (4.22)$$

$$\frac{O_{95}}{D_{85}} < 2 \text{ to } 3 \qquad \text{(Carroll, 1983)} \qquad (4.23)$$

$$\frac{O_{90}}{D_{85}} < 1.3 \text{ to } 1.8 \qquad \text{(Chen and Chen, 1986)} \qquad (4.24)$$

$$\frac{O_{50}}{D_{50}} < 10 \text{ to } 12 \qquad \text{(Chen and Chen, 1986)} \qquad (4.25)$$

The ratio O_{50}/D_{50} ensures that seepage forces within the filter are reasonably small. The reason to choose the upper bound of O_{50}/D_{50} ratio to be 12 is to prevent fines to enter the core. O_{90} was chosen because it can be measured accurately by mercury intrusion method (Chen and Chen, 1986). Kamon (1983) recognized that clay particles are aggregated with average diameter of 50 to 60 microns (0.05 to 0.06 mm). If the average pore size of the filter is less than 50 microns, the clogging of the filter will not take place. Recently, Bergado (1992) conducted laboratory experiments for prefabricated band drain. As preliminary guidelines for soft Bangkok clay, the following criteria are recommended:

$$\frac{O_{90}}{D_{85}} < 2 \text{ to } 3 \qquad (4.26)$$

$$\frac{O_{50}}{D_{50}} < 18 \text{ to } 24 \qquad (4.27)$$

The filter criteria in connection with a prefabricated drain with a PK-124 filter jacket have been satisfied as shown in Fig. 4.29.

4.17 2-D MODELLING OF VERTICAL DRAIN IMPROVED GROUND

Analytical methods developed so far for consolidation analysis of ground improved by vertical drains (PVD) and granular piles (GP) employ the concept of a "unit cell," wherein a circular domain of influence of a single drain or granular pile is analyzed. The assumption is that each unit cell works independently and all strains within the soil mass occur in the vertical direction only (Barron, 1948; Hansbo, 1981). This assumption is strictly valid only for an infinitely wide loaded area and, thus, faces other serious restrictions concerning boundary conditions, taking into account plastic flow in the soil, non-homogenous behavior of subsoils etc. In fact, the assumption of no lateral displacements is not reliable particularly for the case of embankment on soft ground, wherein lateral deformation can occur being one of the important signals indicating the instability of the ground. Therefore, it seems more reasonable that the GP or PVD-ground system should be analyzed as a whole using numerical treatment. The GP or PVD-improved ground is a 3-D problem. However, finite element calculations with discrete modelling of granular piles or vertical drains in 3-D turn out to be very complicated for routine analysis. Thus, 2-D analysis seems more practical.

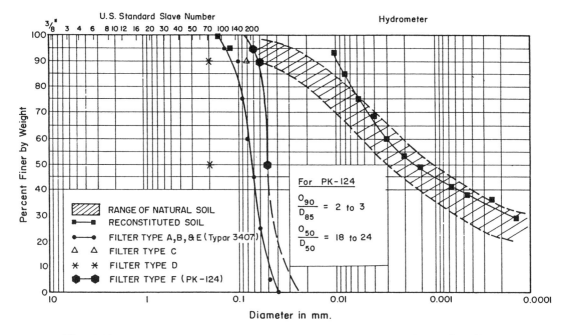

Fig. 4.29 Grain Size Distribution of Original Clay, Slurry and Geotextile Filters

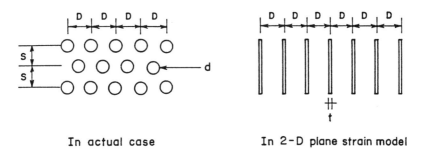

Fig. 4.30 Actual Case and 2-D Plane Model

An "equivalent material" model of improved ground was introduced by Schweiger and Pande (1988) for 2-D plane strain analysis of embankment on soft clay stabilized with stone columns. In this model, the improved zone was treated as an "equivalent material" with the "equivalent parameters" of strength and stiffness. The flow parameters were not considered. Then, only undrained or fully drained cases could be analyzed. Finite element method using elasto-plastic Cam clay model has been used by Asaoka et al. (1991) for analysis of undrained failure of embankment improved by sand compaction piles, wherein the sand compaction piles were transformed into a number of sand walls (see Fig. 4.30). For consolidation analysis, it is necessary to convert the spatial flow in actual case into the laminar one in 2-D plane strain model. One such converted permeability was introduced by Shinsha et al. (1982) for vertical sand drains. Shinsha et al. (1982) based on his assumption that the required time for 50% degree of consolidation in both schemes (in actual case and in 2-D model) are equal. Then, the simple expression was obtained as shown below:

$$k_m / k = (L/D_e)^2 T_{h50} / T_{r50} \qquad (4.28)$$

where L is half the distance between two sand walls in 2-D model, $T_{h50} = 0.197$ is dimensionless time factor at 50% consolidation of laminar flow and T_{r50} is corresponding radial flow in actual case, and k_m and k are horizontal permeability coefficients of 2-D model and actual case, respectively. Comparisons of 3-D with 2-D analysis using Shinsha et al. (1982) method, Cheung et al. (1991) obtained the higher pore pressures in the 2-D model and concluded that this method should be used with caution.

Bergado and Long (1994) have proposed another approach for 2-D modelling of PVD and GP, as shown in Fig. 4.30 and Fig. 4.31. The PVD and GP are transformed into continuous walls with the same spacing as that of the actual case. The following determinations of model parameters are in general form for both vertical drains and granular piles. If a_s is the ratio of granular piles area to the total improved area and the same value of a_s is used for both schemes, then:

$$a_s = tS/DS = t/D \text{ or } t = a_s D \qquad (4.29)$$

where t is the thickness of the walls in 2-D model while D and S are the row spacing and pile spacing of actual case, respectively.

The converted permeability including smear effect is introduced, based on condition of the equal discharge rate in both schemes, with the assumption that the coefficient of permeability is independent on state of seepage flow. In this approach, the permeability of the soil between drain walls in 2-D model is adjusted to make the same discharge between the actual case and 2-D model. The vertical flow in both schemes can be assumed to be the same. Then, only the horizontal flow needs to be converted. Taking the same head boundaries at A and B in Fig. 4.30, then from the condition of the same discharge between both schemes in steady state flows one can get Eq. (4.30), wherein, $n = D/d$, $\alpha = D_e/D$; $S = D$ and $\alpha = 1.05$ for square pattern; $S = 0.866D$ and $\alpha = 1.13$ for triangle pattern.

$$k_m = \frac{\pi(1-a_s)D}{2S\log_e(\alpha n)} k \qquad (4.30)$$

If the smear effect is taken into account, the value of k in Eq. (4.30) can be replaced by k_e, where k_e is the equivalent permeability including smear effects. The value of k_e can be obtained from the condition of the same discharge rate of steady state flows with the same head boundaries (Fig. 4.31) as given below:

$$k_e = \frac{k\log_e(\alpha n)}{\log_e\left[\frac{\alpha D}{d_s}\right] + R_s\log_e\left[\frac{d_s}{d_w}\right]} \qquad (4.31)$$

where $R_s = k/k_s$, k and k_s are horizontal permeabilities in undisturbed and smeared zone, respectively; α, n, D are previously defined and the others are shown in Fig. 4.30. Replacing k in Eq. (4.30) by value of k_e from Eq. (4.31) one can get the value of horizontal converted permeability for the 2-D model, k_m, in terms of the horizontal permeability of unimproved soil, k, as shown below:

$$k_m = \frac{\pi D(1-a_s)k}{2S\left[\log_e\left[\frac{\alpha D}{d_s}\right] + R_s\log_e\left[\frac{d_s}{d_w}\right]\right]} \qquad (4.32)$$

4.18 PREDICTION EQUATION FOR UNDRAINED SHEAR STRENGTH

The increases in undrained shear strength were estimated by SHANSEP (Ladd, 1991) technique as follows:

$$\left(\frac{S_u}{\overline{\sigma}_{vo}}\right)_{OC} = \left(\frac{S_u}{\overline{\sigma}_{vo}}\right)_{NC} OCR^m \qquad (4.33)$$

For Bangkok soft clay:

$$\left(\frac{S_u}{\overline{\sigma}_{vo}}\right)_{NC} = 0.22 \qquad (4.34)$$

$$m = 0.8 \tag{4.35}$$

Therefore, from Eq. (4.33)

$$\left[\frac{S_u}{\overline{\sigma}_{vo}}\right]_{OC} = 0.22 OCR^{0.8} \tag{4.36}$$

In the case of $\sigma_v = \sigma_{vo} + \delta\sigma_v < \sigma_p$, it can be written as:

$$\left[\frac{S_u}{S_{uo}}\right] = \left[\frac{\overline{\sigma}_v}{\overline{\sigma}_{vo}}\right]^{0.2} \tag{4.37}$$

Similarly, for the case of $\sigma_{vo} < \sigma_p < \sigma_v$, it can be expressed as:

$$\left[\frac{S_u}{S_{uo}}\right] = \frac{0.22(\overline{\sigma}_v - \overline{\sigma}_p)}{S_{uo}} + \left[\frac{\overline{\sigma}_p}{\overline{\sigma}_{vo}}\right]^{0.2} \tag{4.38}$$

where:
- S_u = predicted shear strength
- S_{uo} = initial shear strength
- σ_v = effective vertical stress at predicted time
- σ_{vo} = initial effective vertical stress
- σ_p = maximum past pressure
- OCR = overconsolidation ratio

4.19 CASE RECORDS OF SAND DRAINS ON SOFT BANGKOK CLAY

4.19.1 Test Embankment at AIT Campus

According to Akagi (1977a), significant changes in strength and compressibility of soft clay could result due to the installation of displacement sand drains. To verify such assertion, a 2.0 m high full scale test embankment was constructed on soft Bangkok clay foundation improved by 25 closed-end, mandrel-driven sand drains. The sand drains were spaced at 1.2 m in square pattern. The mandrel consisted of 300 mm inside diameter with 7 mm wall thickness, with an expendable bottom disc. Due to the upward forces in the surrounding clay during the withdrawal of the mandrel, the completed sand columns averaged only 5.70 m in length instead of 8.0 m. The test site was located at the AIT campus, 40 km north of Bangkok, Thailand. The subsoil consists of 2 m thick weathered clay underlain by 5 m thick soft clay layer with seams of fine sand and silts. At a depth of 7 m, soft sandy silt and loose sand were encountered followed by a stiff clay layer at 8 m to 9 m depth. The soft clay layer was highly plastic and lightly overconsolidated. The undrained shear strength ranged from 10 to 30 kPa with sensitivity from 3 to 8. The liquid limit is 95 and the plastic limit 30 with average natural water content of 77%. The groundwater table was located at 1 m depth. About 8.5 months after sand drain installation, a 2 m high embankment with 12 x 12 m base and 6 x 6 m top was constructed over sand drained area (Area S). In untreated adjacent area, a similar test

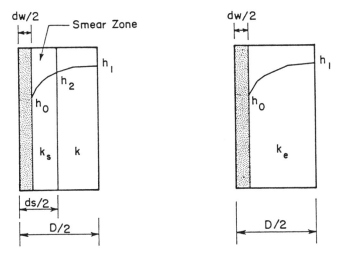

Fig. 4.31 Equivalent Permeability for Smear Effect

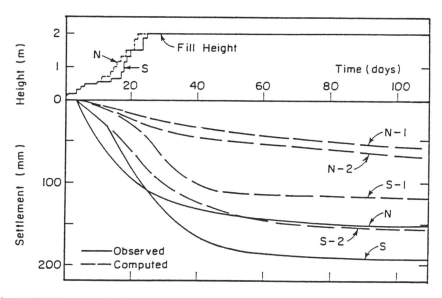

Fig. 4.32 Time vs. Settlement Relationship in Sand Drained Area (S) and Untreated Area (N) (Akagi, 1981)

embankment was constructed (Area N). Figure 4.32 shows the time-settlement data for the first 100 days. It appears that the resulting settlements in Area S was not accelerated significantly enough to justify the sand drain installation in comparison with the corresponding values observed in Area N.

4.19.2 Test Embankment at Pom Prachul Naval Dockyard

The performance of displacement sand drains or sandwicks at the Naval Dockyard site in Pom Prachul, Thailand was investigated by constructing an instrumented test embankment (Balasubramaniam et al. 1980). The embankment was built in 2 stages. It was 90 m long, 33 m wide, and 2.35 high and consisted of 3 sections, namely: no drains, drains with 2.5 m spacing, and drains with 1.5 m spacing. The sand drains consisted of 50 mm diameter sandwicks and were installed into a prebored hole to a depth of 17 m by displacement method. The soil profile is given in Fig. 4.33. The soft Bangkok clay is divided into 3 layers, namely: weathered clay, soft clay and stiff clay. According to the piezometer readings at 10 to 15 m depths, distinct negative drawdown effects were observed in the piezometric levels due to the excessive groundwater pumping of the underlying aquifers. However, below the embankment the drawdown were non-existent, probably due to the recharging action of the sand drains (Balasubramaniam et al. 1980). Down to depths of 6 m and between 12 to 15 m, the surcharge loading just slightly exceeded the maximum past pressures. From 6 to 12 m depth, the surcharge loads just straddle the maximum past pressure. The typical settlement records are shown in Fig. 4.34. It appeared that the major parts of settlement took place in the uppermost 5 m of the soil profile. There was no significant difference in the settlement behavior of the no drain and drain with 2.5 m spacing sections. The section with 1.5 m spacing exhibited larger settlements. The settlements below 7.5 m depth were almost the same for the 3 sections indicating that the sand drains were not effective at deeper depths.

4.19.3 Test Embankment at Nong Ngu Hao

In 1983, three test sections were designed, constructed, and monitored at the proposed Second Bangkok Int'l. Airport at Nong Ngu Hao, Thailand (Moh and Woo, 1987). Non-displacement sand drains of 260 mm diameter at 2.0 m spacing were installed by jet-bailer method under the test sections down to as much as 14.5 m depth. Surcharge loading by embankment fill was used at one test section (Moh and Woo, 1987). The surcharge embankment was built in 2 stages, namely: 2.85 m and 4.2 m high. Vacuum preloading and groundwater lowering by pumping was used at the other test section (Woo et al. 1989). Preloading by vacuum means water and air were sucked from air tight sand drains to create a pressure difference between the surrounding soil and sand drains. The subsoil profile at the site consisted of 5 different strata within the top 35 m, namely: weathered clay, very soft to soft clay, soft to medium clay, stiff clay, and dense sand. The typical soil properties are shown in Fig. 4.35. The groundwater pressure distribution is given in Fig. 4.36 indicating the piezometric drawdown below 11 m depth. To avoid the drawdown effect, sand drains should be limited to 11.0 m. long. Consequently, Moh and Woo (1987) have found that the clay layer between 11 to 15 m depth was found to have little contribution to total settlement. Subsequently, more than 50%

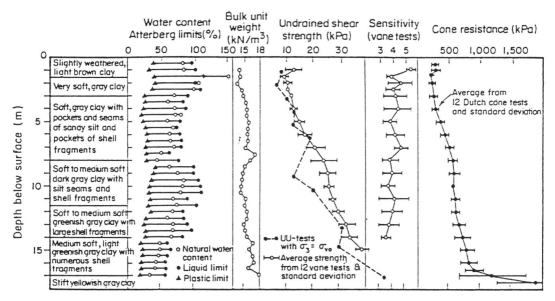

Fig. 4.33 General Properties of Pom Prachul Clay (Balasubramaniam et al. 1980)

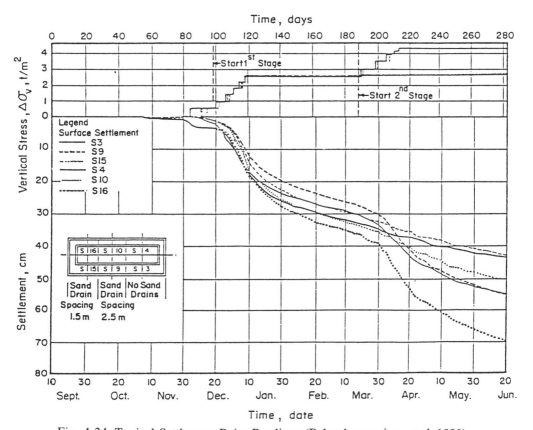

Fig. 4.34 Typical Settlement Point Readings (Balasubramaniam et al. 1980)

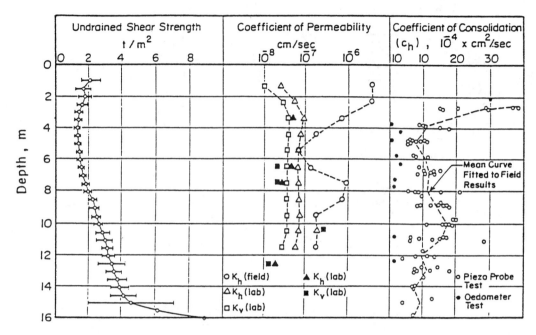

Fig. 4.35 Engineering Properties of the Subsoil at Nong Ngu Hao (Moh and Wu, 1987)

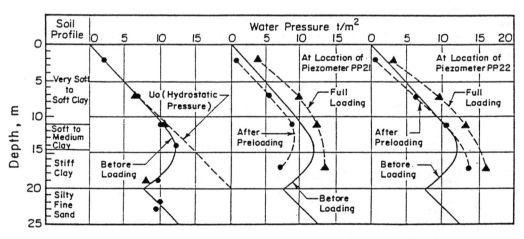

Fig. 4.36 Pore Water Pressure Changes Before and After Preloading (Moh and Woo, 1987)

of the total settlement was observed from compression of the soft clay layer at depths between 3 to 7 m (Woo et al. 1989). For the vacuum loading, the settlements can be separated in 3 stages, namely: dewatering (before sealing), vacuum loading, and cyclic vacuum loading. In general, the ground settled corresponding to various type of loadings indicating the effectiveness of the non-displacement sand drains.

4.20 CASE RECORDS OF PREFABRICATED DRAINS ON SOFT BANGKOK CLAY

4.20.1 Test Embankment Using Mebra Prefabricated Drains

A 4.0 m high test embankment preloading was constructed inside the AIT Campus (approximately 40 km north of Bangkok) on an improved ground with prefabricated vertical (Mebra) drains to study their effectiveness on soft Bangkok clay. The drains were installed in triangular pattern at 1.5 m spacing down to 8.0 m depth (Fig. 4.37). A rectangular-shaped mandrel was used which had inner dimensions of 28.0 mm by 133.0 mm, and outer dimensions of 45.0 mm by 150.0 mm, enough to contain the 3.0 mm by 95.0 mm Mebra prefabricated vertical drains. The embankment plan and section views including the layout of the prefabricated vertical drains are shown in Fig. 4.38. The details of the test embankment have been described by Bergado et al. (1990b). The soil profile consists of the uppermost, 2 m thick, reddish-brown weathered clay underlain by a grayish, soft clay layer down to 8 to 9 m depth. Below the soft clay layer lies the stiff clay layer. The stress history at the site and the surcharge preloading is depicted in Fig. 4.39. Figure 4.40 shows the comparison between the predicted and observed settlements by different methods using back-analyzed parameters (Bergado et al. 1992). The Asaoka (1978) and Skempton and Bjerrum (1957) methods yielded very good predictions. In this test embankment, most of the subsoil compressions were completed in the first 8 months. From back-analysis, values of $k_h/k_s = 10$ and $d_s/d_w = 2.5$ were found to be appropriate. In addition, values of $C_h(field)/C_h(lab) = 4$ and $C_v(field)/C_v(lab) = 12$ were also obtained.

Long (1992) analyzed the test embankment on Mebra prefabricated vertical drains using finite element method (FEM) based on Revised Cam clay model. To analyze the consolidation problem, an FEM program, CON2D, in two-dimensional case was used (Duncan et al. 1981). In this analysis, the isolated vertical drains were converted into 2-D plane strain model as shown in Fig. 4.41. To evaluate the effects of ground subsidence, an axi-symmetric FEM program, CONSAX, was utilized (D'Orazio and Duncan, 1982). To take into account the effect of smear zone, an equivalent permeability, k_e, was obtained from the condition of the same discharge capacity of steady state flow with the same head boundaries.

The predicted consolidation settlements from the FEM analysis are also shown in Fig. 4.40. Excellent agreement with the observed data was obtained. To evaluate the effects of ground subsidence, FEM analyses were performed using 6 m and 8 m long vertical drains. As shown in Fig. 4.41, the total settlements including subsidence effects using 6 m long vertical drains was only about 11% less than the corresponding case using 8 m long vertical drains after 1,200 days. Thus, it is more economical to use only 6 m long vertical drains down to the

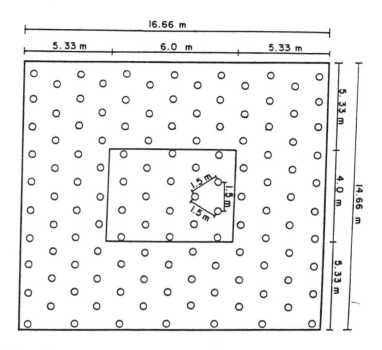

Fig. 4.37 Embankment Plan and Layout of PVD Installation

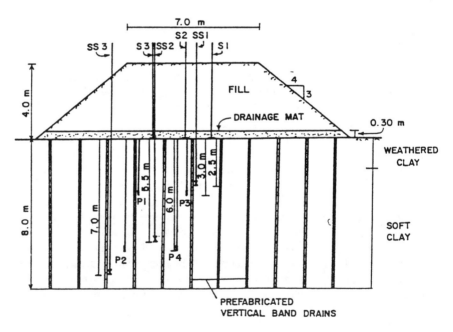

Fig. 4.38 Section View and Layout of Field Instrumentation for Test Embankment on Mebra Prefabricated Drains

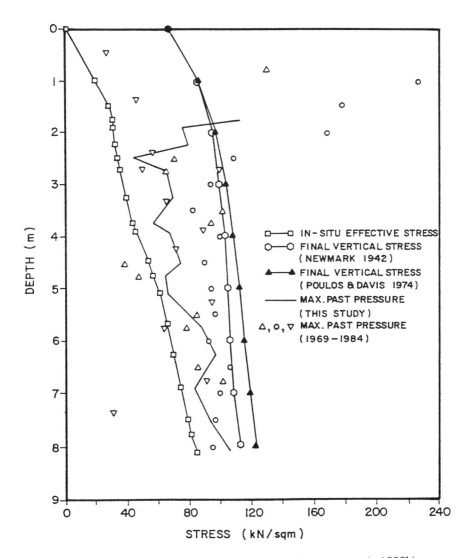

Fig. 4.39 Stress History at the AIT Site (Bergado et al. 1990b)

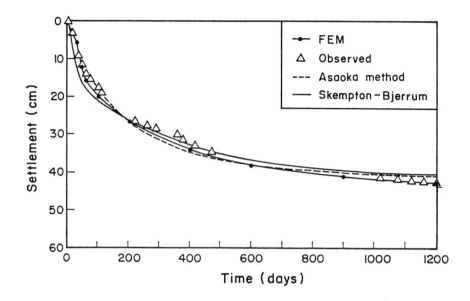

Fig. 4.40 Comparison Between Predicted and Observed Settlement for Test Embankment on Vertical Drains

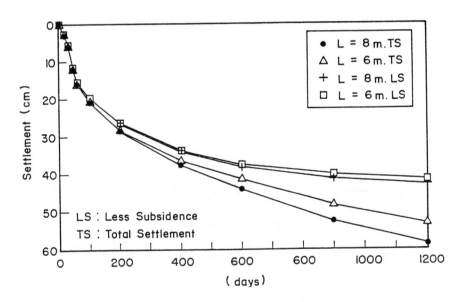

Fig. 4.41 Settlement of Embankment with Different Lengths of Vertical Drains

medium stiff clay layer instead of using 8 m long vertical drains (Bergado and Long, 1994). The observed and predicted piezometric levels of subsiding ground is shown in Fig. 4.42.

4.20.2 Test Embankment Using Alidrain Prefabricated Drains

Bergado et al. (1991) installed prefabricated vertical drains (Alidrains) at 1.2 m spacing in square pattern down to 8.0 m depth at a test site inside the AIT Campus. The installation was carried out using a small mandrel in one-half of the site and a large mandrel in the remaining half to study the smear effects due to the mandrel size. The plan view of the test embankment is shown in Fig. 4.43. The test embankment height was 5 m. Figure 4.44 shows the relationships between k_h/k_h and C_h indicating larger zone of disturbance in the large mandrel area. The results of the back-analysis confirmed the proposal of Hansbo (1987) that $d_s = 2d_m$ and that $k_s = k_v$ in the smeared zone. The time-settlement relationships obtained from full scale field test embankment are shown in Figs. 4.45 and 4.46 for large and small mandrel areas, respectively, together with the settlement predictions given by Mukherjee (1990) using finite element methods (FEM). It can be seen that the performance of the vertical drains is well predicted with smear effect taken into consideration using $k_h/k_s = 10$ and $d_s = 2d_m$. In addition, k_h/k_s ratio was found to affect the settlement rate more than the d_s/d_m ratio. Furthermore, faster settlement rate and slightly higher compression were observed in the small mandrel area than in the large mandrel area.

4.20.3 The Bangkok-Chonburi New Highway Project

This section describes the design as well as the predicted performance of the Prefabricated Vertical Drains (PVD) installed at the Bangkok-Chonburi New Highway Project. In this project, a photograph of PVD installation by displacement method is shown in Fig. 4.47. Figure 4.48 shows the location of the project site and Fig. 4.49 presents the typical section of the embankment with PVD in the soft clay layer.

Section 1-A/1 (Station 0+000 to 5+100)

The general subsoil condition showing the soil profile and the index properties are shown in Fig. 4.50 The index properties for the foundation soils is plotted with depth as illustrated in Fig. 4.51. The subsoil consists of 1 m to 2 m thick of weathered clay, 8 m to 9 m thick of very soft to soft clay, 2 m to 7 m thick of medium stiff clay, and 1 m to 4 m thick of stiff to very stiff clay.

The result of the settlement analyses is plotted in Fig. 4.52 for both PVD lengths (L_{drain}) of 8 m and 10 m. The calculated total primary settlement is about 158 cm. After one year, the predicted settlement is about 114 cm and 123 cm for PVD lengths of 8 m and 10 m, respectively. These computed settlements corresponds to 72% and 78% degree of consolidation. The results also show that there is no significant difference in terms of the settlement magnitudes by using L_{drain} of 8 m and 10 m. The settlement rate, however, differs, in which the consolidation rate using L_{drain} of 10 m is slightly higher by about 6% after one year.

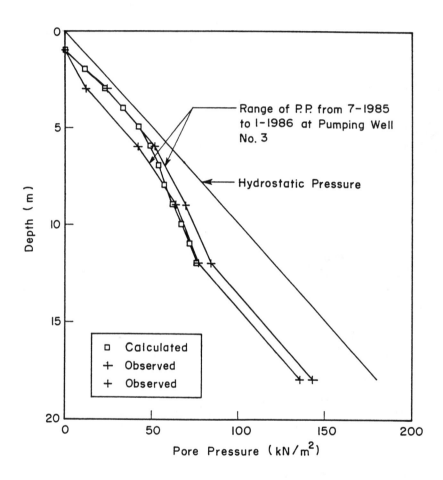

Fig. 4.42 Comparison of Predicted and Observed Pore Pressure Drawdown

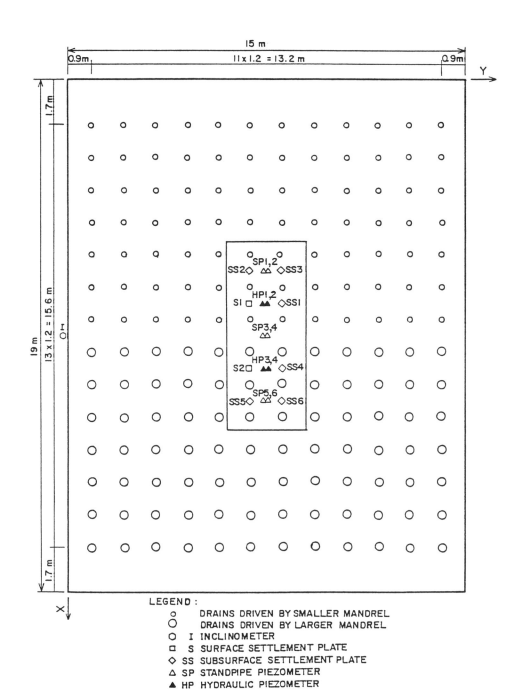

Fig. 4.43 Plan View of Test Embankment on Alidrains

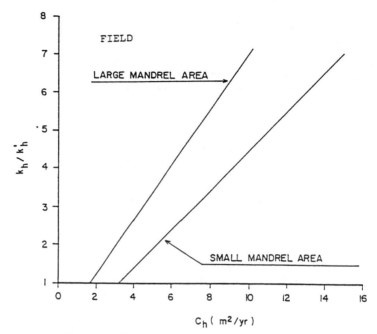

Fig. 4.44 Back-Calculated Sets of K_h/K_h' and C_h Values

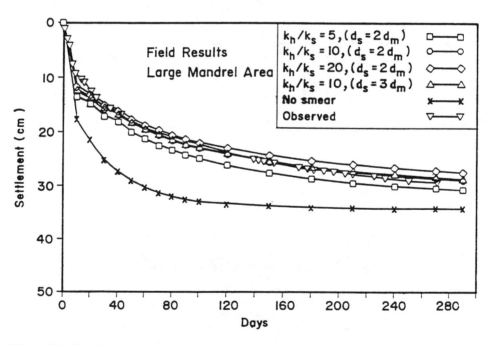

Fig. 4.45 Time-Settlement Relationships at Alidrain Test Embankment: Large Mandrel Area

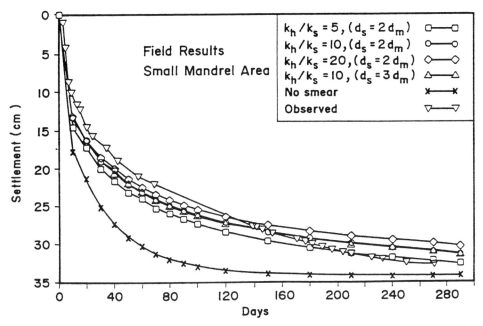

Fig. 4.46 Time-Settlement Relationships of Alidrain Test Embankment: Small Mandrel Area

Fig. 4.47 PVD Installation

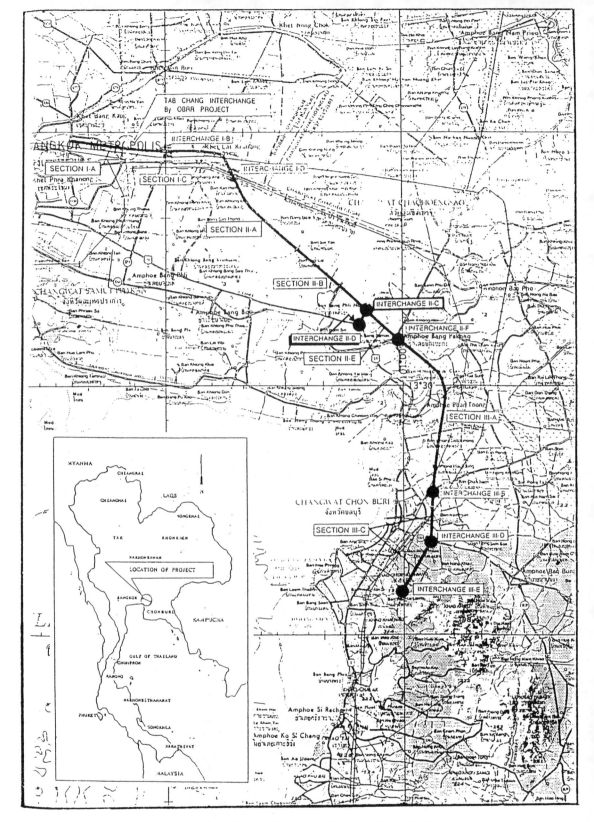

Fig. 4.48 Location Plan of Bangkok-Chonburi New Highway

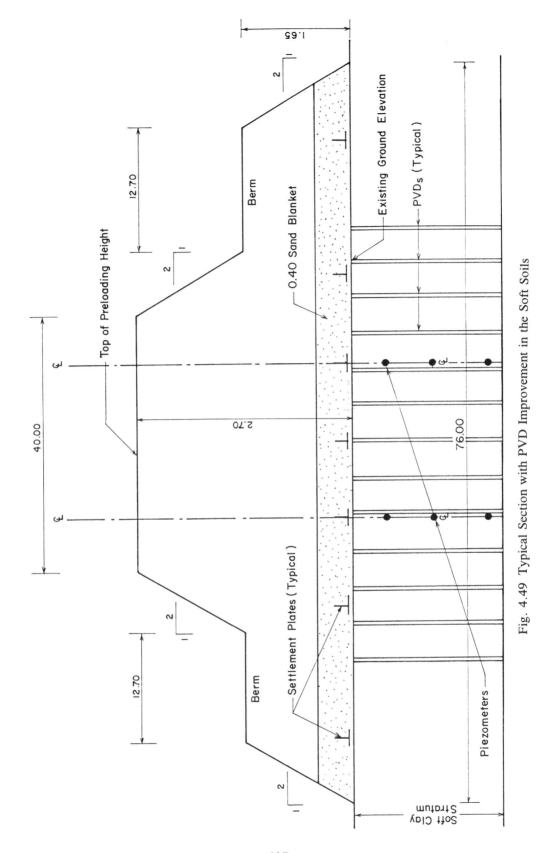

Fig. 4.49 Typical Section with PVD Improvement in the Soft Soils

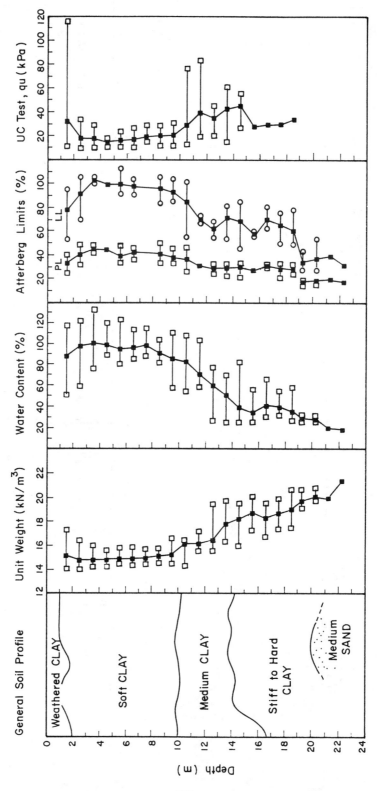

Fig. 4.50 Index Properties of Foundation Soils (Section 1A/1)

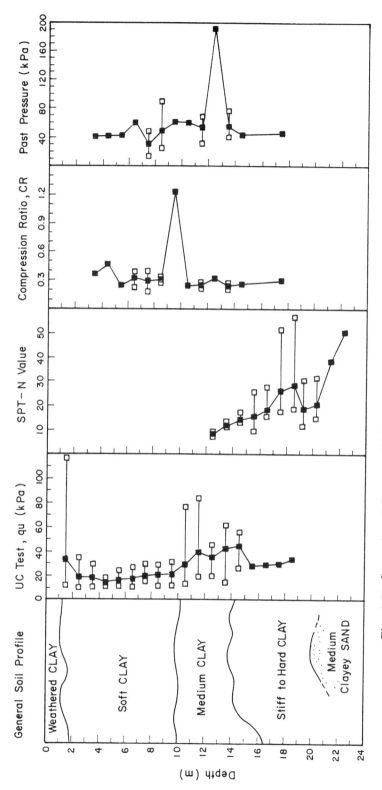

Fig. 4.51 Strength and Compressibility Characteristics of Foundation Soils (Section 1A/1)

The drains were installed in a triangular pattern with a spacing of 1.2 m from center to center. The initial undrained strength was taken from the corrected field vane test results (Fig. 4.53). Bishop Simplified Method was used to analyze the stability of the proposed embankment. Stability calculations were made for both PVD lengths of 8 m and 10 m. The stability analyses were carried out following the construction sequence as presented in Fig. 4.54. Table 4.4 summarizes the results of the settlement and stability analyses.

It was recommended to adopt the length of the drain (L_{drain}) of 8 m. Installation of PVD up to depth of 10 m or more may not be advantageous. The ground improved by 8 m length drains satisfied the stability and settlement criteria with the minimum factor of safety of 1.3 and with the predicted degree of consolidation at 72% after one year. The improvement in undrained strength of soil as presented in Fig. 4.53 exhibits a significant increase after 8 months and one year period. This predicted increase in undrained strength in a short period of time is due to the installation of PVD which accelerates the consolidation process of the soft clay layer. The predicted increase in shear strength was calculated using SHANSEP technique (Ladd, 1991) as demonstrated in Section 4.18.

Section 1 A/2 (Station 5+100 to 10+100)

The generalized soil properties which show the soil profile, index properties, compressibility and strength parameters are demonstrated in Figs. 4.55 and 4.56. Weathered crust (CH) occupies the top most layer until 1.2 to 2 m depth. Very soft to soft clay (CH) ranges at 10 to 15 m depth followed by medium stiff clay (CH) at depth of about 13 to 18 m. Stiff to very stiff clay (CL/CH) ranges from 17 to 24 m and dense to very dense medium to fine silty sand (SM/SC) covers from 17.5 m until the end of boring.

The results of settlement analyses are plotted in Fig. 4.57. The expected total primary settlement is 171 cm. For this section, PVD lengths of 10 m and 12 m were evaluated to ascertain their applicability and effectiveness as determined from the settlement and stability calculations. The average degree of consolidation of 70% is reached at 10 months and 12 months for L_{drain} of 12 m and 10 m, respectively. The stability calculations were made using the corrected field vane tests results. The drains were installed in a triangular pattern at 1.2 m spacing. The most critical case occurs when the embankment reaches 3.8 m of fill thickness (Fig. 4.58). The stability analyses indicate that there is no notable difference in the safety factor using drain lengths of 10 m and 12 m. This observation is based on the fact that the slip surface cannot pass through the layer deeper than 10 m. The results of the settlement and stability calculations are summarized in Table 4.5.

Based on the results of the settlement and stability analyses, a drain length of 10 m was recommended. This proposed drain length satisfies the specified criteria with a factor of safety of 1.25 at critical case condition and with a predicted degree of consolidation of 70% in one

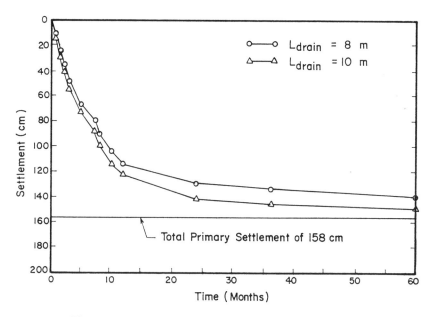

Fig. 4.52 Plot of Settlement Versus Time (Section 1A/1)

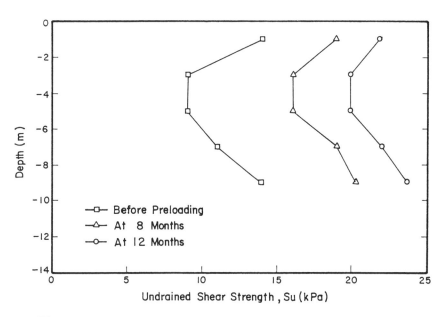

Fig. 4.53 Predicted Improvement of Undrained Strength (Section 1A/1)

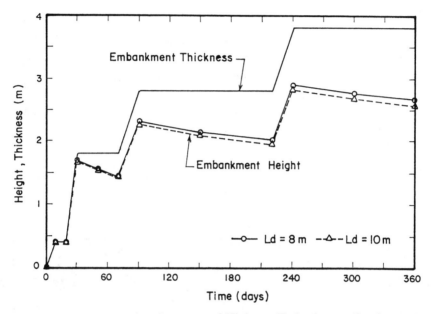

Fig. 4.54 Construction Sequence of Highway Embankment (Section 1A/1)

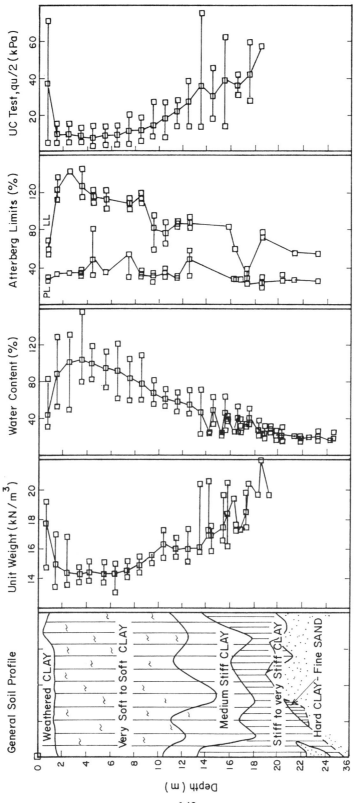

Fig. 4.55 General Soil Profile and Soil Properties (Section 1A/2)

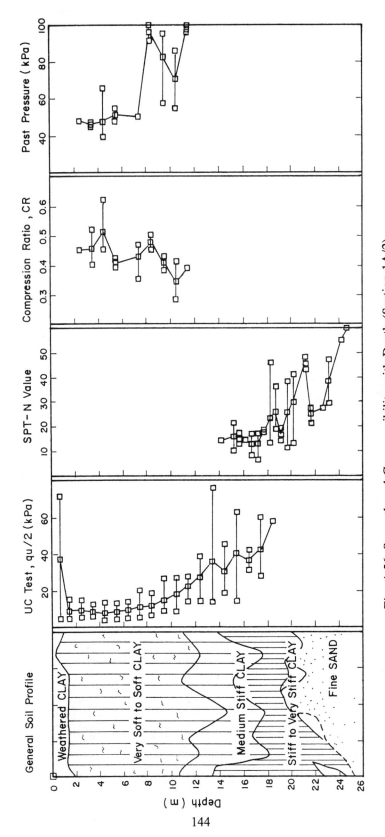

Fig. 4.56 Strength and Compressibility with Depth (Section 1A/2)

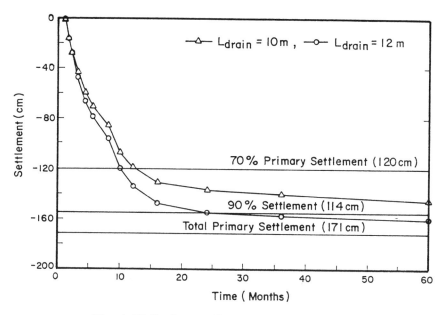

Fig. 4.57 Settlement Versus Time (Section 1A/2)

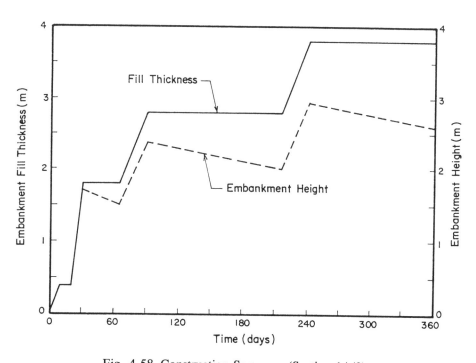

Fig. 4.58 Construction Sequence (Section 1A/2)

Table 4.4   Summary of Results of the Settlement and Stability Analyses (Section 1A/1)

L_d (m)	S (cm)	$S_{1\ year}$ (cm)	$U_{1\ year}$ (%)	$FS_{critical}$	$FS_{1\ year}$
8	158	114	72	1.30	1.43
10	158	123	78	1.34	1.43

Table 4.5   Summary of Results of the Settlement and Stability Analyses (Section 1A/2)

L_d (m)	S (cm)	$S_{1\ year}$ (cm)	t_{70} (months)	$FS_{critical}$	$FS_{1\ year}$
10	171	120	12	1.25	1.40
12	171	134	10	1.25	1.40

Notes:
- L_d : Length of drain.
- S : Total primary settlement.
- $S_{1\ year}$: Settlement after 1 year from start of construction.
- $U_{1\ year}$: Degree of consolidation after 1 year from start of construction.
- t_{70} : Time required for 70% average degree of consolidation.
- $FS_{1\ year}$: Factor of safety after 1 year from start of construction.
- $FS_{critical}$: Factor of safety when the maximum fill thickness of 3.8 m is reached.

year. The predicted improvement in terms of undrained strength by SHANSEP technique (Ladd, 1991) is presented in Fig. 4.59 which shows considerable increase, especially in the soft clay layer. This significant improvement in undrained strength in a short period is obviously caused by the installation of drains which accelerates the consolidation process.

4.20.4 Second Bangkok International Airport Test Embankment

The performance of the prefabricated vertical drains (PVD) installed at Nong Ngu Hao Test Embankment for the Second Bangkok International Airport Project is described in this section. Figure 4.60 exhibits the location plan of the project site. The piezometric drawdown at the site due to excessive groundwater extraction is shown in Fig. 4.61. Figure 4.62 presents the section view and instrumentation layout of the embankment with 12 m length, 1.0 m spaced PVD.

Figure 4.63 shows the general soil condition at the site. The top 2 meters are composed of weathered clay underlain by approximately 16 m thick of very soft to medium clay and followed by 5 m thick of stiff clay. The index properties and Atterberg limits are also shown

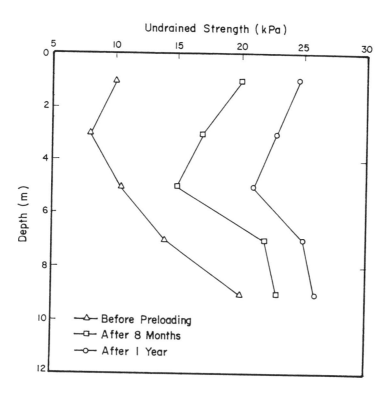

Fig. 4.59 Predicted Improvement in Undrained Strength (Section 1A/2)

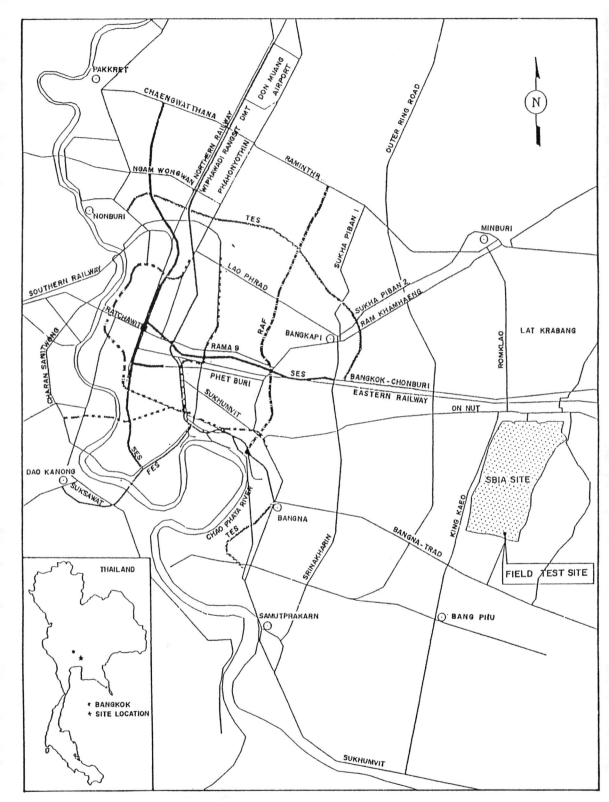

Fig. 4.60 Location Plan of the Second Bangkok International Airport Field Test Site

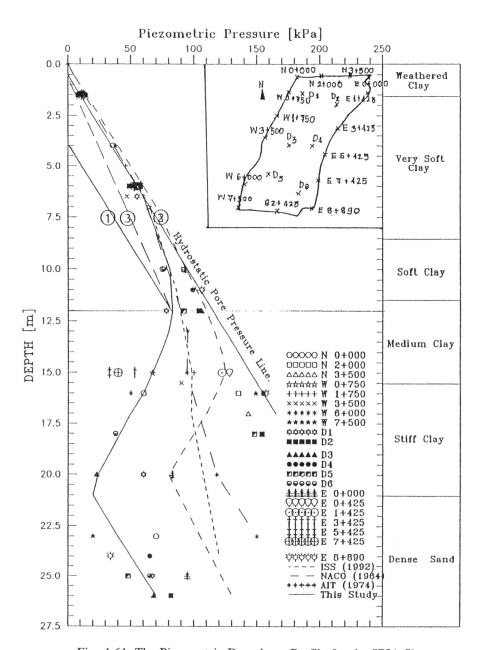

Fig. 4.61 The Piezometric Drawdown Profile for the SBIA Site

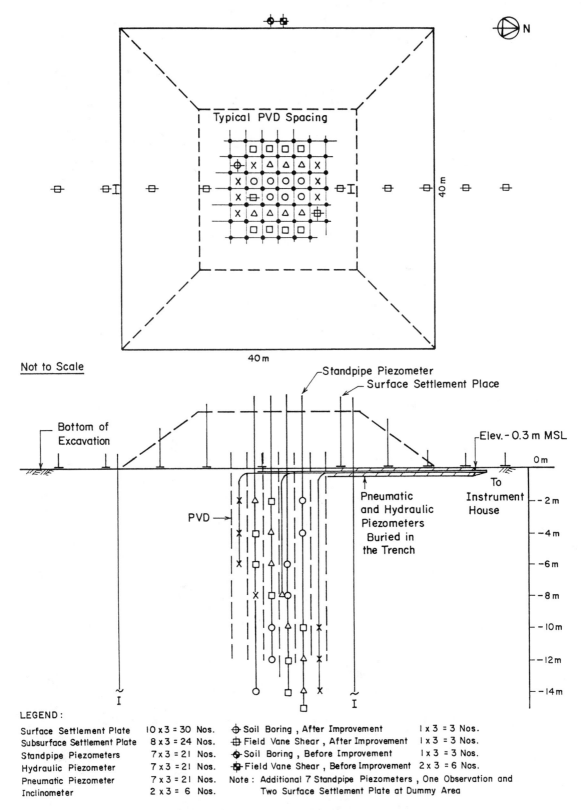

Fig. 4.62 Plan and Section View of Instrumentation

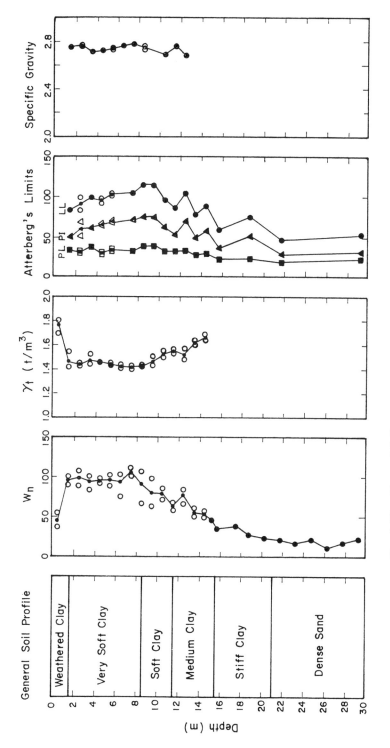

Fig. 4.63 General Soil Profile of Nong Ngu Hao Test Embankment

in Fig. 4.63. The strength and compressibility parameters are plotted with depth in Fig. 4.64. The measured settlement and pore pressure readings are illustrated in Fig. 4.65 and 4.66, respectively. The pore pressure data obtained from pneumatic piezometers show that the pore pressure was not dissipating as would be expected based on consolidation theory. This phenomenon may be due to the creep effect and migration of pore water towards the drained area. The maximum lateral displacement of about 10 cm (Fig. 4.67) could not explain the vertical displacement of 85 cm unless consolidation has taken place. The settlement records, therefore, are reliable although the pore pressure did not dissipate as anticipated. The undrained shear strength increase by SHANSEP technique (Ladd, 1991) showing the predicted and measured values is illustrated in Fig. 4.68. The results show that the predicted values agreed with the measured data. Using the approach of Asaoka (1978) for radial consolidation with PVD, the horizontal coefficient of consolidation, C_h, can be back-calculated from the observed settlements of the test embankments. The back-calculated values are plotted in Fig. 4.69. As shown in Fig. 4.69, the back-calculated values agreed with the results obtained from the piezo probe and piezocone tests. However, the back-calculated values were much higher than the back-calculated C_h values in the 1983 study which did not account for the smear effects.

4.21 EVALUATION OF PREFABRICATED VERTICAL DRAINS (PVD) CONSIDERING SITE CONDITIONS IN JAPAN

Previous studies showed that the performances of PVD are greatly affected by the site condition. Furthermore, the present applications of PVD are only limited to about 20 m depth since there has been no investigation on the performance of PVD installed at deeper depths. This section discusses the results of an extensive investigation on the various factors that influence the rate of consolidation of soil deposit when PVD are installed at deeper clay layer. The types of PVD that were studied are given in Table 4.6.

4.21.1 Effects of PVD Deformation

As the consolidation of the clay subsoil progresses at the site, the phenomena of bending and folding of PVD arises which would have a significant effect on the permeability of PVD. As shown in Fig. 4.70, the magnitude of the effect of folding and bending on the permeability of PVD is greatly dependent on the following factors:

(a) Degree of consolidation.
(b) Long term creep displacement of PD which appears almost at the initial stage of consolidation.
(c) Flow gradient which seems to be high at the time of loading and decreases afterwards as the consolidation progresses.
(d) The presence of air bubbles which decreases the longitudinal permeability capacity of PD.

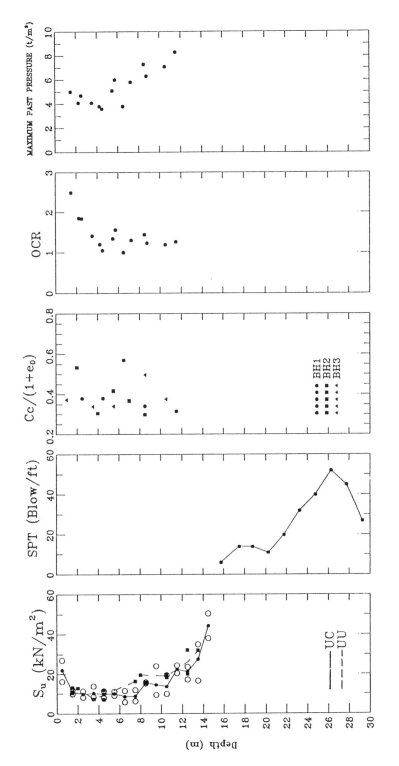

Fig. 4.64 Strength and Compressibility Profile at Nong Ngu Hao Test Site

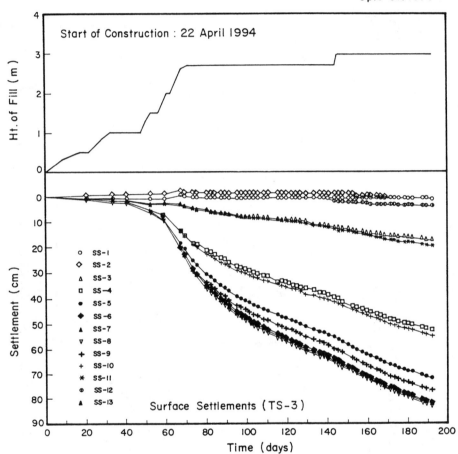

Fig. 4.65 Relationship Between Increasing Embankment Height and the Resulting Surface Settlements

Upto 31/10/94

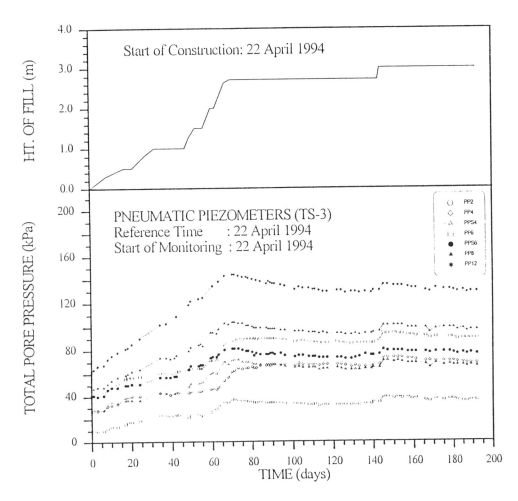

Fig. 4.66 Relationship Between Increasing Embankment Height and the Resulting Pore Pressures

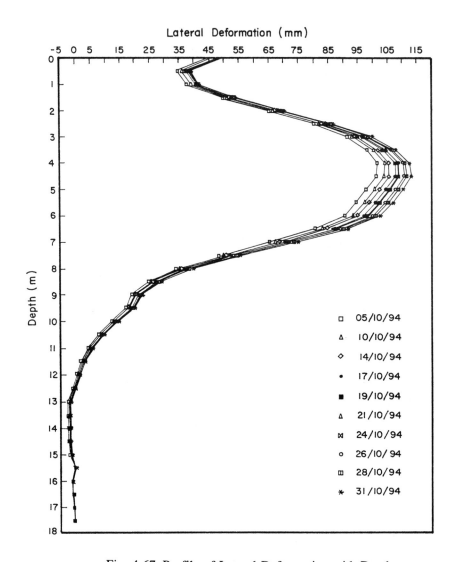

Fig. 4.67 Profile of Lateral Deformation with Depth

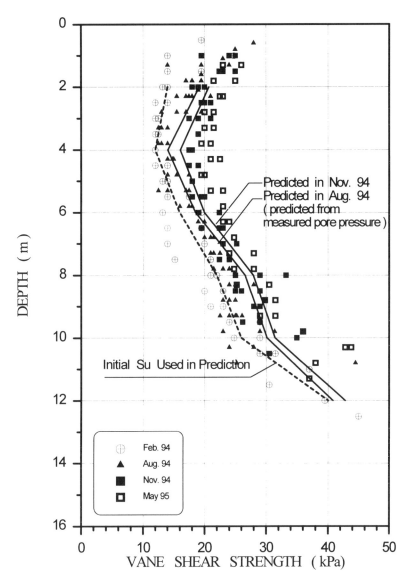

Fig. 4.68 Predicted and Measured Undrained Strength

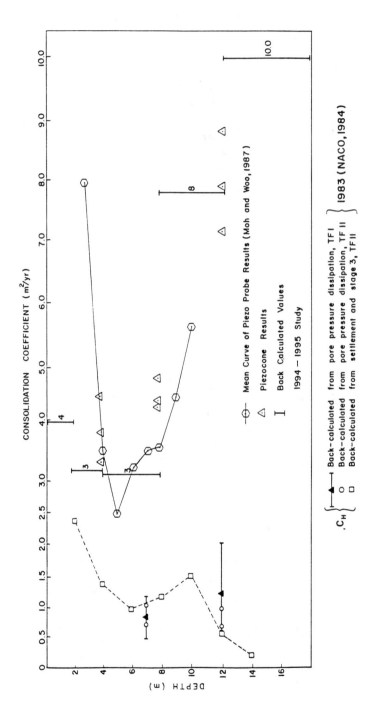

Fig. 4.69 Comparison of C_h Values Computed from TS2 with the Values Obtained from 1983 Test Embankment with Sand Drains and the 1994 Piezocone Tests

Table 4.6 Types of PVD Used in the Experiment

DESIGNATION		GL	MW, MB	CS, CS$_2$	TS	TF
Size (mm)	Thickness	3.4 ± 0.5	3.0 ± 0.5	2.6 ± 0.5	4.6 ± 0.3	7.5 ± 1.0
	Width	95.8 ± 2	100 ± 20	94 ± 2	100 ± 3	100 ± 5
Unit weight (g/m)		100	75	90	100	80
Structural type		Free	Free	Fixed	Free	One body
Material	Filter	Synthetic fiber of cellulose and polyester	Nonwoven fabric made from polypropylene	Spun bonded of polyester	Nonwoven fabric made from polypropylene	Spun bonded nonwoven fabric made from polyethylene
	Core	Polyolefin	Polypropylene	Polyethylene	Polyethylene	
Section diagram						

The plot shown in Fig. 4.70 also demonstrates that when the degree of consolidation reaches 50%, the flow gradient decreases which triggers the bending or folding of PVD. Furthermore, the type of soil at the site has great effect on the longitudinal permeability capacity of PVD (Q_w) and the permeability capacity required at the site (Q_{req}). As illustrated in Fig. 4.70, Q_w always remains higher than Q_{req} all throughout the process of consolidation for clay deposits. In the case of silts, however, Q_w becomes lower than Q_{req} when the degree of consolidation is 50% so the effect of well-resistance cannot be neglected. For silt ground, therefore, the evaluation of the longitudinal permeability of PVD is very important.

The required permeability at the site (Q_{req}) could be estimated using the method proposed by Pradhan et al. (1991), defined as follows:

$$Q_{req} = \frac{\epsilon_f . U_h . FS . L\pi . C_h}{4.T_h.86400} \qquad (4.39)$$

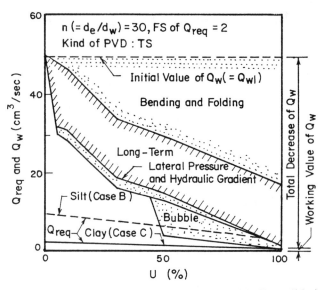

Fig. 4.70 Rate of Decrease of Q_{req} with Consolidation

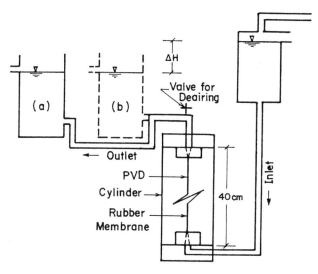

Fig. 4.71 Test Apparatus for Investigating the Influence of Bubbles

where:
- ϵ_f = final vertical strain (0.25H)
- U_h = degree of horizontal (radial) consolidation
- C_h = coefficient of horizontal (radial) consolidation
- L = drain length
- FS = factor of safety (equal to 2)

4.21.2 Factors Affecting Discharge Capacity

The type of equipment that is often used in this kind of testing is similar to the triaxial compression test apparatus, as shown in Fig. 4.71. Both ends of the PVD are connected to the inlet and outlet of water. The test samples are wrapped in a rubber sleeves and the water pressure is applied to the sides. In this testing apparatus, it is necessary to set up the test conditions which suit the actual stress situation acting on PVD at the site.

A. Magnitude of Confining Pressure

The magnitude of confining pressure, σ_3, is estimated based on the PVD depth and the surcharge load as follows:

$$\sigma_3 \leq (L \cdot \acute{\gamma} + h \cdot \gamma_t) K_o \qquad (4.40)$$

where γ is the submerge unit weight of soil, γ_t is the total unit weight of soil, h is the embankment height, and K_o is the coefficient of lateral pressure. As a case in point, with an embankment height of 3 m, PVD depth of 30 m, submerge unit weight of 4.9 kPa, total unit weight of 17.6 kPa and coefficient of lateral pressure of 1.0, the required magnitude of the confining pressure to be applied is approximately 200 kPa. In this case, therefore, it is advisable to use an equipment which could load about twice as the approximated confining pressure, that is, 400 kPa.

The effect on the discharge capacity of the PVD (Q_w) by varying the confining pressure (σ_3) is exhibited in Fig. 4.72. The results show that Q_w decreases with increasing confining pressure. Reports on some previous research show that the relationship between Q_w and σ_3 is exponential. However, from a practical viewpoint, it is considered reasonable to assume a linear relationship between Q_w and σ_3 over the range of σ_3 from 50 to 400 kPa.

The length of PVD test samples lies between 10 to 300 mm in many cases. The factors to be considered in deciding the length of the test samples are the total head, the head loss of the apparatus (0.6 cm in the case of the equipment shown in Fig. 4.71), the hydraulic gradient and the influence of bubbles. For tests which are aimed to investigate the effect of bending and folding of PVD, an allowance in length should be provided to warrant forced deformation at some designated sections of the PVD. Considering the above mentioned factors, PVD test samples of 40 to 50 cm long is recommended.

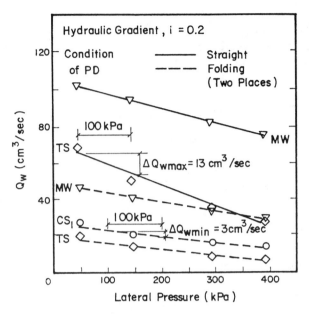

Fig. 4.72   Variation of Q_w with Lateral Pressure

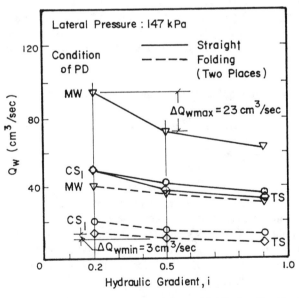

Fig. 4.73   Variation of Q_w with Hydraulic Gradient

B. Hydraulic Gradient

Pradhan et al. (1991) reported that although the hydraulic gradient adopted in a longitudinal permeability tests is the same, the resulting Q_w varies depending on the hydraulic head difference. Hence, it is important to execute the laboratory tests adopting the flow gradient suitable for the site condition. To calculate the flow gradient that is applicable at the site, the following formula can be used:

$$i = \frac{\Delta H}{(L/2)} \tag{4.41}$$

where ΔH = head of water = $\Delta p/\gamma_w$. Nakanodo et al. (1991), by varying the head distribution of the PVD at the site, found out that the flow gradient ranges from 0.03 to 0.8. Figure 4.73 shows the variation of Q_w with the flow gradient and it shows that Q_w is decreasing with increasing flow gradient. In the laboratory, the flow of water inside the PVD may changed from laminar flow to random flow, hence, the Darcy Rule may not be applied. Moreover, Q_w becomes almost independent of the flow gradient when $i > 0.5$. It is quite reasonable to assume that the flow of water from the soil with low permeability into the PVD is laminar. It is therefore recommended to adopt i ranging from 0.2 to 0.5 in laboratory permeability tests.

C. Influence due to Bending of PVD

When the compression strain of the clay reaches more than 20%, PVD will be deformed greatly. The compression strain of the clay due to consolidation is considered high at the shallower depth. Figure 4.74 shows how the pore ratio is varied in the direction of the depth when the embankment load of 35.2 kPa is applied. The strain at each layer is shown in Fig. 4.75. Figure 4.75 illustrates that the effect due to the bending of PVD is great at shallower depth and it becomes lesser at deeper layers. The strain of each layer corresponding to the consolidation quantity is shown in Fig. 4.76, of which the embankment load has been applied for 300 days on PVD-improved Ariake clay, in the Saga plains area of Japan.

D. Influence due to Bubbles

The equipment used to investigate the longitudinal permeability could be classified into 2 types as shown in positions a and b in Fig. 4.71. At first, experiments were conducted by using the equipment shown in position a of Fig. 4.71. It was observed that the permeability was gradually reduced and finally becomes zero. This high reduction in permeability was found to be caused by bubbles inside the PVD or in the apparatus. An experiment was also conducted using the equipment as shown in position a of Fig. 4.71 under the condition of forced deformation at one location with a head difference (ΔH) of 8 cm and a flow gradient of 0.2. The results shown in Fig. 4.77 demonstrate that Q_w is gradually reduced and finally becomes zero after the lapse of 50 hours if bubbles were not removed. On the other hand, if bubbles were removed from inside the hose, Q_w is restored 60% to 70%. When the cylinder is located above the top of the equipment as shown in position b of Fig. 4.71, there will be no influence

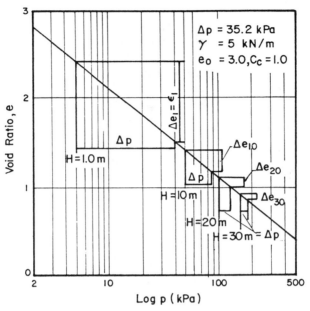

Fig. 4.74 Change in Void Ratio due to Fill Loading at Each Depth of Soft Ground

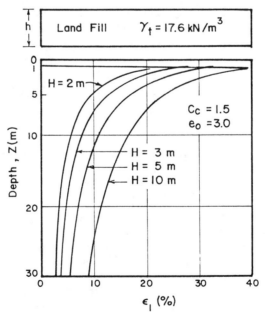

Fig. 4.75 Distribution of Hydraulic Gradient with Depth

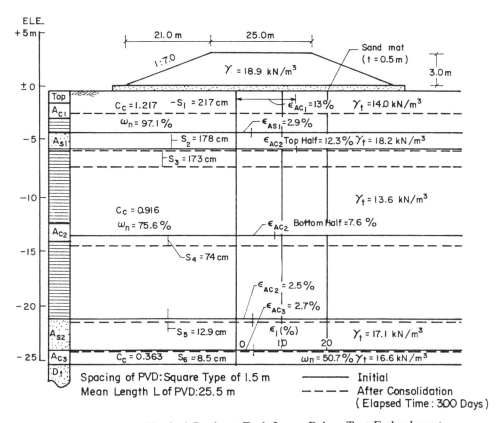

Fig. 4.76 Vertical Strain at Each Layer Below Test Embankment After 300 Days

due to bubbles. It is, therefore, necessary to perform the test as shown in position b of Fig. 4.71 in order that bubbles will not affect the results. The observations as discussed above imply that the malfunction of PVD may be brought about by the presence of bubbles through the dissolved air in the ground during the consolidation process.

Immediately after the maximum embankment height is achieved, there will be a forced release of bubbles inside the PVD due to fairly high imposed pressure. However, as the pore pressure dissipates and when the deformation of the PVD becomes large, the effects of bubbles cannot be neglected.

E. Long Term Permeability Tests

It is a general practice to measure Q_w on a short-term basis. Considering that the stage embankment loading method is followed at the site, it is also important to evaluate the variation of Q_w over a long period of time corresponding to any stage of the loading process. Figure 4.78 shows the test results over a period of 8 days under a condition of forced deformation at one location for 3 different types of PVD. The bubbles were completely removed and the results became stable in 1140 to 5000 minutes after the first Q_w had been reduced to 50% to 70%. It was observed that the Q_w was reduced due to reduced flow section as a result of the embedment of filter into the PVD core. It is therefore advantageous to measure Q_w after three days in order that the effect of reduced flow section due to filter embedment to the core will be taken into consideration.

4.21.3 Effects of Smear Zone

A. Measurement Method

The effect of smear zones (diameter, ds) brought about by the installation of PVD influence greatly the consolidation rate. So far, there is no standard procedure to measure the effect of smear zones at the site. The following steps to investigate the effect of smear zone at the site were followed. The square rod with a cross section of 12 cm x 12 cm and 4 m long (model mandrel with equivalent diameter, dw) was pushed into the clay ground and subsequently pulled up. The ground around the model mandrel was then excavated 2 m and a pit was dug 1 m away from the model mandrel. Three (3) soil samples were collected by inserting the sampler horizontally toward the mandrel. The samples were taken at constant intervals and these are used for one dimensional compression tests, measurement of cone indices by the portable cone penetrometer (denoted by PCP with the cross section area of 0.385 cm^2 and the tip angle of 30°), and for standard consolidation tests. Thus, the investigation was designed so as to evaluate the variations of lateral permeability coefficients and the strength distribution.

B. Range of Smear Zone

Figure 4.79 shows a decreasing trend of q_c obtained from PCP in the region around the mandrel. The vertical axis represents the cone index of the smear zone, q_{cs}, divided by that of

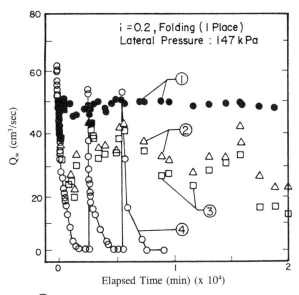

①: Perfect Removal of Bubble in the Hose and the PVD
②: Removal of Bubble in the Hose
③: Flow with Bubble in the Hose and the PVD
④: Reduction of Flow to Zero due to the Presence of Bubbles in the Hose and the PVD

Fig. 4.77   Influence of Bubbles on Q_w

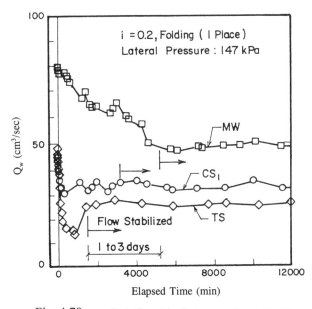

Fig. 4.78   Relationship Between Q_w with Time

the undisturbed clay, qco, in order to normalize the cone index of the smear zone. It is intended to show the degree of disturbance in the smear zone in accordance to the value of q_{cs}/q_{co}. The effect due to uneven characteristics of the ground was reduced by multiplying q_{cs}/q_{co} by e_{os}/e_o, where e_{os} is the pore ratio measured at the site and e_o is the pore ratio of the undisturbed clay. The horizontal axis, on the other hand, represents the ratio of the distance from the center of the mandrel, d, to the diameter of the mandrel, dw. The region where $q_{cs}/q_{co}=0.75$ almost agrees with the region of q_{cs}/q_{co} which has been calculated based upon the experiments of pulling the mandrel up and down as reported in the previous papers (Miura et al. 1993). This region, therefore, is defined as the smear zone. At this region, ds/dm becomes 5 to 7. It was also found out that the smear zone obtained by the experiments at the site is more widely distributed than the result of the previous laboratory tests (ds/dm=2.5 to 3.0). The region of disturbance (ds/dm) measured at the site varies from 14 to 15. Furthermore, a value of ds/dm=15 was obtained by the one-dimensional compression tests. It is considered that the influence of soil disturbance is due to the heavy equipment at the site. Although the region of disturbance is greatly varied as a result of the experiments, it is considered reasonable to adopt the ratio, q_{cs}/q_{co}, to be 0.75.

C. Variation of Lateral Permeability Coefficient

In order to evaluate the variation of the lateral permeability coefficients inside the smear zone, the standard consolidation tests were performed. Kobayashi et al. (1990) compare the experimental and analyzed results with the use of improved lateral consolidation container and proposed the method to determine the horizontal consolidation coefficients of clays. The values of horizontal coefficient, K_h, however, are different from the results obtained by Kobayashi et al. (1990). Figure 4.80 demonstrates the relationship between K_h and d/dm. K_h in the completely disturbed zone (approximately 1.5 dm) is decreased to 1/5 of the K_h value for undisturbed clay. The degree of disturbance decreases as the distance from the mandrel increases and the effect of the disturbance due to the mandrel is negligible when ds/dm=14. A range of values of ks/kc from 0.75 to 0.83 were obtained by PCP which corresponds to ds/dm ranging from 5 to 7 and q_{cs}/q_{co} of about 0.75. Moreover, it was found out the rate of decrease of the permeability coefficient and strength as a result of clay disturbance is almost the same.

4.22 LABORATORY AND FIELD TESTS FOR GROUND IMPROVEMENT OF SAGA AIRPORT PROJECT, JAPAN

The Saga Plains has been formed by natural deposition and artificial land reclamation over a period of about 2,000 years and is situated in low lying plains facing the Bay of Ariake whose maximum difference between the ebb and flood tides extend to 6.5 m. The Ariake clay is marine clay which is very susceptible to large compression. Moreover, differential settlements also cause serious problems. When embankment or excavation work is conducted, slope failures such as landslides often take place. Thus, careful consideration should be carried out in designing foundations and the associated ground improvement works.

The Saga Airport Project utilizes vertical drains as a means to improve the ground. A

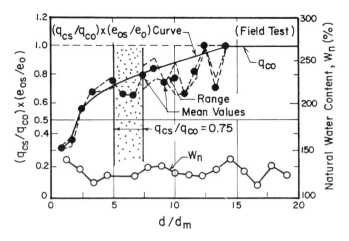

Fig. 4.79 Distribution of q_c around the Mandrel

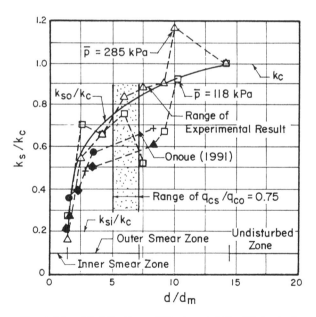

Fig. 4.80 Distribution of Horizontal Coefficient of Permeability, k_s, in Smear Zone

full test embankment was constructed prior to the actual construction work. Three ground improvement techniques were evaluated and compared, namely: sand-packed drains (SD), prefabricated vertical drains (PVD), and preloading (without drains).

4.22.1 Overview of Saga Airport Project

The Saga Airport is located about 13 km south of Saga City, Japan. It is to be constructed on a reclaimed land situated at the most inner part of the Bay of Ariake (Fig. 4.81). The area is an average of 0 to 1 m above sea level. This airport in Japan is the first to be constructed on a land that is located below sea level (middle level between the ebb and flood tides). The total area is approximately 110.10 hectares which comprises the runway, the landing zone, the guiding pathway and the four berths in the apron, as shown in Fig. 4.81.

The subsoil of the airport construction site mostly consists of soft Ariake clay. The clay thickness extends up to 15 to 30 meters. This clay is bluish-gray clay to silty clay. Sand seams of about 1 to 3 m thick are also present in the weak soil layers. Underlying these layers are gravel which had been formed about 2 million to 10 thousand years ago. Figure 4.82 shows the estimated cross section of the earth layers along the airport construction site. The B layer is the topmost layer consisting of a clay or silty clay layer 0.5 to 1.0 m thick. The AC_1 layer comprises the upper soft clay to silty clay layer. The AC_2 layer consists of intermediate soft clay layer from 5 to 15 m thick. The AC_3 layer composed of bottom soft clay layer with interbedding of sand layers. The D_s layer consists of coarse to medium sand layer with medium to dense gravel with SPT N-values of 10 to 40. The layers AC_1 and AC_2 posed the main problem for the airport construction with natural water content varying from 75 to 92%, liquid limit from 68% to 78% and plasticity index from 33% to 45%.

4.22.2 Consolidation Characteristics of Ariake Clay

The coefficient of consolidation, Cv, derived from the standard consolidation tests performed on samples taken from the construction site ranges from 7.30 to 65.70 m²/year. Figure 4.83 presents the relationship between the plasticity index (I_p) versus both the compression index ($\lambda=0.435C_c$) and the expansion index ($\kappa=0.435C_s$) for Ariake clay together with the results from various types of clays which were evaluated by Kamei et al. (1984). The results as illustrated in Fig. 4.83 show higher compression characteristics of Ariake clay at normal consolidated state.

In order to determine the secondary coefficient of consolidation (C_α), long term consolidation tests were conducted using the standard consolidometer. The results revealed that C_α ranges from 0.016 to 0.018 for AC_2 layer and 0.002 to 0.022 for AC_3 layer. Figure 4.84 further implies that the C_α value does not depend on the embankment pressure nor the thickness of the test samples. Instead, it is dependent on the stress increment ratio and the duration of loading. The relationship between C_α and C_c of Ariake clay is in close agreement with the findings of Mesri (1973) with C_α/C_c ranging from 0.03 to 0.05 for AC_1 and from 0.03 to 0.04 for AC_2 layer.

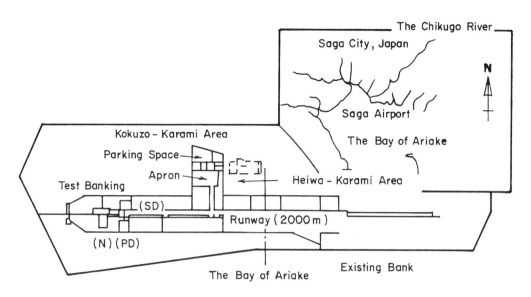

Fig. 4.81   Location of Saga Airport

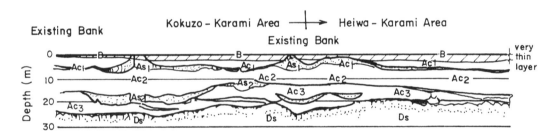

B: Surface Layer; Ac_1, Ac_2, Ac_3: Alluvial Soil Layer;
As_1, As_2: Sand Layer; Ds: Diluvial Sand Soil Layer

Fig. 4.82   Estimated Cross Section of Soil Profile of Airport Construction Site

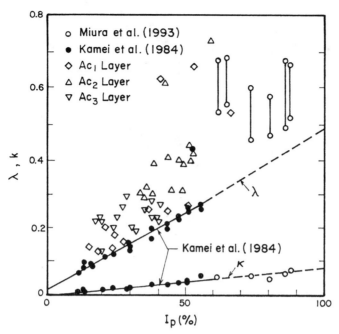

Fig. 4.83　Relationship with λ and κ with I_p

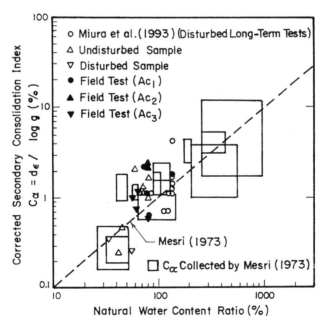

Fig. 4.84　Settlement Curve by Large-Scale Consolidation Test Equipment in Experiment I

4.22.3 PVD Drainage Performance Evaluation by Large-Scale Consolidation Test Equipment

In any ground improvement work which utilizes prefabricated vertical drains, basic investigations are needed on the discharge reduction due to bending or folding of drains and influence of well-resistance. The investigation in this study was carried out using large-scale consolidation test equipment as well as consolidation apparatus.

A. <u>Experimental Method</u>

The different types of PVD that were used in this study are tabulated in Table 4.6.

1) Cs - which is composed of glued polyester synthetic resin textiles as filter with the polyethylene resin as cores.

2) G - the filter of which is made of cellulose and polyester synthetic textiles and the core is made from polyolefine resin.

3) MD - in which the filter is made from unwoven fabric while the core is manufactured from polypropylene resin.

4) TS - made of unwoven fabric polypropylene textile filter and polyethylene resin core.

The whole experimental investigation was carried out in two phases, namely: Experiments I and II. The following are the soil parameters adopted for both Experiment I and Experiment II.

Parameters	Experiment I	Experiment II
Natural water content (%)	140	134
Liquid limit (%)	128	97
Plasticity index	80	54
Specific gravity	2.62	2.62
Coefficient of consolidation (m^2/yr)	7.66 (p=0.5kg/cm^2)	5.91 (p=3 kg/cm^2)

In Experiment I, the PVD width was modified to 5 cm. The SD samples were prepared by filling sand into the filters with equal perimeter area as PVD samples. This relationship in terms of perimeter area between PVD and SD samples was desired in order to acquire an accurate comparison between the test results from both samples. The PVD or SD samples of 5 cm width were set up at the center of the core padding in the large-scale consolidometer of 45 cm in diameter and 90 cm high. The Ariake clay was then placed inside the apparatus up to 85 cm deep and a vertical force of 0.5 kg/cm^2 was applied through a rubber balloon membrane.

In Experiment II, the Ariake clay was placed up to 100 cm deep inside the steel cylinder with diameter and height of 120 cm and 130 cm, respectively. It was preconsolidated by applying a vertical pressure of 0.5 kg/cm² through a 30 cm long piston. After 50 days, the piston was removed and two types of PVD (CS and TF) were inserted at the center of the cylinder using an iron-made model mandrel. The model mandrel has a cross section of 10.8 cm by 2.3 cm. The sample was then consolidated under a pressure of 1.0 kg/cm². After the primary consolidation was completed, the pressure was increased to 3 kg/cm².

B. Experimental Results and Investigations

Figure 4.85 presents the settlement curve of the drains obtained from Experiment I. The consolidation settlement characteristics are more or less the same for all types of drains that were used. The final settlement quantity, S_f, as well as the 50% degree of consolidation, were calculated using the hyperbolic method. The horizontal coefficient of consolidation, C_h, was determined by applying these expressions defined as follows:

$$F(n) = \frac{n^2}{(n-1)} \times \frac{\ln[n-(3n^2-1)]}{4n^2} \tag{4.42}$$

$$C_h = d_e^2 \times \frac{F(n)}{8} \times \ln\frac{1}{(1-U_h)} \tag{4.43}$$

The C_h value corresponding to t_{50} was 0.86 m²/yr for SD and it ranges from 0.55 to 0.86 m²/yr for PVD. The fact that the back-calculated values of C_h for both PVD and SD are almost the same reveals that the drainage capability of both materials is more or less similar.

For Experiment II, the consolidation curves using CS and TF drains are shown in Fig. 4.86. Figure 4.86 demonstrates that the rate of consolidation for both CS and TF drains are almost the same. However, the final settlement magnitude for CS drain is 2 cm higher than that of TF drain. The degree of consolidation reached 97% after 1030 days for the PVD samples. The horizontal coefficient of consolidation, C_h, calculated from the settlement curve under the first-stage loading was 1.35 m²/yr for CS drain and 1.25 m²/yr for TF drain. Under the second-stage loading, C_h was 2.20 m²/yr and 2.59 m²/yr for CS and TF drains, respectively. The vertical strain reached at about 40% under the vertical consolidation pressure of 3 kg/cm² and it was found that the drainage function of the drains was not disrupted even at such high deformation.

After the lapse of 1030 days, the soil sample was extracted using a thin-wall sampler. Samples were taken in both vertical and orthogonal directions. Standard consolidation tests were run using these samples and it gave a C_v value of 11.42 m²/yr and C_h value of 19.34 m²/yr under the consolidation pressure of 3 kg/cm². This increase in C_h value is due to strengthening of the clay particles resulting from long-term consolidation under high pressure.

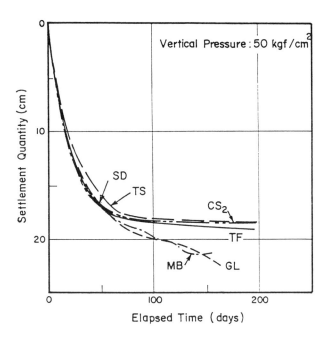

Fig. 4.85 Settlement Curve of Drains from Experiment I

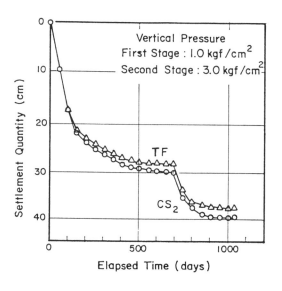

Fig. 4.86 Settlement Curve by Large-Scale Consolidation Test Equipment in Experiment II

Table 4.7 Final Settlement Magnitudes by Hyperbolic Method

		S_o (cm)	$1/\beta$ (cm)	R^2	S_f (cm)	S_t (cm)	S_t/S_f (%)
PVD	CS$_2$	179.9	65.3	0.985	245.2	231.0	94.2
	MB	177.5	70.1	0.924	247.6	228.6	92.3
SD		165.8	72.7	0.998	238.5	221.6	92.9
N		95.8	110.1	0.982	205.9	159.7	77.6

Note: S_t is the settlement at 175 days after completion of embankment construction

4.22.4 Test Embankment

The cross section of the test embankment is shown in Fig. 4.87. The subsoil properties are as follows:

Liquidity index	1.05 to 1.20
Natural water content	greater than liquid limit
Sensitivity ratio	greater than 16
C_v (m^2/yr)	10.45 to 36.50
C_c	0.5 to 1.5

The tests were carried out in four sections, namely: one section with sand-packed SD, two sections with PVD, and one unimproved section. The sand mat (sand with permeability coefficient, k, > 1x10^{-2} cm/s) 50 cm deep was laid at the base of the test embankment of each section. The embankment width is 71.0 m at the base. The crest is 25 m square shape with slope of 1:7. Ten layers of decomposed granite soil were laid 3 meters deep as embankment fill. The rate of loading was 3 cm/day and it was completed after 100 days. The drains penetrated below the Ac$_3$ layer since the designed depth of improvement was 25.5 m from the embankment base.

The settlement curves for the four sections are shown in Fig. 4.88. After 175 days from construction completion of the embankment, the vertical settlement was 221.6 cm for SD section, 229.8 cm for PVD section, and 159.7 cm for the unimproved section. Table 4.7 presents the final settlement magnitudes at each section which was calculated by hyperbolic method.

Figure 4.89 exhibits the degree of consolidation at each layer of PVD (Cs$_2$) and unimproved section estimated under the following conditions:

a) immediately after the construction completion of the embankment,
b) 50 days after completion of embankment construction, and
c) 100 days after completion of embankment construction.

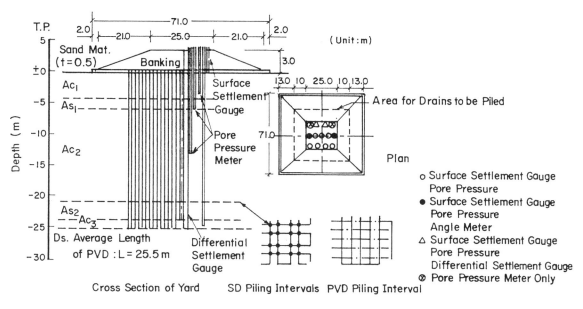

Fig. 4.87 Detailed Test Embankment of Soil Improvement Section

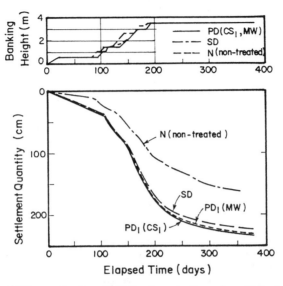

Fig. 4.88 Surface Soil Settlement Quantity of Each Section

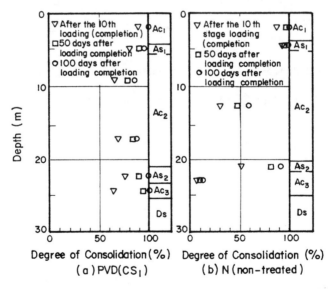

Fig. 4.89 Relationship Between Degree of Consolidation with Depth

After 100 days from completion of embankment construction, the degree of consolidation (U) at each layer for PVD (Cs_2) section reached 85%. The value of U at unimproved section is very much lower with 60% and 10% for Ac_2 and Ac_3 layers, respectively. Likewise, the behavior of the rate of consolidation for both PVD (MWs) and SD sections are similar to PVD (Cs_2) section. It was ascertained, therefore, that the consolidation settlement is almost uniform with depth by using vertical drains.

The relationship between horizontal displacement with depth is illustrated in Fig. 4.90. A maximum of 4 to 7 cm of horizontal displacement took place for all sections. For PVD section, there was no significant horizontal displacement that occurred 100 days after completion of the embankment. On the other hand, the horizontal displacement for the unimproved section progressed to notable magnitude even after 100 days from embankment completion. The end of primary consolidation was already attained 100 days after embankment construction for PVD section while the primary settlement was still in progress for unimproved section.

Figure 4.91 presents the behavior of the excess pore pressure with time. The most significant difference in terms of pore pressure behavior between the improved and unimproved section can be observed at the middle of the Ac_2 layer. For the improved section, the pore pressure increased during the construction period and it dissipated significantly thereafter. The pore pressure at the unimproved section, on the other hand, showed considerable increase during the embankment construction period but later showed very little dissipation. This contrast in the behavior of pore pressure between the improved and unimproved section has therefore confirmed the effectiveness of vertical drains as ground improvement technique.

4.23 CONCLUSIONS

Vertical drains are used to accelerate the consolidation of soft clay foundations. The state-of-practice of using vertical drains has been outlined in this paper. The theory of radial consolidation and its solution have been presented including the effects of well-resistance and smear. The influence of the different flow parameters of the soil have been examined. The diameter of the smeared zone can be assumed to be twice the equivalent diameter of the mandrel and the horizontal permeability coefficient in the smeared zone is approximately equal to the corresponding values in the vertical direction. A design chart has been suggested for preliminary design of PVD for convenience. Moreover, evaluations of PVD performance considering actual conditions have been made including the effects of PVD deformation, factors affecting discharge capacity and effects of smear zone. Finally, the applications of vertical drains in numerous case studies in soft Bangkok clay in Thailand and in soft Ariake clay in Japan have been presented.

4.24 REFERENCES

Akagi, T. (1977a), Effect of mandrel-driven sand drains on strength, Proc. 8th Intl. Conf. Soil Mech. and Found. Eng'g., Tokyo, Vol. 1, pp. 3-6.

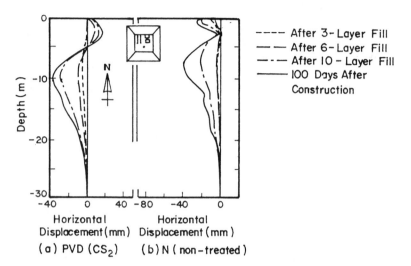

Fig. 4.90　Relationship Between Subsoil Horizontal Displacement with Depth

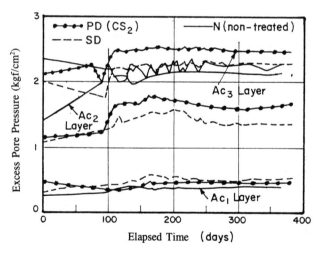

Fig. 4.91　Relationship Between Dissipation Characteristics with Time at Layers Ac_1, Ac_2, and Ac_3

Akagi, T. (1977b), Effect of displacement type sand drains on strength and compressibility of Soft Clays, Publications Dept. of Civil Eng'g., Tokyo Univ., Japan, 403 pp.

Akagi, T. (1979), Consolidation caused by mandrel-driven sand drains, Proc. 6th Asian Regional Conf. Soil Mech. and Found. Eng'g., Singapore, Vol. 1, pp. 125-128.

Akagi, T. (1981), Effects of mandrel-driven sand drains on soft clay, Proc. 10th Intl. Conf. Soil Mech. and Found. Eng'g., Stockholm, Vol. 1, 1981, pp. 581-584.

Albakri, W.H., Othman K., How, K.T., and Chan, P.C.(1990), Vertical drain embankment trial at Sungai Juru, Proc. Sem. N-S Expressway, Kuala Lumpur, Malaysia, pp. 195-205.

Asaoka, A. (1978), Observational procedure for settlement prediction, Soils and Foundations, Vol. 18, No. 4, pp. 87-101.

Asaoka, A., Matsuo, M., and Kodaka, T. (1991), Undrained bearing capacity of clay with sand piles, Proc. 9th Asian Reg. Conf. on Soil Mech. and Found. Eng'g., Bangkok, Thailand, Vol. 1, pp. 467-470.

Balasubramaniam, A.S., Brenner, R.P., Mallawaaratchy, G.U., and Kuvijitjaru, S. (1980), Performance of sand drains in a test embankment in soft Bangkok clay, Proc. 6th Southeast Asian Conf. Soil Eng'g., Taipei, pp. 447-468.

Barron, R.A. (1948), Consolidation of fine-grained soils by drain wells, Trans. ASCE No. 2346, pp. 718-754.

Belloni, L., Bondioli, S., and Sembenelli, P.(1979), Building permanent structures on swampy soils for Kalayaan Pumped Storage Plant, Philippines, Proc. Asian Reg. Conf. Soil Mech. and Found. Eng'g., Singapore, Vol. 2, pp. 215-225.

Bergado, D.T. (1992), Investigations of effectiveness of Flodrain vertical drains on 'undisturbed' soft Bangkok clay using laboratory model test, Report submitted to General Engineering Ltd., Bangkok, Thailand, and Geosynthetics Department Nylex (Malaysia) Berhad, Malaysia.

Bergado, D.T., Miura, N., Singh, N., and Panichayatum, B. (1988), Improvement of soft Bangkok clay using vertical band drains based on full-scale test, Proc. Int. Conf. Eng'g. Problems of Reg. Soils, Beijing, China, pp. 379-384.

Bergado, D.T., Ahmed, S., Sampaco, C.L., and Balasubramaniam, A.S. (1990a), Settlements of Bangna-Bangpakong Highway on soft Bangkok clay, J. Geotech. Eng'g. Div., ASCE, Vol. 116, No. 1, pp. 136-155.

Bergado, D.T., Singh, N., Sim, S.H., Panichayatum, B., Sampaco, C.L., and Balasubramaniam, A.S.(1990b), Improvement of soft Bangkok clay using vertical geotextile band drains compared with granular piles, Geotextiles and Geomembranes J., Vol 9, No.3, pp. 203-231.

Bergado, D.T., Asakami, H., Alfaro, M.C., and Balasubramaniam, A.S. (1991), Smear effects of vertical drains on soft Bangkok clay, J. Geotech. Eng'g. Div., ASCE, Vol. 117, No. 10 pp. 1509-1530.

Bergado, D.T., Enriquez, A.S., Sampaco, C.L., Alfaro, M.C., and Balasubramaniam, A.S. (1992), Inverse analysis of geotechnical parameters on improved soft Bangkok clay, J. Geotech. Eng'g. Div., ASCE, Vol. 118, No.7, pp. 1012-1030.

Bergado, D.T., Alfaro, M.C., and Balasubramaniam, A.S. (1993a), Improvement of soft Bangkok clay using vertical drains, Geotextiles and Geomembranes J., Vol. 12, No. 7, pp. 615-664.

Bergado, D.T., Mukherjee, K., Alfaro, M.C., and Balasubramaniam, A.S. (1993b), Prediction of vertical band drain performance by finite element method, Geotextiles and Geomembranes J., Vol. 12, No. 6, pp. 567-586.

Bergado, D.T., and Long, P.V. (1994), Numerical analysis of embankment on subsiding ground improved by vertical drains, Proc. Int'l. Conf. Soil Mech. and Found. Eng'g., New Delhi, pp. 1361-1366.

Brenner, R.P., Balasubramaniam, A.S., Chotivittayathanin, R., and Pananookooln (1979), Pore Pressures from pile driving in Bangkok clay, Proc. 6th Asian Reg. Conf. Soil Mech. and Found. Eng'g., Singapore, Vol. 1, pp. 133-136.

Brenner, R.P., and Prebaharan, N. (1983), Analysis of sandwick performance in soft Bangkok clay, Proc. 8th European Conf. Soil Mech. and Found. Eng'g., Helsinki, pp. 579-586.

Carillo, N. (1942), Simple two-and three-dimensional cases in the theory of consolidation of soils, J. of Mathematics and Physics, Vol. 21, No. 1, pp. 1-5.

Carroll, R.G. Jr. (1983), Geotextile filter criteria, TRR 916, Engineering Fabrics in Transportation Construction, Washington DC., pp. 46-53.

Casagrande, L., and Poulos, S. (1969), On effectiveness of sand drains, Canadian Geotech. J., Vol. 6, No. 3, pp. 236-287.

Cedergreen, H. (1972), Seepage control in earth dams, In Embankment Dam Eng'g. - Casagrande Volume, John Wiley and Sons, pp. 21-45.

Chen, R.H., and Chen, C.N., (1986), Permeability characteristics of prefabricated vertical drains, Proc. 3rd Intl. Conf. on Geotextiles, Vienna, Austria, pp. 785-790.

Cheung, Y.K., and Lee, P.K. (1991), Some remarks on two and three dimensional consolidation analysis of sand-drained ground, Computers and Geotechnics, Vol. 12, pp. 73-87.

Choa, V., Karunaratne, G.P., Ramaswamy, S.D., Vijiaratnam, A., and Lee, S.L. (1979), Pilot test for soil stabilization at Changi Airport, Proc. 6th Asian Reg. Conf. Soil Mech. and Found. Eng'g., Singapore, Vol. 1, pp. 141-144.

Choa, V., Karunaratne, G.P., Ramaswamy, S.D., Vijiaratnam, A., and Lee, S.L. (1981), Drain performance in Changi marine clay, Proc. 10th Intl. Conf. Soil Mech. and Found. Eng'g., Stockholm, Vol. 3, pp. 623-626.

Chou, N.S., Chou, K.T., Lee, C.C., and Tsai, K.W. (1980), Preloading by water testing laminated sand drains for a 65,000 ton raw water tank in Taiwan, Proc. 6th Southeast Asian Reg. Conf. Soil Eng'g., Taipei, pp. 485-492.

Duncan, J.M., D'Ozario, T.B., Chang, C.S., Wong, K.S., and Namiq, L.I. (1981), CON2D: A finite element computer program for analysis of consolidation, Report No. UCB/GT/81-01 to US Army Engineers Waterways Experiment Station, Vickburg, Mississippi, and Nikken-Sekkei, Japan, Univ. of California, Berkeley.

D'Orazio, T.B., and Duncan,J.M. (1982), CONSAX: A computer program for axisymmetric finite element analysis of consolidation, Report No. UCB/GT/82-01 to Nikken-Sekkei, Japan, Univ. of California, Berkeley, Calif.

Hansbo, S. (1979), Consolidation of clay by bandshaped prefabricated drains. Ground Eng'g., Vol. 12, No. 5, pp. 16-25.

Hansbo, S. (1981), Consolidation of fine-grained soils by prefabricated drains, Proc. l0th Intl. Conf. Soil Mech. and Found. Eng'g., Stockholm, Vol. 3, pp. 12-22.

Hansbo, S. (1987), Design aspects of vertical drains and lime column installations, Proc. 9th Southeast Asian Geotech. Conf., Bangkok, Thailand, Vol. 2, pp. 8-12.

Holtz, R.D., and Holm, G. (1973), Excavation and sampling around some sand drains at Ska-Edeby, Sweden, Proc. Nordic Geotech. Meeting, Trondheim, NGI, Oslo.

Holtz, R.D. (1987), Preloading with prefabricated vertical strip drain, Geotextiles and Geomembranes J., Vol. 6, pp. l09-131.

Jamiolkowski, M., and Lancellotta, R. (1981), Consolidation by vertical drains-uncertainties involved in prediction of settlement rates, Panel Discussion, Proc. 10th Intl. Conf. Soil Mech. and Found. Eng'g., Stockholm.

Jamiolkowski, M., Lancellotta, R., and Wolski, W. (1983), Precompression and speeding up consolidation, General Report, Special Session 6, Proc. 8th Europe Conf. Soil Mech. and Found. Eng'g., Helsinki, Vol. 3, pp. 1201-1226.

Jamiolkowski, M., Ladd, C.C., Germaine, J.T., and Lancellotta, R. (1985), New developments in field and laboratory testing of soils, Theme Lecture, Proc. llth Intl. Conf. Soil Mech. and Found. Eng'g., San Francisco.

Janbu, N., Bjerrum, L., and Kjaernsli, B. (1956), Veileduing ved lesning av fundamenteringsoppgaver, NGI Publ. No. l6.

Kamei, K., Hiratsuka, T., and Nakamizo, A. (1984), Drainage and shearing characteristics of normal consolidated viscous clay (In Japanese), J. of Soil Eng'g. Society, Vol. 19, pp. 390-392.

Kamon, M. (1983), Design and performance criteria for methods of precompression and speeding up of consolidation, other than vertical drains, Specialty Session 6, Proc. 8th Europian Conf. Soil Mech. and Found. Eng'g., Helsinki.

Kellner, L., Bally, R.T., and Matter, S. (1983), Some aspects concerning retaining capacity of geotextiles, Proc. 2nd Intl. Conf. on Geotextiles, Las Vegas, Nevada, pp. 85-90.

Kobayashi, M., Minami, J., and Tsuchida, T. (1990), Determination method of horizontal consolidation coefficient of clay, Tech. Report, Research Center of Harbor Eng'g. of Transport Ministry, Vol. 29, No.2, pp. 63-83.

Ladd, C.C. (1991), Stability evaluation during staged construction, J. Geotech. Eng'g. Div., ASCE, Vol. 117, No.4, pp. 540-615.

Lee, S.L., Karunaratne, G.P., Yong, K.Y., and Ramaswamy, S.D. (1989), Performance of fibredrain in consolidation of soft soils, Proc. 12th Int. Conf. Soil Mech. and Found. Eng'g., Brazil.

Long, P.V. (1992), Numerical analysis of embankments improved by granular piles and vertical drains on subsiding ground, M. Eng'g. Thesis No. GT-91-10, A.I.T., Bangkok, Thailand.

Mesri, G. (1973), Coefficient of secondary compression, J. Soil Mech. and Found. Div., ASCE, Vol. 99, No. SM1, pp. 123-147.

Miura, N., Park, Y.M., and Madhav, M.R. (1993), Basic research on drainage performance of plastic-board drains, Japanese Civil Eng'g. Society, No. 481, III-25, pp. 31-40.

Moh, Z.C., and Woo, S.M. (1987), Preconsolidation of Bangkok clay by non-displacement sand drains and surcharge, Proc. 9th Southeast Asian Geotech. Conf., Vol. 2, pp. 8-184.

Mukherjee K. (1990), Prediction of field and laboratory performance of vertical drains by numerical modelling, M. Eng. Thesis No. GT-89-3, Asian Institute of Technology, Bangkok, Thailand.

NACO (1984), The Second International Airport, Master plan study, Design and construction phasing, Final Detailed Geotechnical Report, Vol. II, Test Sections, Their Performances and Evaluations, Airports Authority of Thailand.

Newmark, N. (1942), Influence charts for computations of stresses in elastic foundations, Bull. No. 338, Univ. of Illinois, Urbana, Illinois.

Nicholls, R.A. (1981), Deep vertical drain installation, Proc. 10th Intl. Conf. Soil Mech. and Found. Eng'g., Stockholm.

Onoue, A., Ting, N.H., Germaine, M., and Whitman, R.V. (1991), Permeability of disturbed zone around vertical drains, Proc. 1991 ASCE Geotech. Eng'g. Congress, Vol. 2, pp. 879-890.

Pilot, G. (1981), Methods of improving the engineering properties of soft clay, In Soft Clay Engineering, Elsevier Publ. Co., New York, pp. 637-691.

Poulos, H.G. (1967), Stresses and displacements in an elastic layer underlain by rough rigid base, Geotechnique, Vol. 17, No. 4, pp. 378-410.

Poulos, H.G., and Davis, E.H. (1974), Elastic solutions for soil and rock mechanics, John Wiley and Sons, Inc., New York.

Pradhan, T.B.S., Kamon, M., and Suwa, S. (1991), Design method of the evaluation of discharge capacity of prefabricated vertical drain, Proc. 9th Asian Reg. Conf. Soil Mech. and Found. Eng'g., Bangkok, Thailand.

Rahman, J., Suppiah, A., Yong, K.W., and Shahrizaila, Z. (1990), Performance of stage constructed embankment on soft clay with vertical drains, Proc. Sem. Geotech. Aspects of N-S Expressway, Kuala Lumpur, pp. 245-253.

Rixner, J.J., Kraemer, S.R., and Smith, A.D. (1986), Prefabricated vertical drains, Vol. 1 (Eng'g. Guidelines), Federal Highway Administration, Report No. FHWA-RD-86/168, Washington D.C.

Schober, W., and Teindl, H. (1979), Filter criteria for geotextiles, In Design Parameters in Geotechnical Eng'g., BGS, London, England, Vol. 2, pp. 121-129.

Schweiger, H.F., and Pande, G.N. (1988), Numerical analysis of a road embankment constructed on soft clay stabilized with stone columns, J. of Numerical Methods in Geomechanics, pp. 1329-1334.

Sherard, J.L., and Dunnigan, L.P. (1989), Critical filters for impervious soils, J. Geotech. Eng'g. Div., ASCE, Vol. 115, No. 7, pp. 927-947.

Shinsha, H., Hara, H., Abe, T., and Tanaka, A. (1982), Consolidation settlement and lateral displacement of soft ground improved by sand drains, Tsuchi-to-Kiso, Japan Society Soil Mech. and Found. Eng'g., Vol. 30, No. 2, pp. 7-12.

Skempton, A.W., and Bjerrum, L. (1957), A contribution to the settlement analysis of foundations on clay, Geotechnique, London, Vol. 7, pp. 168-178.

Suzuki, S., and Yamada, M. (1990), Construction of an artificial island for the Kansai Intl. Airport, Proc. Sem. Eng'g for Coastal Dev., IEM and Kosai Club, Malaysia, pp. 511-520.

Takai, T., Oikawa, K., and Higuchi, Y. (1989), Construction of an artificial island for the Kansai Intl. Airport, Proc. 1989 Sem. on Coastal Dev., Bangkok, Kosai Club, pp. 511-520.

Tanimoto, K., Ukita, Y., and Suematsu, N. (1979), Stabilization of reclaimed land by sand drains, Proc. 6th Asian Reg. Conf. Soil Mech. and Found. Eng'g., Singapore, pp. 187-190.

Terzaghi, K. (1929), Effect of minor details on the safety of dams, Bull. American Inst. of Mining Engrs., Tech. Publ. No. 215.

Terzaghi, K. (1943), Theoretical soil mechanics, John Wiley and Sons, New York.

Tominaga, M., Sakaki, T., Hashimoto, M., and Paulino, C.C. (1979), Embankment tests in reclamation area, Manila Bay, Proc. 6th Asian Reg. Conf. Soil Mech. and Found. Eng'g., Singapore, pp. 195-198.

Volders, R. (1984), Soil improvement by preloading and vertical drains at Port Klang Power Station Storage Facilities, Malaysia.

Vreeken, C., van den Berg, F., and Loxham, M. (1983), The effect of clay-drain interface erosion on the performance of band-shaped vertical drain, Proc. 8th Europ. Conf. Soil Mech. and Found. Eng'g., Helsinki, pp. 713-716.

Woo, S.M., Cheng, L.K., Dawn, J.D., and Ou, C.D. (1988), Use of vertical drains for soil improvement at bridge approaches, Proc. 8th Southeast Asian Geotech. Conf., Kuala Lumpur, pp. 2-23 to 2-30.

Woo, S.M., Moh, Z.C., Van Weele, A.F., Chotivittayathanin, R., and Transkarahart, T. (1989), Preconsolidation of soft Bangkok clay by vacuum loading combined with non-displacement sand drains, Proc. 12th Intl. Conf. Soil Mech. and Found. Eng'g., Brazil.

Chapter 5

GRANULAR PILES

5.1 GENERAL

The Central Plain of Thailand is situated on a flat deltaic-marine deposit with a north-south dimension of about 300 km and an east-west width of about 200 km. The presence of thick deposits of soft Bangkok clay and the effects of ground subsidence due to excessive pumping of groundwater create foundation problems to earth structures and infrastructures, especially at approach embankments (transition units) to bridges and viaducts. Basically, the problem lies in the occurrence of differential movements between the pile-supported bridges and viaducts and the ground-supported approach embankments. To mitigate such a natural hazard, granular piles are proposed as the most appropriate and viable foundation treatment.

Granular piles are composed of compacted sand or gravel inserted into the soft clay foundation by displacement method. The term "granular piles" used in this paper refers to the component of compacted gravel and/or sand piles. It also refers to those known as stone columns. The ground improved by compacted granular piles is termed as composite ground. When loaded, the pile deforms by bulging into the subsoil strata and distributes the stresses at the upper portion of the soil profile rather than transferring the stresses into the deeper layers, thus causing the soil to support it. As a result, the strength and bearing capacity of the composite ground can be increased and the compressibility reduced. In addition, lesser stress concentration is developed on the granular piles. Since the component materials are granular and have higher permeability, granular piles could also accelerate the consolidation settlements and minimize the post construction settlements.

5.2 METHODS OF GRANULAR PILE CONSTRUCTION

Various methods for installation of granular piles have been used all over the world depending on their proven applicability and availability of equipment in the locality. The following common methods will be briefly described with their corresponding references.

5.2.1 Vibro-Compaction Method

The vibro-compaction method is used to improve the density of cohesionless, granular soils using a vibroflot which sinks in the ground under its own weight and with the assistance of water and vibration (Baumann and Bauer, 1974; Engelhardt and Kirsch, 1977). After reaching the predetermined depth, the vibroflot is then withdrawn gradually from the ground with subsequent addition of granular backfill thereby causing compaction. The process is repeated in stages forming a compacted column of granular piles. Figure 5.1 illustrates the steps in vibro-compaction process. The range of grain size distribution of soils suitable for this method is shown in Fig. 5.2. The fines content is limited to not more than 18% passing No. 200 U.S. standard sieves.

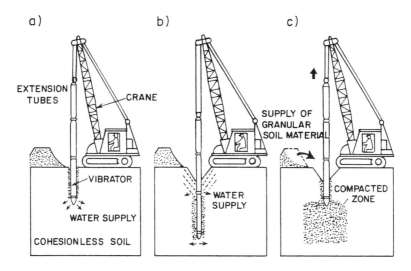

Fig. 5.1 The Vibro-Compaction Process (Baumann and Bauer, 1974)

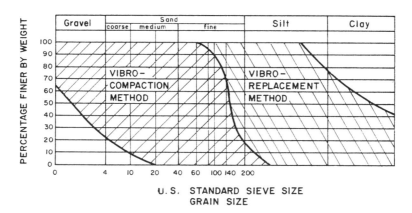

Fig. 5.2 Range of Soils Suitable for Vibro-Compaction Methods (Baumann and Bauer, 1974)

5.2.2 Vibro-Compozer Method

This method is popularized in Japan and is used for stabilizing soft clays in the presence of high groundwater level (Aboshi et al. 1979; Aboshi and Suematsu, 1985; Barksdale, 1981). The installation procedures are illustrated in Fig. 5.3. The resulting pile is usually termed as the sand compaction pile. The sand compaction piles are constructed by driving the casing pipe to the desired depth using a heavy and vertical vibratory hammer located at the top of the pipe. The casing is filled with a specified volume of sand and the casing is then repeatedly extracted and partially redriven using the vibratory hammer starting from the bottom. The process is repeated until a fully penetrating compacted granular pile is constructed.

5.2.3 Cased-Borehole Method

In this method, the piles are constructed by ramming granular materials in the prebored holes in stages using a heavy falling weight (usually of 15 to 20 kN) from a height of 1.0 to 1.5 m (Datye and Nagaraju, 1975; Datye, 1978; Datye and Nagaraju, 1981; Bergado et al. 1984; Ranjan, 1989). The method is a good substitute for vibrator compaction considering its low cost. However, disturbance and subsequent remolding by the ramming operation may limit its applicability to sensitive soils. The method is useful in developing countries utilizing only an indigenous equipment in contrast to the methods described above which require special equipments and trained personnel (Rao, 1982; Ranjan and Rao, 1983). The installation process is illustrated in Fig. 5.4.

Another method of granular pile construction is the vibro-replacement method. This method is discussed in detail in Chapter 3 (Deep Compaction).

5.3 ENGINEERING BEHAVIOR OF COMPOSITE GROUND

The performance of composite ground is best investigated in terms of ultimate bearing capacity, settlement, and general stability. In the following sections, basic relationships of the composite ground as well as failure mechanisms of granular piles on homogeneous soft clay are first described and the ultimate bearing capacity, settlement, and stability of the composite ground based on experimental and analytical studies are then presented.

5.3.1 Basic Relationships

The tributary area of the soil surrounding each granular pile is closely approximated by an equivalent circular area. For an equilateral triangular pattern of granular piles, the equivalent circle has an effective diameter of:

$$D_e = 1.05S \tag{5.1}$$

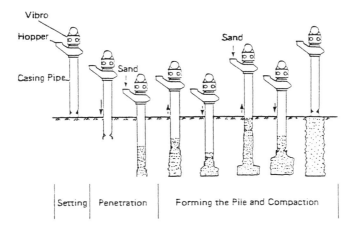

Fig. 5.3   The Vibro-Composer Method (Aboshi and Suematsu, 1985)

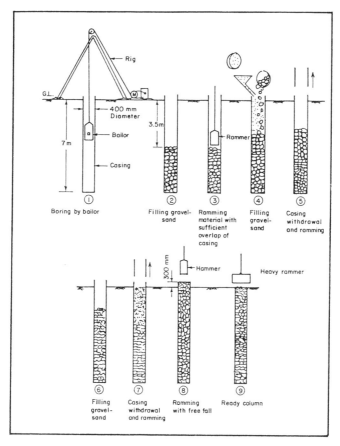

Fig. 5.4   The Cased-Borehole Method (Datye and Nagaraju, 1975)

while for a square pattern,

$$D_e = 1.13S \tag{5.2}$$

where S is the center to center spacing of granular piles. The equilateral triangular pattern gives the most dense packing of granular piles in a given area. The resulting cylinder of composite ground with diameter D_e enclosing the tributary soil and one granular pile is known as the unit cell.

Figure 5.5 illustrates the area replacement factor as well as the stress concentration in the granular pile. The area replacement ratio is defined as the ratio of the granular pile area over the whole area of the equivalent cylindrical unit within the unit cell and expressed as:

$$a_s = \frac{A_s}{A_s + A_c} \tag{5.3}$$

where A_s is the horizontal area of a granular pile, and A_c is the horizontal area of the clayey ground surrounding the pile. The area replacement ratio can also be expressed in terms of the diameter (D) and spacing (S) of the granular pile as follows:

$$a_s = c_1 \left(\frac{D}{S}\right)^2 \tag{5.4}$$

where c_1 is a constant depending upon the pattern of granular piles used; for the square pattern $c_1 = \pi/4$ and for the equilateral triangular pattern $c_1 = \pi/(2\sqrt{3})$.

When the composite ground is loaded, studies indicated that concentration of stress occurs in the granular pile accompanied by the reduction in stress which occurs in the surrounding less stiff clayey soil (Fig. 5.5). This can be explained by the fact that, when loaded, the vertical settlement of the granular pile and the surrounding soil is approximately the same, causing the occurrence of stress concentration in the granular pile which is stiffer than the surrounding cohesive or loose cohesionless soil. The distribution of vertical stress within the unit cell can be expressed by a stress concentration factor, n, defined as:

$$n = \frac{\sigma_s}{\sigma_c} \tag{5.5}$$

where σ_s is the stress in the granular pile, and σ_c is the stress in the surrounding cohesive soil. The magnitude of stress concentration also depend on the relative stiffness of the granular pile and the surrounding soil. The variation of stress concentration factor with area replacement ratio compiled by Barksdale and Bachus (1983) ranged from 2 to 5. Meanwhile, Aboshi et al. (1979) and Bergado et al. (1987) obtained a higher value of as much as 9. The higher stress concentration factor obtained by Bergado et al. (1987) was probably due to the high rigidity of the plates used during the load tests. From full scale test embankment observations on soft Bangkok clay at the low area replacement ratio of 0.06, the stress concentration factor of 2 was obtained and was found to decrease to 1.45 with the increasing applied load (Bergado et al. 1988). The average stress, σ, over the unit cell area corresponding to a given area replacement ratio, a_s, is expressed as:

$$\sigma = \sigma_s a_s + \sigma_c(1-a_s) \qquad (5.6)$$

The stresses in the pile and the clay using the stress concentration factor are:

$$\sigma_s = \frac{n\sigma}{[1+(n-1)a_s]} = \mu_s \sigma \qquad (5.7)$$

$$\sigma_c = \frac{\sigma}{[1+(n-1)a_s]} = \mu_c \sigma \qquad (5.8)$$

where μ_s and μ_c are the ratio of stress in the pile and clay, respectively, to the average stress over the unit cell area.

5.3.2 Failure Mechanisms

In practice, granular piles are usually constructed fully penetrating a soft soil layer overlying a firm stratum. It may be constructed also as floating piles with their tips embedded within the soft clay layer. Granular piles may fail individually or as a group. The failure mechanisms for a single pile are illustrated in Figs. 5.6, respectively, indicating the possible failures as: a) bulging, b) general shear, and c) sliding.

5.4 ULTIMATE BEARING CAPACITY OF SINGLE, ISOLATED GRANULAR PILE

For single, isolated granular piles, the most probable failure mechanism is bulging failure. This mechanism develops whether the tip of the pile is floating in the soft soil or fully penetrating and bearing on a firm layer. The lateral confining stress which supports the granular pile is usually taken as the ultimate passive resistance which the surrounding soil can mobilize as the pile bulges outward. Most of the approaches in predicting the ultimate bearing capacity of a single, isolated granular pile has been developed based on the above assumption. Table 5.1 tabulates the different methods to estimate the ultimate bearing capacity corresponding to bulging, general shear, and sliding modes of failure as presented by Aboshi and Suematsu

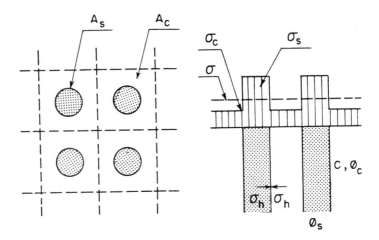

Fig. 5.5 Diagram of Composite Ground

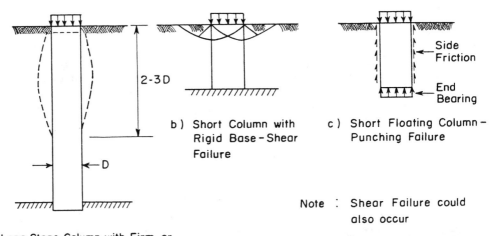

a) Long Stone Column with Firm or Floating Support-Bulging Failure

b) Short Column with Rigid Base-Shear Failure

c) Short Floating Column - Punching Failure

Note : Shear Failure could also occur

Fig. 5.6 Failure Mechanisms of a Single Granular Pile in a Homogeneous Soft Layer (Barksdale and Bachus, 1983)

Table 5.1 Estimation of Ultimate Bearing Capacity (Aboshi and Suematsu, 1985)

MODE OF FAILURE	DERIVED FORMULA	REFERENCE
Bulging	$q_{ult} = (\gamma_c Z K_{pc} + 2C_o\sqrt{K_{pc}}) \dfrac{1 + \sin\phi_s}{1 - \sin\phi_s}$	Greenwood (1970)
	$q_{ult} = (F'_c C_o + F'_q Q_o) \dfrac{1 + \sin\phi_s}{1 - \sin\phi_s}$	Vesic (1972) Datye and Nagaraju (1975)
	$q_{ult} = (\sigma_{ro} + 4C_o) \dfrac{1 + \sin\phi_s}{1 - \sin\phi_s}$	Hughes and Withers (1974)
	$q_{ult} = \dfrac{1 + \sin\phi_s}{1 - \sin\phi_s}(4C_o + \sigma_{ro} + K_o q_s)(W/B)^2 + [1-(W/B)^2]q_s$	Madhav et al. (1979)
General Shear	$q_{ult} = C_o N_c + (1/2)\gamma_c B N_\gamma + \gamma_c D_f N_q$	Madhav and Vitkar (1978)
	$q_{ult} = 2A_s(K_{pc}q_o + 2C_o\sqrt{K_{pc}}) + (1/K_{as})[3d_s K_{pc}\gamma_c(1-(3d_s/2L))]$	Wong (1975)
	$q_{ult} = (1/2)\gamma_c B \tan^3\psi + 2C_o \tan^2\psi + 2(1-a_s)C_o \tan\psi$ $\psi = 45° + \dfrac{\tan^{-1}(\mu_s a_s \tan\phi_s)}{2}$	Barksdale and Bachus (1983)
Sliding Surface	$\tau = (1-a_s)C_o + (\gamma_s z + \mu_s \sigma_z)a_s \tan\phi_s \cos^2\theta$ $\mu_s = \dfrac{n}{1 + (n-1)a_s}$	Aboshi et al. (1979)

Note: Refer to APPENDIX for Notations

(1985). A relationship between ultimate bearing capacity and area replacement ratio is shown in Fig. 5.7. The relationship between internal friction angle of granular material, strength of the surrounding clay, and the ultimate bearing capacity of single granular pile is shown in Fig. 5.8.

5.5 ULTIMATE BEARING CAPACITY OF GRANULAR PILE GROUPS

The common method for estimating the ultimate bearing capacity of granular pile groups assumed that the angle of internal friction in the surrounding cohesive soil and the cohesion in the granular pile are negligible. Furthermore, the full strength of both the granular pile and cohesive soil has been mobilized. The pile group is also assumed to be loaded by rigid foundation. The ultimate bearing capacity of granular pile groups as suggested by Barksdale and Bachus (1983) is determined by approximating the failure surface with two straight rupture lines as shown in Fig. 5.9. Assuming the ultimate vertical stress, q_{ult}, and the ultimate lateral stress, σ_3, to be the principal stresses, then the equilibrium of the wedge requires:

$$q_{ult} = \sigma_3 \tan^2\beta + 2c_{avg}\tan\beta \tag{5.9}$$

where:

$$\sigma_3 = \frac{\gamma_c B \tan\beta}{2} + 2c \tag{5.10}$$

$$\beta = 45 + \frac{\phi_{avg}}{2} \tag{5.11}$$

$$\phi_{avg} = \tan^{-1}(\mu_s a_s \tan\phi_s) \tag{5.12}$$

$$c_{avg} = (1 - a_s) c \tag{5.13}$$

where γ_c = saturated or wet unit weight of the cohesive soil; B = foundation width; ß = failure surface inclination; c = undrained shear strength within the unreinforced cohesive soil; ϕ_s = angle of internal friction of the granular soil; ϕ_{avg} = composite angle of internal friction; c_{avg} = composite cohesion on the shear surface.

The development of the above approach did not consider the possibility of a local bulging failure of the individual pile. Hence, the approach is only applicable for firm and stronger cohesive soils having an undrained strengths greater than 30-40 kN/m². However, it is useful for approximately determining the relative effects on ultimate bearing capacity design variables such as pile diameter, spacing, gain in shear strength due to consolidation, and angle of internal friction.

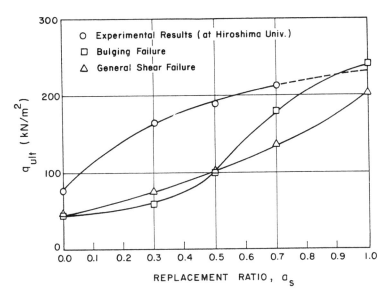

Fig. 5.7 Relationship Between Ultimate Bearing Capacity and Area Replacement Ratio (Aboshi and Suematsu, 1985)

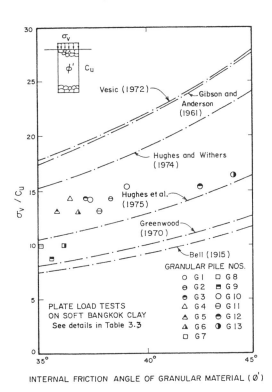

Fig. 5.8 Relationship Between Internal Friction Angle of Granular Material, Strength of Surrounding Clay and Ultimate Bearing Capacity of Single Granular Pile (Bergado and Lam, 1987)

For the case of the soft and very soft cohesive soils, the pile group capacity is predicted using the capacity of a single, isolated pile located within a group and to be multiplied by the number of piles (Barksdale and Bachus, 1983). The ultimate bearing capacity for a single, isolated pile in this case is expressed as:

$$q_{ult} = cN_c' \tag{5.14}$$

where N_c = composite bearing capacity factor for the granular pile which ranges from 18 to 22. For the soft Bangkok clay, N_c ranges from 15 to 18 using an initial pile diameter of 25.4 cm with the gravel compacted by a 0.16 ton hammer dropping 0.70 m (Bergado and Lam, 1987).

5.6 SETTLEMENT OF THE COMPOSITE GROUND

Most of the approaches in estimating settlement of the composite ground assumed an infinitely wide, loaded area reinforced with granular piles having a constant diameter and spacing. For this loading condition and geometry, the unit cell idealization is assumed to be valid. The model of a unit cell loaded by a rigid plate is analogous to a one-dimensional consolidation test. Thus, the unit cell is confined by a rigid frictionless wall and the vertical strains at any horizontal level are uniform. Different methods for estimating the settlement of the composite ground are summarized in Table 5.2. The settlement reduction ratio is expressed as:

$$R = \frac{S_t}{S_o} \tag{5.15}$$

where S_t = settlement of the composite ground and S_o = settlement of the unimproved ground. In the case of the equilibrium method, estimation of the settlement of the composite ground is expressed as:

$$S_t = m_v(\mu_c \sigma) \tag{5.16}$$

where m_v = modulus of volume compressibility and H = thickness of layer. The settlement reduction ratio is also expressed as a function of the area replacement ratio (a_s), angle of internal friction of the granular materials (ϕ_s), stress concentration factor, etc. Figure 5.10 shows the relationships between the settlement reduction ratio and the aforementioned parameters based on different methods together with the results from the work of Bergado et al. (1987) on soft Bangkok clay.

5.7 SLOPE STABILITY OF THE COMPOSITE GROUND

Granular piles could be used to increase the stability of slopes and embankments constructed over soft cohesive ground. The method of stability analysis on a composite ground performed exactly in the same manner as for a normal slope stability problem except that stress concentration is considered. When circular rotational failure is expected, the simplified method

Table 5.2 Estimation of Settlement of Composite Ground (Aboshi and Suematsu, 1985)

METHODS	CONTENTS	REFERENCES
Equilibrium Method	$S_t = m_v(\mu_c \sigma)H$ $R = \mu_c = \dfrac{1}{1 + (n-1)a_s}$	Aboshi et al. (1979)
Priebe Method	$\dfrac{1}{R} = 1 + a_s \left[\dfrac{1/2 + f(\mu, a_s)}{(K_A)_s f(\mu, a_s)} - 1 \right]$ $f(\mu, a_s) = \left[\dfrac{1-\mu^2}{1-\mu-2\mu^2} \right] \left[\dfrac{(1-2\mu)(1-a_s)}{1-2\mu+a_s} \right] \quad (K_A)_s = \tan^2\left(45°- \dfrac{\phi_s}{2}\right)$	Priebe (1976)
Granular Wall Method	$S_t = RH(1-\mu^2)\left(1 - \dfrac{\mu^2}{1-\mu^2}\right) \dfrac{\sigma}{E}$ $R = f(a_s, \phi_s, \mu, \sigma/E)$	Van Impe and De Beer (1983)
Incremental Method	$\varepsilon_v = (1-a_s) \dfrac{C_c}{1+e_0} \log_{10}\left[\dfrac{(P_o)_{vc}+\Delta P}{(P_o)_{vc}}\right]$ $\Delta P = \dfrac{(\Delta P)^*_{vc}}{1+2K_O}\left[1 + K + K_O \begin{pmatrix} K \text{ if } K>1 \\ 1 \text{ if } K\leq 1 \end{pmatrix}\right]$ $K = K_O + \dfrac{1}{\varepsilon_v}\left[\sqrt{\dfrac{1}{1-\varepsilon_v}} - 1\right] \dfrac{\sqrt{a_s}}{1-\sqrt{a_s}}$ $(\Delta P)^*_{vc} = \dfrac{(\Delta P)^*_v + (P_o)_{vc}a_s - K_O(P_o)_{vc}a_s\tan^2(45+\phi_s/2)}{KFa_s \tan^2(45+\phi_s/2)}$ $R_p = \dfrac{\varepsilon_v}{\dfrac{C_c}{1+e_0}\log_{10}\left[\dfrac{(P_o)_{vc}+(\Delta P)^*_v}{(P_o)_{vc}}\right]}$	Goughnour (1983) Baumann and Bauer (1974) Hughes et al. (1975)
Finite Element Method	$[K_E](\Delta\sigma^{(m-1)}) = ((\Delta F_E) + [K_c^{(m)}](\Delta\sigma^{(m)}) + (\Delta F_{DN}^{(m)})$	alaam and Poulos (1983)

Note: Refer to APPENDIX for Notations

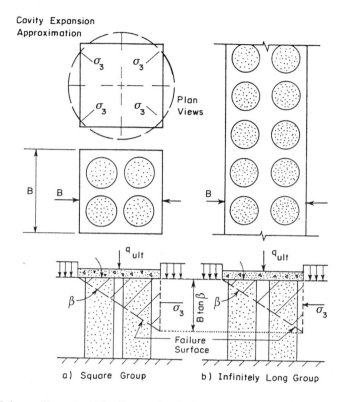

Fig. 5.9 Granular Pile Group Analysis (Barksdale and Bachus, 1983)

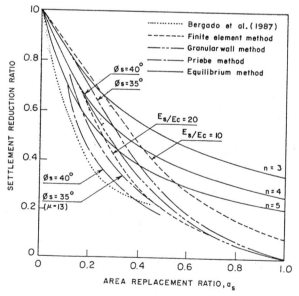

Fig. 5.10 Comparison of Estimating Settlement Reduction of Improved Ground (Aboshi and Suematsu, 1985)

of slices is recommended (Fredlund and Krahn, 1977; Whitman and Bailey, 1967). The method of stability analysis is extended, as an approximation, to evaluate the stability over large areas improved with granular piles and imposed with heavy loads. Stability analysis are usually carried out with the implementation of computer programs. The three general techniques used in stability analysis of the composite ground consist of the profile method, the average shear strength method, and the lumped parameter method, described as follows (see Barksdale and Bachus, 1983):

5.7.1 Profile Method

In the profile method, each row of the granular piles is converted into an equivalent, continuous strip with width, w, as shown in Fig. 5.11. Each strip of granular and cohesive soils is then analyzed using its actual geometry and material properties. For an economical design, the stress concentration developed in the piles must be taken into consideration. The stress concentration in the granular pile results in an increase in resisting shear force. The effect of the stress concentration is being handled by placing thin, fictitious strips of soil above the in-situ soil and granular piles at the embankment interface (see Fig. 5.11). The weight of the fictitious strips of soil placed above the granular piles is relatively large to cause the desired stress concentration when added to the stress caused by the embankment. The weight of the fictitious soil placed above the in-situ soil must be negative to give proper reduction in stress when added to that caused by the embankment. The fictitious strips placed above the in-situ soil and granular piles would have no shear strength, and their weights are respectively expressed as:

$$\gamma_f^c = \frac{(\mu_c - 1)\gamma_1 H'}{\overline{T}} \tag{5.17}$$

$$\gamma_f^s = \frac{(\mu_s - 1)\gamma_1 H'}{\overline{T}} \tag{5.18}$$

where μ_c and μ_s are the stress concentration factors of the in-situ soil and granular piles, respectively, and the other terms are defined as indicated in Fig. 5.11. It must be noted that limits should be imposed on the radius and/or grid size of circle centers so that the critical circle should not be controlled by the weak, fictitious interface layer.

5.7.2 Average Shear Strength Method

The average shear strength method is widely used in stability analysis for sand compaction piles (Aboshi et al. 1979; Barksdale, 1981). The method considers the weighted average material properties of the materials within the unit cell (Fig. 5.12). The soil having the fictitious weighted material properties is then used in stability analysis. Since average properties can be readily calculated, this approach is appealing for both hand and computer calculations.

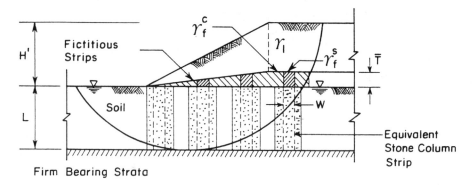

Fig. 5.11 Granular Pile Strip Idealization and Fictitious Soil Layer for Slope Stability Analysis (Barksdale and Bachus, 1983)

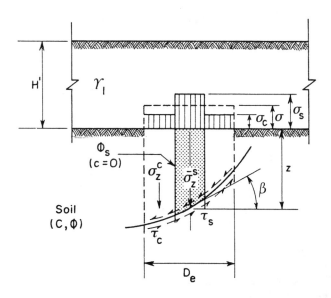

Fig. 5.12 Average Stress Method of Stability Analysis (Barksdale and Bachus, 1983)

200

However, average properties cannot be generally used in standard computer programs when stress concentration in the granular piles is considered in the analysis.

When stress concentration is considered, hand calculation is preferred. Within the unit cell, the granular pile has only internal friction, ϕ_s, and the surrounding soil is undrained but has cohesion, c, and internal friction, ϕ_c. The state of stresses within the unit cell is also shown in Fig. 5.13. The effective stresses in the granular pile and the total stress in the surrounding soil are respectively expressed as:

$$\sigma_z^s = \gamma_s z + \sigma \mu_s \quad (5.19)$$

$$\sigma_z^c = \gamma_c z + \sigma \mu_c \quad (5.20)$$

where γ_s = bouyant weight of the granular materials; γ_c = saturated unit weight of the surrounding soil; z = depth below the ground surface; σ = stress due to embankment loading, and the other terms are already defined. The shear strength of the granular pile and the surrounding cohesive soil are:

$$\tau_s = \left(\sigma_z^s \cos^2\beta\right)\tan\phi_s \quad (5.21)$$

$$\tau_c = c + \left(\sigma_z^c \cos^2\beta\right)\tan\phi_c \quad (5.22)$$

where ß = inclination of the shear surface with respect to the horizontal. The average weighted shear strength within the area tributary to the granular pile is:

$$\tau = (1-a_s)\tau_c + a_s\tau_s \quad (5.23)$$

The weighted average unit weight, γ_{avg}, within the composite ground used in calculating the driving moment is:

$$\gamma_{avg} = \gamma_s a_s + \gamma_c a_c \quad (5.24)$$

where γ_s and γ_c are the defined previously. In this approach, the weighted shear strength and unit weight are calculated for each row of granular piles and then used in conventional hand calculation.

When stress concentration is not considered, as in the case of some landslide problems, a standard computer analysis employing average strengths and unit weights, can be performed using a conventional computer program. Neglecting the cohesion in the granular materials and the stress concentration, the shear strength parameters for use in the average shear strength method are:

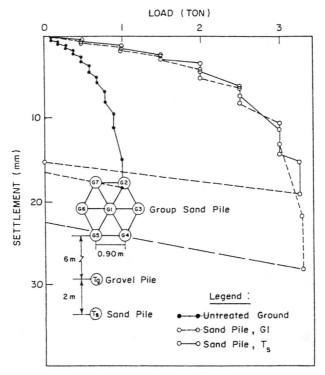

Fig. 5.13 Load-Settlement Relationship of Granular Piles from Full-Scale Load Tests (Bergado et al., 1984)

$$c_{avg} = c(a_s) \tag{5.25}$$

$$(\tan\phi)_{avg} = \frac{\gamma_s a_s \tan\phi_s + \gamma_c a_c \tan\phi_c}{\gamma_{avg}} \tag{5.26}$$

where γ_{avg} is given by Eq. (5.24) using bouyant weight for γ_s and saturated weight for γ_c. DiMaggio (1978) reported that the use of $(\tan\phi)_{avg}$ based just on the area ratio is not correct as can be demonstrated by considering the case when $\phi_c = 0$. If averages based on the area were used, then:

$$(\tan\phi)_{avg} = a_s \tan\phi_s \tag{5.27}$$

The above expression would be appropriate to use if $\gamma_{avg} = \gamma_s$, but incorrect if γ_{avg} is used to calculate the driving moment.

5.7.3 Lumped Parameter Method

The lumped parameter method can be used to determine the safety factor of selected trial circles by either hand calculations or with the aid of a computer. The general approach is described by Chambosse and Dobson (1982). The safety factor of the composite ground is calculated by:

$$SF = \frac{(RM + \Delta RM)}{(DM + \Delta DM)} \tag{5.28}$$

where RM = resisting moment, DM = driving moment, ΔRM = excess resisting moment due to granular piles, ΔDM = excess driving moment due to granular piles. The DM and RM are first calculated for the condition of unimproved ground. Then ΔRM and ΔDM are added to the previously calculated moments, RM and DM, respectively. The approach is generally suited for hand calculations. The use of computer programs is possible only when adding ΔRM and ΔDM which could be calculated by hand.

5.8 RATE OF PRIMARY CONSOLIDATION SETTLEMENT

Previous studies assumed that granular piles could accelerate the consolidation process in the same way as sand drains. In a cohesive soil reinforced with granular piles, pore water moves toward the pile in a curved path having both vertical and radial components of flow. The average degree of primary consolidation could be handled by considering the vertical and radial consolidation effects separately, as expressed by the following equation (Carillo, 1942):

$$U = 1 - (1-U_z)(1-U_r) \tag{5.29}$$

where U = average degree of consolidation of the cohesive layer considering both vertical and radial drainage, U_z = degree of consolidation considering only vertical flow, and U_r = degree of consolidation considering radial flow. The degree of consolidation in the vertical direction is calculated by Terzaghi's one-dimensional theory, while that in the radial direction is calculated by Barron's theory. The primary consolidation settlement at any time, t, is expressed as:

$$S_c(t) = U(S_{cf}) \tag{5.30}$$

where $S_c(t)$ = primary consolidation settlement at any time, t; S_{cf} = final primary consolidation settlement.

5.9 STRENGTH INCREASE OF CLAY DUE TO CONSOLIDATION

The rate of construction of embankments on ground improved with granular piles is frequently controlled to allow the shear strength to increase so that the required safety factor against instability is maintained. The undrained shear strength of a normally consolidated clay has been found to increase linearly with effective overburden pressure (Leonards, 1962). Consolidation results in an increase in effective stress due to the dissipation of pore pressure. For a cohesive soil having a linear increase in shear strength with effective stress, the increase in undrained shear strength, $\Delta c(t)$, with time due to consolidation can be expressed as:

$$c(t) = k_1[\sigma \mu_c][U(t)] \tag{5.31}$$

where $k_1 = c/\sigma$, constant of proportionality defining the linear increase in shear strength with effective stress; $U(t)$ = degree of consolidation of the clay at any time, t, and the other terms are defined previously. The ratio of the measured increase in undrained shear strength to the calculated ones is a function of area replacement ratio and stress concentration factor.

5.10 SECONDARY SETTLEMENT

Secondary settlement is important in organic soils and some soft clays, especially with the presence of free draining granular piles wherein primary consolidation is thought to occur in a short period of time. The prediction for secondary settlement is based on the work of Mesri (1973) and can be calculated by:

$$S_s = C_a H \log_{10} \frac{t_2}{t_1} \tag{5.32}$$

where S_s = secondary compression of the layer; C_α = physical constant evaluated by continuing a one-dimensional consolidation test past the end of primary consolidation for a suitable load increment; H = thickness of compressible layer; t_1 = time at the beginning of secondary compression (the time corresponding to 90% of primary consolidation is sometimes used); t_2 = time at which the value of secondary settlement is desired.

5.11 FULL-SCALE LOAD TESTS ON GRANULAR PILES

Initial full-scale load tests on granular piles were performed to study the feasibility of improving the soft Bangkok clay (see Bergado et al. 1984). The results indicated that the granular piles increased the bearing capacity more than 3 to 4 times that of the untreated ground. Further, the adjacent piles acted independently provided that the pile spacing is 3 times the pile diameter or greater, as shown in Fig. 5.13. An investigation of the behavior of granular piles with different densities and different proportions of gravel and sand on soft Bangkok clay was carried out by Bergado and Lam (1987). Table 5.3 shows that for the same granular materials, the ultimate bearing capacity increases with number of blows per layer because of the increase in the densities and friction angle. Using different proportions of gravel and sand shown in Table 5.3, the resulting load-settlement curves are compared in Fig. 5.14 and indicated a higher ultimate pile capacity for pure gravel. The average deformed shape of the granular piles is typically bulging type. It was observed that the maximum bulge occurred near the top of the pile and ranged from 10 cm to 30 cm below the ground surface. With an initial pile diameter of 30 cm, the measurements of bulge are in close agreement with the observations of Hughes et al. (1975), wherein the maximum bulge occurred near the ground surface at a depth approximately equal to one-half to one pile diameter. The results of full-scale load tests on granular piles with different sizes of plates were reported by Bergado et al. (1987). As expected, the settlement decreased as the size of the plate increased for the same amount of total load. It was also indicated that as the ratio D_e/D increased, the settlement of the treated ground approaches the settlement of untreated ground. This implies that beyond $D_e/D = 4.0$, the settlement of treated ground is almost the same as the settlement of untreated ground. It should be noted that D_e is directly related to the spacing of granular piles depending on the pattern used. The variations of stress concentration in granular piles and clay with D_e/D are shown in Fig. 5.15.

5.12 TEST EMBANKMENT ON GRANULAR PILES

A full-scale test embankment 2.4 m high was constructed by Sim (1986) on a granular pile-improved foundation, and was raised to a height of 4.0 m by Panichayatum (1987) to provide a meaningful basis of comparison with the performance of the nearby 4.0 m high test embankment constructed on Mebra vertical drain-improved foundation by Singh (1986). The performances of these two embankments have been already reported by Bergado et al. (1988,1990a). An evaluation by back-analysis of the geotechnical parameters of the foundation soil improved by granular piles and vertical drains was carried out by Enriquez (1989).

Table 5.3 Properties of Granular Piles (Bergado and Lam, 1987)

Group	1			2			3			4		5	
No. of pile	G1	G2	G3	G4	G5	G6	G7	G8	G9	G10	G11	G12	G13
Proportion of sand in volume		1.0			1.0			1.0			0.3		0.0
Proportion of gravel in volume		0.0			0.0			0.0			1.0		1.0
Blows per compacted layer		20			15			10			15		15
In-situ average density (t/m)	1.73	1.71	1.66	1.64	1.51	1.67	1.47	1.53	1.50	1.91	1.96	1.76	1.79
Average		1.70 t/m^3			1.61 t/m^3			1.50 t/m^3			1.94 t/m^3		1.74 t/m^3
Friction angle (degree)	39.1	38.4	37.2	37.0	36.0	37.6	35.1	36.2	35.6	37.4	37.9	42.5	44.7
Average		38.2°			36.9°			35.6°			37.7°		43.3°
Ultimate Load (tons)	3.50	3.25	3.25	3.25	3.00	3.00	2.25	2.25	2.00	3.25	3.00	3.50	3.75
Average		3.33 tons			3.08 tons			2.17 tons			3.13 tons		3.63 tons

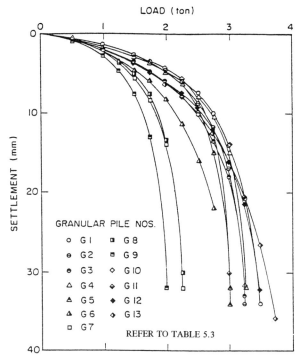

Fig. 5.14 Comparison of Load-Settlement Curves Between Piles with Different Number of Blows per each Compacted Layer During Installation (Bergado and Lam, 1987)

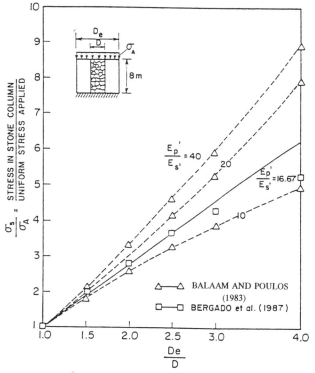

Fig. 5.15 Variation of Vertical Stress in Granular Pile with De/D (Bergado et al. 1987)

The site is located at the Asian Institute of Technology (AIT) campus, about 42 km north of Bangkok at the Central (Chao Phraya) Plain of Thailand. The soil profile at the site, together with the index and strength properties of the soil, is given in Fig. 5.16. The soil profile consists of 2 to 3 m of weathered marine clay underlain by 6 to 8 m thick of soft marine clay and followed by a stratum 5 to 6 m thick of stiff clay layer. The groundwater table fluctuated with the season and varied from 0.5 to 2.0 m below the ground level.

The test embankment had a first stage height of 2.4 m and subsequently was raised to a second stage height of 4.0 m after 345 days. The plan and cross-section of the embankment including the layout of the monitoring instruments are illustrated in Fig. 5.17. The monitoring program included surface and subsurface settlement measurements, pore pressures, vertical earth pressures, and lateral movement. The instrumentations installed consisted of 3 surface settlement plates, 4 subsurface settlement gauges at 1.5 m, 3.0 m, 6.0 m, and 8.0 m depths, 8 closed hydraulic piezometers, 2 earth pressure cells, an SIS Geotechnica C412 type inclinometer casing with readings provided by the SIS Geotechnica sensor, and three lateral movement stakes. The embankment was compacted in layers by a light vibrating plate tamper and was found to have an average density of 18.0 kN/m^3.

The Cased Borehole Method was employed in constructing the granular piles (Bergado et al. 1984; Bergado and Lam, 1987). The compaction was done by means of dropping the 1.6 kN hammer at 0.6 m falling height to the steel disc which was placed on the surface of the granular material. Each layer of the pile was compacted with 15 blows per layer, with a compacted thickness of about 0.6 m. The friction angles of the compacted granular piles obtained from direct shear tests varied from 39 to 45 degrees with compacted densities ranging from 17 to 18.1 kN/m^3. The compacted granular piles were arranged in a triangular pattern with a spacing of 1.5 m. The piles were 30 cm in diameter and have a length of 8.0 m, fully penetrating the soft clay layer. The granular materials consist of whitish-gray, poorly graded crushed limestones with a maximum size of 20 mm. A drainage blanket of 0.25 m thick consisting of clean sand was laid on top of the compacted granular piles.

5.13 PERFORMANCE OF GRANULAR PILES UNDER EMBANKMENT LOADING

Based on full-scale load tests on granular piles, Bergado et al. (1984, 1987) and Bergado and Lam (1987) reported that the bearing capacity of the soft Bangkok clay using granular piles increased by up to 4 times, the total settlements reduced by at least 30%, and the slope stability safety factor increased by at least 25%. The comparative study of the performance of test embankments on granular piles and on vertical drains indicated that the embankment on granular piles settled about 40% less than the embankment on vertical drains, as shown in Fig. 5.18. The study confirms the idea that granular piles function as a reinforcement to the clay rather than drains (Bergado et al. 1988), attributed perhaps to the larger zone of disturbance (smeared zone) surrounding the pile caused by the method of installation.

As mentioned in the previous chapters, using the finite element method (FEM), Long (1992) analyzed the consolidation due to the embankment load by 2-D model and evaluated the

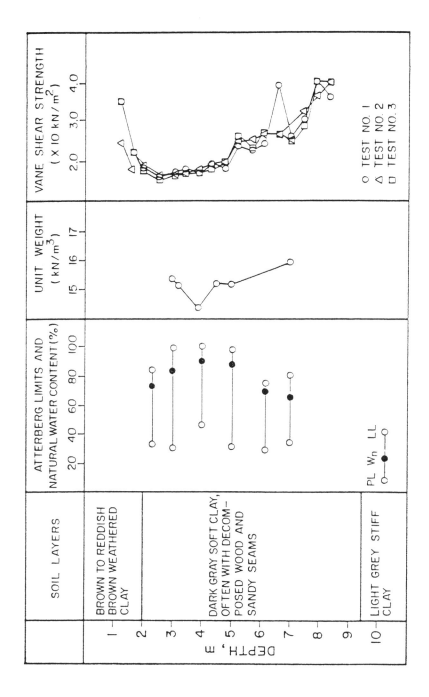

Fig. 5.16 Soil Profile and Properties at AIT Campus

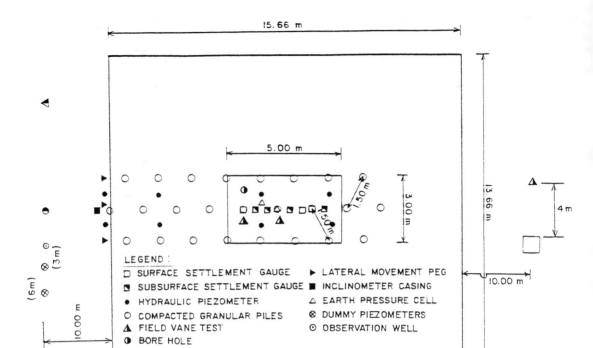

(a)

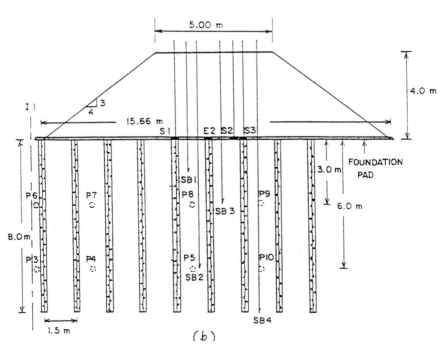

(b)

Fig. 5.17 Plan and Embankment Cross-Section Including Locations of Instrumentations, Field Tests and Field Sampling

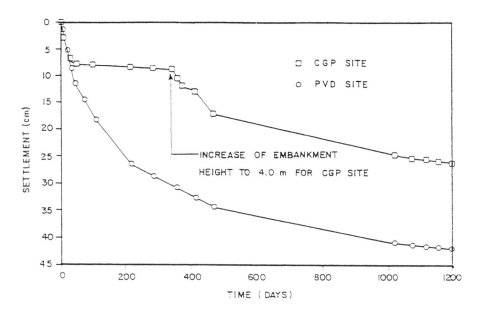

Fig. 5.18 Comparison of Maximum Surface Settlements at Granular Piles and Vertical Drains Test Embankments (Bergado et al. 1990a)

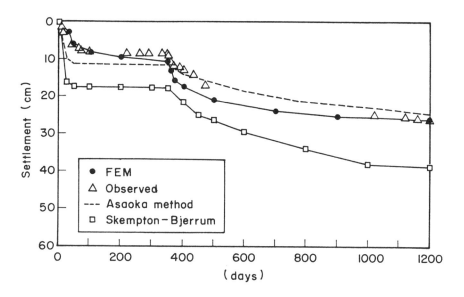

Fig. 5.19 Comparison of Primary Settlement at Embankment with Granular Piles

effects of ground subsidence caused by pore pressure drawdown (see Fig. 4.37) by axi-symmetric model. For the two-dimensional case, the FEM program CON2D was used (Duncan et al. 1981). In the case of the axi-symmetric model, the FEM program CONSAX was utilized (D'Orazio and Duncan, 1982). For 2-D consolidation analysis of the test embankment, the granular piles were converted into continuous granular walls with the same spacing and area replacement ratio as that of the actual case (see Fig. 4.31). The converted permeability including the smear zone is introduced based on the condition of equal discharge with the assumption that the coefficient of permeability is independent on the state of flow (see Fig. 4.32). The subsidence effects were evaluated assuming that the pore pressure drawdown due to deep well pumping is uniform throughout the improved ground and that the axi-symmetric model is valid.

The predicted and actual consolidation settlements are shown in Fig. 5.19. The results of FEM and Asaoka (1978) method agreed with the observed data. The results from the Skempton and Bjerrum (1957) method overestimated the settlements. Figure 5.20 shows the predicted settlement at different pile lengths. The total subsidence including ground subsidence effects were almost the same using 6 m and 8 m length of piles. Therefore, it is more economical to use 6 m length of granular piles down to the medium stiff clay layer (Bergado and Long, 1992).

5.14 MODEL INFRASTRUCTURE PROJECTS ON SOFT GROUND

A test excavation was constructed to study the mechanical behavior and applicability of different soil improvement methods to excavated slopes on soft Bangkok clay (Warinsisak, 1991). The soft clay properties at the site are given in Fig. 5.21. The test excavations consists of two unimproved and two improved slopes, as shown in Fig. 5.22. One of the improved slopes was treated with sand compaction piles. The instrumentation consisted of 6 inclinometers, 1 extensometer, 7 settlement gauges, 6 piezometers, and a water stand pipe. The excavation was carried first using the dry method using water jets.

A simplified compozer method was employed in the construction of sand compaction piles and was carried out before the test excavation. The steps in the construction are illustrated in Fig. 5.23. A 3.6 ton (35.3 kN) hammer was used to drive the 0.40 m diameter casing pipe into the soft ground to the designated bottom. The desired volume of sand was loaded into the bucket with a hopper using a backhoe and poured into the casing as the latter was being pulled up. The hammer, dropped from a height of about 10 cm, was used to compact the sand until the casing sunk 1.0 m making a sand compaction pile of 50 cm in length for every cycle of compaction. A photograph of the simplified compozer method equipment for installation of sand compaction piles is shown in Fig. 5.24.

During the excavation process, a crack was observed along the shoulder about 7.0 m from the top of the slope. The horizontal and vertical separation of the crack measured about 250 mm and 100 mm, respectively. The excessive horizontal displacement during the excavation process as shown in Fig. 5.25 could also indicate slope failure. The slope when the excavation

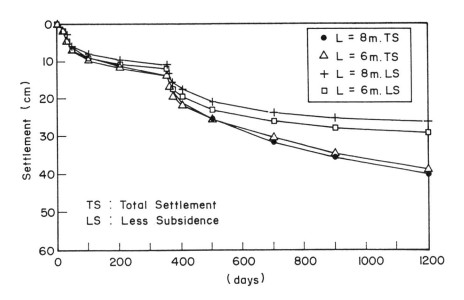

Fig. 5.20 Settlement with Differential Lengths of Granular Piles

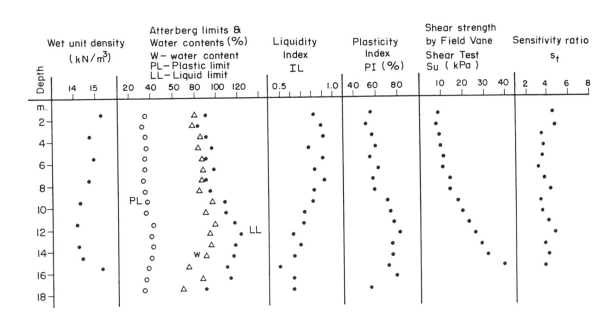

Fig. 5.21 Engineering Properties of Soft Bangkok Clay at Bang Bo

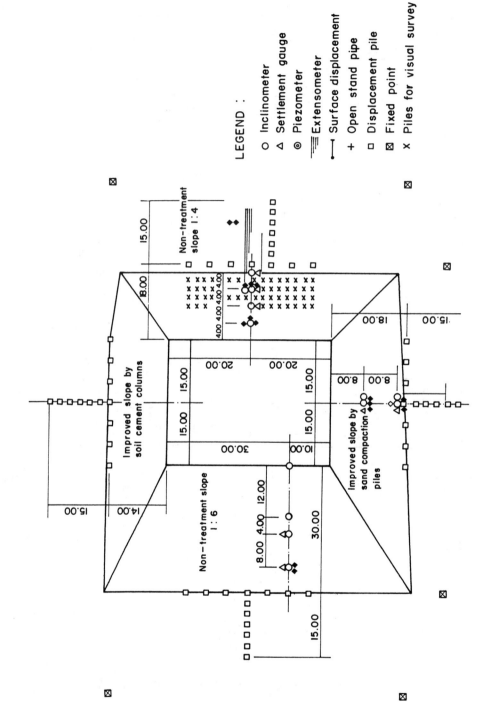

Fig. 5.22 Plan of Excavated Slope Facility

1. A sand bag is put on the spot where the sand compaction pile is to be constructed
2. Pile the casing pipe into the ground with the stone hammer
3. Pour the sand in the casing 2.0 m in height
4. Pull up the casing 1.5 m giving compressed air
5. Compact the casing with stone hammer until the casing sinks 1.0 m
6. Repeat the procedure 3 to 5

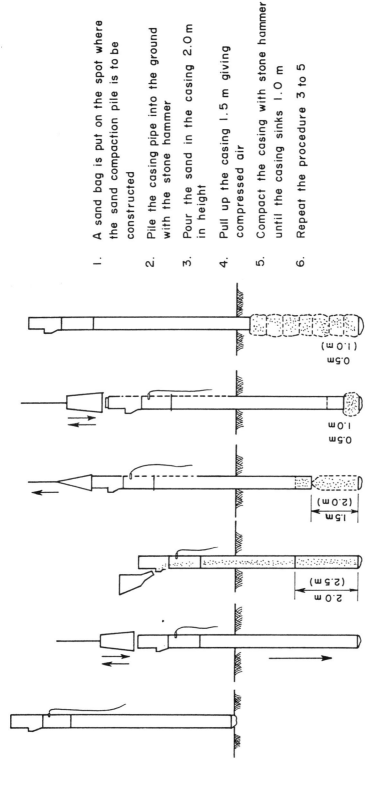

Fig. 5.23 Installation of Sand Compaction Piles

Fig. 5.24 Construction of Sand Compaction Pile on Soft Bangkok Clay By Sand Compaction Method

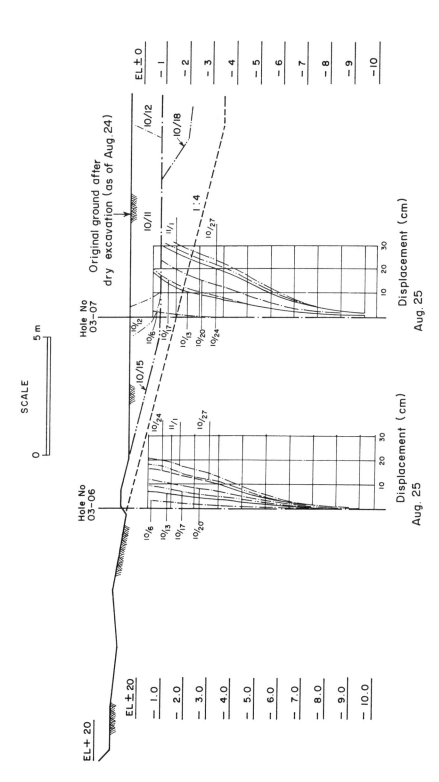

Fig. 5.25 Observed Lateral Displacements at Bang Bo Site

was stopped had a gradient of 1V:6H with maximum depth of 3.5 m.

5.15 EVALUATION OF LOCALLY AVAILABLE MATERIALS FOR SAND COMPACTION PILES (SCP)

In Thailand and in the Southeast Asian Region, the problem of the SCP method is the selection and identification of alternative types of sand suitable for SCP construction which are less costly, though may be of less quality, to be competitive. Clean sands are quite expensive. While clayey to silty sands cost half as much. Based on previous works of Nutalaya et al. (1984), Tjakrawiralaksana (1980) and Selvakumar (1977), the probable sources of sandy materials in the Lower Central Plain of Thailand (Fig. 5.26) can easily be identified. The extent of these deposits varies in depth, from 2 to 15 m, and covers areas of several square kilometers. The areas investigated were located within 100 km from Bangkok considering the most economical maximum distance of hauling of the materials for SCP to be competitive.

In order to evaluate the physical and engineering properties of the materials for SCP, sand samples were collected from the Ayutthaya, Chonburi, Kampengsen, and Ratchaburi areas. The materials obtained from former three locations were sandy soils, whereas the latter was laterite. The possibility of using lateritic residual soil as SCP construction material is considered since the material is readily available and cheap in Thailand.

Table 5.4 summarizes the laboratory test performed in order to determine physical and engineering properties of the materials (Bersabe, 1992). The grain size distributions of the four materials are shown in Fig. 5.27 with the suitable range of gradation for SCP materials which is used in Japan and the U.S. as reported by Fudo Construction (1991) and indicated by the broken lines.

Constant volume direct shear and triaxial (CID) tests were conducted to determine the strength properties and stress-strain behavior of the samples. Nine CID tests were performed for each type of sample corresponding to 3 different relative densities (60%, 70%, and 90%) and 3 different confining stresses (5, 10, and 15 tsm). Except lateritic soil, the other three samples were used for direct shear test, 9 tests per each sample using the same relative density levels used in CID tests and under three different normal stresses (5, 10, and 15 tsm). To achieve the desired relative densities in the sample preparation, a combination of equal-volume and undercompaction, as reported by Dennis (1988), was employed.

Table 5.4 also shows the results of the direct shear and triaxial (CID) tests along with the permeabilities of the samples. The Ayutthaya and Kampengsen sands indicated a significant increase in the angle of internal friction with increasing relative densities under consolidated drained conditions. The laterite showed only a slight increase in internal friction angle which was probably caused by grain crushing during the shearing. This difference is not so prominent in Chonburi material due to its fine grained nature (18% passing the No. 200 US Standard sieve). Generally, laboratory experiments indicate that the shearing strength and internal friction angle increase with increase in relative density of the sample. However, according to Vesic and

Clough (1968), at higher pressures, particles of cohesionless soils tend to crush each other thereby reducing the angle of shearing resistance. The behavior of the lateritic soil agrees with this. Table 5.4 also shows that the permeability of the samples decreases with the increase of relative density.

Large scale SCP model tests were performed by Leong (1992) and Cahulogan (1993) in the Asian Institute of Technology's Geotechnical Engineering Experimentation Site (Fig. 5.28). The schematic diagram of the model test system is illustrated in Fig. 5.29. A 0.90 diameter, 2.5 m high SCP steel mold was used to carry out model tests at three different relative densities (approximately 60%, 70%, and 90%). Each sample was tested in three set-ups corresponding to the desired relative density. A molding water content of about 5% was maintained as the sample was placed and compaction was made by layers using a compaction vibration and a tamper. The saturation process was provided through a pressure tank filled with water while carbon dioxide gas was used to saturate the sample. The samples were consolidated at three different pressure levels by using dead loads which were also used to apply confining pressure. Standard Penetration Tests (SPT) and Cone Penetration Tests (CPT) were performed for each set-up. Constant head permeability tests were also performed in every set-up to observe the drainage behavior.

Table 5.4 Internal Friction Angles and Permeabilities of the Samples

SAMPLE	Relative Density (%)	Friction Angle (ϕ) deg. (CID)	Friction Angle (ϕ) deg. (Direct Shear)	Permeability (x 10^{-4}) (cm/sec)
Ayutthaya	90	38.7	26.6	0.87
	70	34.6		4.02
	60	32.0	24.5	5.30
Chonburi	90	35.5	33.4	6.24
	70	35.0	31.0	12.80
	60	34.1	28.8	14.68
Kampengsen	90	39.0	30.1	2.50
	70	36.5	26.6	7.97
	60	34.6	25.6	16.5
Laterite	90	37.1		2.66
	70	36.2		8.35
	60	34.9		17.53

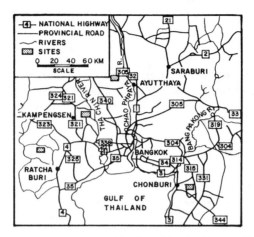

Fig. 5.26 Map of Areas Explored in the Lower Central Plain of Thailand

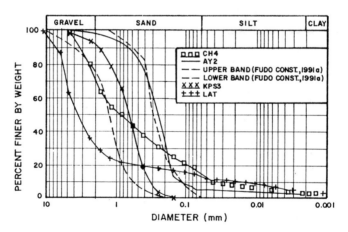

Fig. 5.27 Grain Size Distribution Curves

Fig. 5.28 SCP Model Tests in AIT

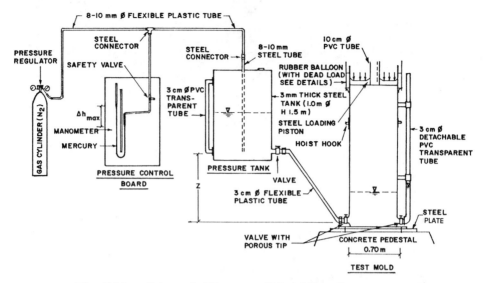

Fig. 5.29   Schematic Diagram of Model Test System

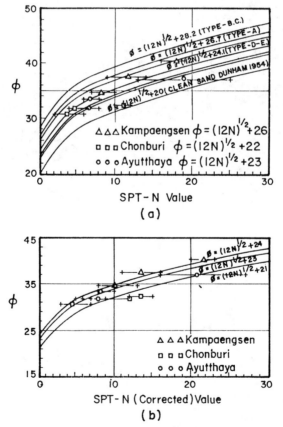

Fig. 5.30   Relationship Between Internal Friction Angle and SPT-N Values

The SPT-N values of the samples used in the model tests at 3 different relative densities (60%, 70%, and 90%) and three different vertical stresses are summarized in Table 5.4. Two sets of tests were conducted for each specimen for better comparison of the results. Raw SPT-N values from the model tests were corrected using the empirical equation of Gibbs and Holtz (1957). The SPT-N values significantly increased corresponding to the increase in relative density. Correlations of the internal friction angle with SPT-N values from the model tests indicated that the materials from Ayuthaya and Kampengsen followed the general equation of Dunham (1954) together with the SCP materials in Japan indicated as Type A to E in Fig. 5.30a,b.

The Dutch cone test results in the model tests showed good correlation between the internal friction angle and cone resistance which generally agree with the graphical correlation recommended by Meyerhoff (1974) and Robertson and Campanella (1983). Figures 5.31 and 5.32 show these agreements, respectively.

Based on the preliminary results of the laboratory and model tests, Ayuthaya and Kampengsen materials are recommended as suitable SCP construction materials. Their fine contents (percentage passing the No. 200 U.S. standard sieve) must be limited to about 8% for them to function well. However, the Ayuthaya material is recommended to undergo reliable filter characteristics in as much as it has finer grains than the Kampengsen material. Chonburi material can also be a good SCP construction material if it will undergo washing which will limit its fine contents to about 5% to 8%.

The lateritic soil showed quite acceptable properties needed for SCP design. However, uncertainties may arise, especially that its swelling potential was not considered. No further tests and investigations were made.

Since SPT has been employed widely to determine the actual performance of SCP, the following mathematical equations derived from the model test results can be used to assess the internal friction angles of the materials from SPT-N values as follows:

From raw SPT-N values:

Kampengsen:

$$\phi = (12N)^{1/2} + 23.3 \quad (5.33)$$

Chonburi:

$$\phi = (12N)^{1/2} + 22 \quad (5.34)$$

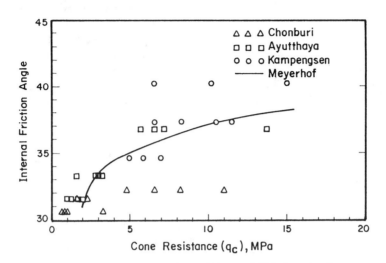

Fig. 5.31 Correlation of Internal Friction Angle and Cone Resistance

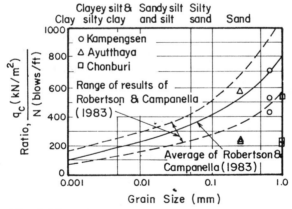

Fig. 5.32 Variation of q_c/N Ratio with Mean Grain Size for Sand Materials

Ayuthaya:

$$\phi = (12N)^{1/2} + 22.8 \tag{5.35}$$

The regression equations using corrected SPT-N values are as follows:

From Corrected SPT-N values:

Kampengsen:

$$\phi = (12\acute{N})^{1/2} + 26 \tag{5.36}$$

Chonburi:

$$\phi = (12\acute{N})^{1/2} + 23.4 \tag{5.37}$$

Ayuthaya:

$$\phi = (12\acute{N})^{1/2} + 22.4 \tag{5.38}$$

These correlations will facilitate the design and quality control of SCP construction.

5.16 GRANULAR PILES IN COMBINATION WITH OTHER SOIL IMPROVEMENT TECHNIQUES

At Ebetsu in Hokkaido, Japan, two full scale test embankments were constructed with different soil stabilization methods in addition to the test embankment without treatment (Aboshi and Suematsu, 1985). One of the test embankments was stabilized with sand drains and steel sheet reinforcements while the other was stabilized with sand compaction piles. At an embankment height of 3.5 m, the test embankment without treatment collapsed with substantial deformations on the subsoil and cracks in the embankment fill. While the embankments treated with sand drains and steel reinforcements and with sand compaction piles were completed up to 8 m high without failures. The test embankment stabilized by sand compaction piles yielded a stress concentration factor of 3 which is in close agreement with the observations of Bergado et al. (1988) on soft Bangkok clay improved with granular piles having a stress concentration factor ranging from 2 to 5. Based on the results of these test embankments, it is envisioned that the use of granular piles and steel grids reinforcements could be a good combination for embankments of soft Bangkok clay. The steel grids reinforcements help minimize the lateral spreading of the embankment and provide steeper side slopes or even vertical sides. The granular piles provide the reduction of settlements as well as the increase in strength and bearing capacity of the soft ground foundation. An example for such an application scheme is on transition units

Table 5.5 Raw and Corrected SPT-N Values from Model Tests

Sample	γ_t	Vertical Pressure (6.2 tsm)				Vertical Pressure (8.5 tsm)				Vertical Pressure (10 tsm)			
		N		N'		N		N'		N		N'	
		Set 1	Set 2	Set 1	Set 2	Set 1	Set 2	Set 1	Set 2	Set 1	Set 2	Set 1	Set 2
Ayuthaya	60	5	6	6	8	8	8	9	9	4	5	4	5
	70	6	8	8	10	7	8	8	9	8	9	9	10
	90	15	13	18	16	20	25	22	27	15	33	10	14
Choburi	60	2	3	3	4	4	6	4	7	4	5	4	5
	70	5	5	6	6	5	6	5	7	5	5	5	5
	90	10	12	13	15	13	11	14	12	14	12	15	13
Kampengsen	60	8	10	10	13	8	7	9	8	10	5	11	5
	70	9	13	11	16	14	11	15	12	15	9	16	10
	90	19	16	23	20	25	20	27	-	23	25	18	20

to bridges and viaducts as shown in Fig. 5.33. As an alternative of using expensive ideal materials for embankment fill such as sand, savings in the construction costs can be realized by using cheaper, locally-available, cohesive-frictional soil with more than 18% having particle size diameter lower than 0.74 mm. Extensive research has been done on steel grids reinforcements with poor quality backfills consisting of weathered clay, lateritic soils and clayey sand (see Bergado et al. 1990a). However, the combination scheme of granular piles and steel grids reinforcements to improve the ground and embankment fill, respectively, must be studied through full scale field prototype so that their effectiveness on soft and subsiding Bangkok clay would be proven and the actual reduction of lateral spreading can be measured.

5.17 CONCLUSIONS

The current state-of-the-art on granular pile scheme has been discussed but there are still some loopholes which need to be studied for better understanding of this ground improvement method. Among these are the influences of the method of construction, characteristics of pile materials, better estimation of stress concentration factors, stress distribution with depth and time, improvement factors as well as the effect of overlying surcharge consisting of mechanically stabilized (reinforced) embankment. It was indicated that there is a need for additional research to improve the design methods and develop a complete understanding of the mechanics of granular pile behavior. More full scale and model tests with extensive instrumentations in combination of finite element model studies should be carried out to shed light to the uncertainties and improve confidence, especially when the scheme is applied to the soft and subsiding ground.

Already, a test embankment on granular piles has been constructed at AIT to evaluate and investigate the applicability of such a scheme on the soft and subsiding Bangkok clay. The performance of this test embankment has been described in the preceding section. The improvement factors on bearing capacity, settlement and gain in strength have been determined and indicated substantial values. Based on the results of these studies as well as in Japan, it was suggested to use granular piles in the subsoil in combination with mechanically stabilized earth (MSE) embankments using grid reinforcements as one alternative for ground improvement at the approach embankments to bridges and viaducts. The implementation of this ground improvement scheme was postponed mainly due to the lack of available data regarding its actual performance. Therefore, it is deemed necessary to construct a full scale field prototype so that the effectiveness of granular piles, possibly combined with other soil improvement techniques, as an alternative scheme for ground improvement on approach embankments to bridges and viaducts would be proven in the soft and subsiding environment of Bangkok. Finally, the use of cheaper materials for SCP construction has been found to be viable and economical alternative.

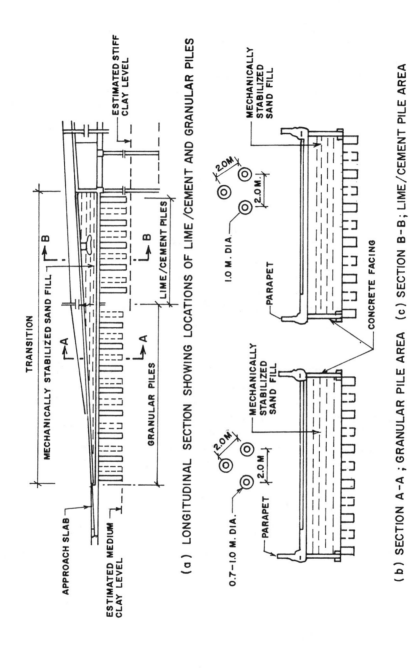

Fig.5.33 Granular Piles with Mechanically Stabilized Embankment

5.18 REFERENCES

Aboshi, H., Ichimoto, E., Enoki, M., and Harada, K. (1979), The Compozer - A Method to Improve Characteristics of Soft Clays by Inclusion of Large Diameter Sand Columns, Proc. Intl. Conf. on Soil Reinforcement: Reinforced Earth and Other Techniques, Paris, Vol. 1, pp. 211-216.

Aboshi, H., and Suematsu, N. (1985), Sand Compaction Pile Method: State-of-the-Art Paper, Proc. 3rd Intl. Geotechnical Seminar on Soil Improvement Methods, Nanyang Technological Institute, Singapore.

Akagi, T. (1979), Effect of Displacement Type Sand Drains on Strength and Compressibility of Soft Clays, Tokyo University Publication, Japan.

Aoyama, M., Nakamura, N., Kuwabara, M., and Nuzo, M. (1990), Some Examples of Field Tests for Soil Improvement Methods in Japan, Proc. Symp. Dev. of Laboratory and Field Tests in Geotech. Eng. Practice, Bangkok, Thailand.

Asaoka, A. (1978), Observational Procedure for Settlement Prediction, Soils and Foundations, Vol. 18, No. 4, pp. 87-101.

Balaam, N.P., and Poulos, H.G. (1983), The Behaviour of Foundations Supported by Clays Stabilized by Stone Columns, The University of Sydney Research Report No. R424, Sydney, Australia.

Barksdale, R.D. (1981), Site Improvement in Japan Using Sand Compaction Piles, Georgia Institute of Technology, Atlanta.

Barksdale, R.D., and Bachus, R.C. (1983), Design and Construction of Stone Columns, Report No. FHWA/RD-83/026, National Technical Information Service, Springfield, Virginia.

Baumann, V., and Bauer, G.E.A. (1974), The Performance of Foundation on Various Soils Stabilized by Vibrocompaction Method, Canadian Geotechnical Journal, Vol. 11, pp. 509-530.

Bell, A.L. (1915), The lateral pressure and the resistance of clay and supporting power of clay foundations, Proc. Inst. of Civil Engrs.

Bergado, D.T., Rantucci, G., and Widodo, S. (1984), Full Scale Load Tests on Granular Piles and Sand Drains in the Soft Bangkok Clay, Proc. Intl. Conf. on In-situ Soil and Rock Reinforcement, Paris, pp. 111-118.

Bergado, D.T., and Lam, F.L. (1987), Full Scale Load Test of Granular Piles with Different Densities and Different Proportions of Gravel and Sand in the Soft Bangkok Clay, Soils and Foundations Journal, Vol. 27, No. 1, pp. 86-93.

Bergado, D.T., and Long P.V. (1992), Numerical Analysis of Embankment on Subsiding Ground Improved by Granular Piles, Proc. Hanoi Intl. Geotech. Conf. (NTFE92), Hanoi, Vietnam.

Bergado, D.T., Sim, S.H., and Kalvade, S. (1987), Improvement of Soft Bangkok Clay Using Granular Piles in Subsiding Environment, Proc. 5th Intl. Geotechnical Seminar on Case Histories in Soft Clay, Singapore, pp. 219-226.

Bergado, D.T., Miura, N., Panichayatum, B., and Sampaco, C.L. (1988), Reinforcement of Soft Bangkok Clay Using Granular Piles, Proc. Intl. Geotechnical Symp. on Theory and Practice of Earth Reinforcement, Fukuoka, Japan, pp. 179-184.

Bergado, D.T., Singh, N., Sim, S.H., Panichayatum, B., Sampaco, C.L., and Balasubramaniam, A.S. (1990a), Improvement of Soft Bangkok Clay Using Vertical Drains Compared with Granular Piles, Geotextiles and Geomembranes Journal, Vol. 9, pp. 203-231.

Bergado, D.T., Sampaco, C.L., Alfaro, M.C., Shivashankar, R., and Balasubramaniam, A.S.(1990b), Welded Wire Wall and Embankment System With Poor Quality Backfill on Soft Ground, 4th Progress Report Submitted to the USAID, Thailand.

Bersabe, N.D. (1992), Engineering properties of locally-available clayey to silty sand and their applicability as sand compaction pile material, M. Eng'g. Thesis No. GT-91-20, Asian Institute of Technology, Bangkok, Thailand.

Cahulogan, R.H. (1993), Correlation of Strength from Model and Laboratory Tests of Locally Available Silty Sand and its Applicability as Sand Compaction Pile Material, M. Eng'g. Thesis, Asian Institute of Technology, Bangkok, Thailand.

Carillo, N. (1942), Simple two-and three-dimensional cases in the theory of consolidation of soils, J. of Mathematics and Physics, Vol. 21, No. 1, pp. 1-5.

Chambosse, G., and Dobson, T. (1982), Stone Columns I - Estimation of Bearing Capacity and Expected Settlement in Cohesive Soils, GKN Keller, Inc., Tampa, Florida.

Datye, K.R., and Nagaraju, S.S. (1975), Installation and Testing of Rammed Stone Columns, Proc. IGS Specialty Session, 5th Asian Regional Conf. on Soil Mech. and Found. Engineering, Bangalore, India, pp. 101-104.

Datye, K.R. (1978), Special Construction Techniques, Proc. IGS Conf.on Geotech. Eng'g, New Delhi, India, pp.30-44.

Datye, K.R., and Nagaraju, S.S. (1981), Design Approach and Field Control for Stone Columns, Proc. 10th Intl. Conf. on Soil Mech. Found. Eng'g., Stockholm.

DiMaggio, J.A. (1978), Stone Columns - A Foundation Treatment (In situ Stabilization of Cohesive Soils), Demonstration Project No. 4-6, Federal Highway Administration, Region 15, Arlington, Virginia, U.S.A.

Duncan, J.M., D'Orazio, T.B., Chang, C.S., Wong, K.S., and Namiq, L.I. (1981), CON2D: A Finite Element Computer Program for Analysis of Consolidation, Report No. UCB/GT/81-01 to U.S. Army Corps of Engineers, Waterways Experiment Station, Vicksburg, Mississipi, U.S.A.

Dunham, J.W. (1954), Pile foundations for buildings, J. Soil Mech. and Found. Div., ASCE, pp. 385-1 to 385-21.

D'Orazio, T.B., and Duncan, J.M. (1982), CONSAX: A Computer Program for Axisymmetric Finite Element Analysis of Consolidation, Report No. UCB/GT/82-01, University of California, Berkeley, Calif.

Engelhardt, K., and Kirsch, K. (1977), Soil Improvement by Deep Vibration Technique, Proc. 5th Southeast Asian Conference on Soil Engineering, Bangkok, Thailand.

Enriquez, A.S. (1989), Inverse Analysis of Settlement Data on Improved Soft Clays by Test Embankments and Laboratory Test, M. Eng. Thesis, Asian Institute of Technology, Bangkok, Thailand.

Fredlund, D.G., and Krahn, J. (1977), Comparison of Slope Stability Methods of Analysis, Canadian Geotechnical Journal, Vol. 14, No. 3, pp. 429-439.

Fudo Construction (1991a), Data Concerning Specification of Sandy Materials for Sand Compaction Pile on Ground Improvement Works, Japan.

Gibbs, H.J., and Holtz, W.G. (1957), Research on Determining Density of Sand by Spoon Penetration Test, Proc. 4th International Conference on Soil Mechanics and Foundation Engineering, Vol. 1, pp. 35-39.

Gibson, R.E., and Anderson, W.F. (1961), In-situ measurements of soil properties with the pressuremeter, Civil Eng'g. and Public Review, Vol. 56, No. 658.

Goughnour, R.R. (1983), Settlement of Vertically Loaded Stone Columns in Soft Ground, Proc. Specialty Session - 8th European Conf. on Soil Mech. and Found. Eng'g., Vol. 2, Helsinki.

Greenwood, D.A. (1970), Mechanical Improvement of Soils Below Ground Surface, Proc. Ground Eng'g. Conference, Institute of Civil Eng'g., pp. 9-20.

Hughes, J.M.O., and Withers, N.J. (1974), Reinforcing Soft Cohesive Soil with Stone Columns, Ground Engineering, Vol. 7, No. 3, pp. 42-49.

Hughes, J.M.O., Withers, N.J., and Greenwood, D.A. (1975), A Field Trial of Reinforcing Effects of Stone Columns in Soil, Geotechnique, Vol. 25, No. 1, pp. 31-44.

Leonards, G.A. (1962), Foundation engineering, McGraw-Hill, Inc., New York.

Leong, K.H. (1992), Correlation of Strength and Density of Locally Available Silty Sand Based on Model Tests for Sand Compaction Piles, M. Eng'g. Thesis, Asian Institute of Technology, Bangkok, Thailand

Long, P.V. (1992), Numerical Analysis of Embankments Improved by Granular Piles and Vertical Drains on Subsiding Ground, M. Eng'g. Thesis No. 91-10, GTE Division, Asian Institute of Technology, Bangkok, Thailand.

Madhav, M.R., and Vitkar, R.P. (1978), Strip Footing on Weak Clay Stabilized with Granular Trench or Pile, Canadian Geotechnical Journal, Vol. 15, No. 4, pp. 605-609.

Madhav, M.R., Iyengar, N.G.R., Vitkar, R.P., and Nandia, A. (1979), Increased Bearing Capacity and Reduced Settlements Due to Inclusions in Soil, Proc. Intl. Conf. on Soil Reinforcement: Reinforced Earth and Other Techniques, Vol. 2, pp. 239-333.

Mesri, G. (1973), Coefficient of Secondary Compression, Journal of the Soil Mech. and Found. Eng'g. Div., ASCE, Vol. 99, SM1. Panichayatum, B. (1987), Comparison of the Performance of Embankments on Granular Piles and Vertical Drains with Probabilistic Slope Assessment, M. Eng'g.. Thesis, Asian Institute of Technology, Bangkok, Thailand.

Meyerhof, G.G. (1974), Penetration testing outside Europe, General Report, Proc. Europ. Symp. on Penetration Testing, Stockholm, Vol. 2.1.

Nutalaya, P., Rau, J., and Sodsee, S. (1984), Surficial geology of the Lower Central Plain, Thailand, Proc. Conf. on Appl. of Geol. and Natl. Dev., Bangkok, Thailand.

Panichayatum, B. (1987), Comparison of the performance of embankments on granular piles and vertical drains with probabilistic slope assessment, M. Eng'g. Thesis No. GT-86-8, Asian Institute of Technology, Bangkok, Thailand.

Priebe, H. (1976), Estimating Settlements in a Gravel Column Consolidated Soil, Die Bautechnik 53, pp. 160-162 (in German).

Rao, B.G (1982), Behavior of Skirted Granular Piles, Doctoral Thesis, Civil Eng'g. Dept., University of Roorkee, Roorkee, India.

Ranjan, G., and Rao, B.G. (1983), Skirted Granular Piles for Ground Improvement, Proc. VIII European Conf. on Soil Mech. and Found. Eng'g., Helsinki.

Ranjan, G. (1989), Ground Treated with Granular Piles and Its Response Under Load, Indian Geotechnical Journal, Vol. 19, No. 1, pp. 1-85.

Robertson, P.K., and Campanella, R.G. (1983), Interpretation of cone penetration test, Part I: sand, Canadian Geotech. J., Vol. 20, No. 4, pp. 718-733.

Selvakumar, S. (1977), Analysis of quarternary terrace levels of the Chao Phraya-Mae Khlong Basins, Thailand, M. Eng'g. Thesis, Asian Institute of Technology, Bangkok, Thailand.

Sim, S.H. (1986), Stability and Settlement of AIT Test Embankment Supported by Granular Piles on Subsiding Ground, M. Eng'g.. Thesis, Asian Institute of Technology, Bangkok, Thailand.

Singh, N. (1986), Soil Improvement by Preloading and Plastic Band Drains in Subsiding Ground at AIT Campus, M. Eng'g. Thesis, Asian Institute of Technology, Bangkok, Thailand.

Skempton, A.W., and Bjerrum, L. (1975), A Contribution to the Settlement Analysis of Foundations on Clay, Geotechnique, Vol. 7, pp. 168-178.

Tjakrawiralaksana, R. (1980), Study on sands as construction materials in the East Chao Phraya River, Special Study SSPR No. 80-4, Asian Institute of Technology, Bangkok, Thailand.

Van Impe, W., and De Beer, E. (1983), Improvement of Settlement Behavior of Soft Layers by Means of Stone Columns, Proc. 8th European Conf. on Soil Mech. and Found. Eng'g., Helsinki, Vol. 1, pp. 309-312.

Vesic, A.S., (1972), Expansion of Cavities in Infinite Soil Mass, Journal of Soil Mech. and Found. Eng'g. Div., ASCE, Vol. 98, No. SM3, pp. 265-290.

Vesic, A.S., and Clough, G.W. (1968), Behavior of Granular Materials Under High Stress, Journal of Soil Mechanics Foundation Division, ASCE 94, SM3, pp. 661-666.

Warinsisak, Y. (1991), On the Results of a Large-Scaled Experimental Excavated Slopes in Soft Bangkok, M. Eng'g. Thesis, AIT, Bangkok, Thailand.

Whitman, R.V., and Bailey, W.A. (1967), Use of Microcomputers for Slope Stability Analysis, Journal of the Soil Mech. and Found. Eng'g. Div., ASCE, Vol. 93, No. SM4, pp. 475-498.

Wong, H.Y. (1975), Vibroflotation - Its Effect on Weak Cohesive Soils, Civil Engineering (London), No. 824, pp. 44-67.

APPENDIX
(Notations for Tables 5.1 and 5.2)

A_s	: cross-sectional area of sand pile	n	: stress concentration factor
B	: width of loaded area	q_o	: overburden pressure
C_c	: compression index of clay	q_s	: bearing capacity of soft soil expressed as $(2/3)C_o N_c$
C_o	: original undrained shear strength of clay	q_{ult}	: ultimate bearing capacity
D_f	: depth of foundation	γ_s, γ_c	: unit weight of sand and clay, respectively
E	: modulus of elasticity	$\{\Delta F_{DN}\}$	: vector of incremental nodal forces at the dual nodes along the pile clay interface
F'_c, F'_q	: cavity expansion factors		
H	: thickness of layer		
K	: earth pressure coefficient applying to the load increments	$\{\Delta F_E\}$	: vector of incremental nodal forces due to applied tractions (usually applied along the top of the sand pile)
K_{as}	: active earth pressure coefficient of sand pile		
$[K_c^{(m)}]\{\Delta\sigma^{(m)}\}$	: vector of corrections due to yielding of the pile and/or clay; these corrections are treated as an additional external load	$(\Delta P)^*_{vc}$	: effective vertical stress increase in the clay averaged over the horizontal projected area of clay
$[K_E]$	: elastic stiffness matrix		
K_o	: coefficient of earth pressure at rest	$(\Delta P)^*_v$	: effective vertical stress increase in the clay averaged over the horizontal projected area of unit cell
K_{pc}	: soil coefficient of passive earth pressure		
L	: length of sand pile		
N_c, N_γ, N_q	: dimensionless factors which depend on properties of soil and pile material and area replacement ratio	$\{\Delta\sigma^{(m+1)}\}$	: vector incremental deflections
		ε_v	: vertical strain (same for sand and clay)
$(P_o)_{vc}$	: initial vertical stress in the clay	θ	: vertical angle of sliding surface at each sand pile
		μ	: Poisson's ratio
R	: settlement reduction factor	μ_c	: reduction in stress coefficient of clay
S_t	: settlement of composite foundation		
		σ	: vertical stress
W	: width of equivalent granular pile strip	σ_{ro}	: initial radial stress along the granular pile
Z	: depth from surface of composite foundation	σ_z	: overburden pressure at depth z
a_s	: area replacement ratio	τ	: shear resistance of composite foundation
d_s	: pile diameter		
e_o	: initial void ratio	ϕ_s	: angle of internal friction
m	: iteration number	Ψ	: angle between the assumed failure surface and foundation
m_v	: modulus of volume compressibility		

Chapter 6

LIME/CEMENT STABILIZATION

6.1 GENERAL

Chemical admixture stabilization has been extensively used in both shallow and deep stabilization in order to improve inherent properties of the soil such as strength and deformation behavior. An increment in strength, a reduction in compressibility, an improvement of the swelling or squeezing characteristics and increasing the durability of soil are the main aims of the admixture stabilization. Lime or cement have commonly been used as chemical admixtures for soil stabilization. The lime/cement mixing method has been used to improve the properties of soils since olden times. These processes were developed in the 1970's simultaneously in Sweden and Japan. However, cement columns, using cement powder, were preliminarily reported to be successfully executed in practice in 1980 and 1982 (DJM Research Group, 1984; Chida, 1982; Miura et al. 1986). The method adopted is known as the DJM (Dry Jet Mixing) Method. Since the mid-80's, lime and cement have been utilized increasingly as stabilizing agents. Deep stabilization of soft soils with lime and/or cement stabilized columns has been the subject of research in Sweden, Japan, and other countries for some time. The Swedish Geotechnical Institute together with Linden-Alimak AB and Professor Bengt Broms have done extensive work on the usage of lime column technique for foundations and earth works both for embankments and excavations in soft clays. Modern application of this method for deep mixing of in-situ soils started in the late 1970's in Japan (Terashi et al. 1979; Kawasaki et al. 1981; Suzuki, 1982). The deep mixing method (DMM) originally was developed to improve the soft ground for port and harbor structures. DMM is now applied to foundation of structures built on land such as embankments, buildings, and storage tanks. However, as suggested by Broms (1986), in Southeast Asia, it is preferable to use cement instead of lime, because of the following reasons: the low cost of cement compared to lime; the difficulty of storing unslaked lime in a hot and humid climate; the greater strength which can be obtained with cement while there is a limit of maximum strength that can be obtained with lime.

Lime and cement treatment has been extensively used for road construction purposes resulting in increased bearing capacity of soft subgrade, enabling a reduction in the thickness of the base course. Treatment with cement or lime has been mainly used in the field of highways, railroads and airport constructions in order to improve the mechanical properties of the bearing layers. The use of lime or cement stabilization has more recently been extended to greater depth in which lime or cement columns act as a type of soil reinforcement. Layers of lime or cement stabilized clays, with their high strength and high modulus, can also function as a rigid crust which is useful in spreading the applied loads to the subsoil. As such, additional applications of the technique have been realized. These present applications include the use of cement or lime column to improve the stability of slopes, trenches, and deep excavations; to increase the bearing capacity and reduce the total and differential settlements below lightly loaded structures; to reduce the negative skin friction on structural piles, to prevent sliding failure of embankments; to reduce the vibrations from traffic loads, blasting, and pile driving; and also to accelerate the consolidation settlements under embankments. Because of the proven versatility of cement and lime stabilization, the methods have gained a wider acceptance in different countries of the world and more recently in Southeast Asia. Some applications of lime/cement columns are illustrated in Fig. 6.1.

6.2 FUNDAMENTAL CONCEPTS OF SOIL-CEMENT STABILIZATION

A Portland cement particle is a heterogeneous substance, containing minute tricalcium silicate (C_3S), dicalcium silicate (C_2S), tricalcium aluminate (C_3A), and a solid solution described as tetracalcium alumino-ferrite (C_4A) [Lea, 1956]. These four main constituents are major strength producing compounds. When the pore water of the soil encounters with the cement, hydration of the cement occurs rapidly and the major hydration (primary cementitious) products are hydrated calcium silicates (C_2SH_x, $C_3S_2H_x$), hydrated calcium aluminates (C_3AH_x, C_4AH_x), and hydrated lime $Ca(OH)_2$. The first two of the hydration products listed above are the main cementitious products formed and the hydrated lime is deposited as a separate crystalline solid phase. These cement particles bind the adjacent cement grains together during hardening and form a hardened skeleton matrix, which encloses unaltered soil particles. The silicate and aluminate phases are internally mixed, so it is most likely that none is completely crystalline. Part of the $Ca(OH)_2$ may also be mixed with other hydrated phases, being only partially crystalline. In addition, the hydration of cement leads to a rise of pH value of the pore water, which is caused by the dissociation of the hydrated lime. The strong bases dissolve the soil silica and alumina (which are inherently acidic) from both the clay minerals and amorphous materials on the clay particle surfaces, in a manner similar to the reaction between a weak acid and strong base. The hydrous silica and alumina will then gradually react with the calcium ions liberated from the hydrolysis of cement to form insoluble compounds (secondary cementitious products), which harden when cured to stabilize the soil. This secondary reaction is known as the pozzolanic reaction. The composition of hydrated cements is still not clearly defined by a chemical formula, so considerable variations are feasible. The compounds in the Portland Cement are transformed on the addition of water as follows:

$2(3CaO.SiO_2)$ + $6H_2O$ = $3CaO.2SiO_2.3H_2O$ + $3Ca(OH)_2$ --------(6.1)
(tricalcium silicate)   (water)   (tobermorite gel)   (calcium hydroxide)

$2(2CaO.SiO_2)$ + $4H_2O$ = $3CaO.2SiO_2.3H_2O$ + $Ca(OH)_2$ --------(6.2)
(bicalcium silicate)   (water)   (tobermorite gel)   (calcium hydroxide)

$4CaO.Al_2O_3.Fe_2O_3$ + $10H_2O$ + $2Ca(OH)_2$ = $6CaO.Al_2O_3.Fe_2O_3.12H_2O$ --------(6.3)
(tetracalciumaluminoferite)   (calcium aluminoferrite hydrate)

$3CaO.Al_2O_3$ + $12H_2O$ + $Ca(OH)_2$ = $3CaO.Al_2O_3.Ca(OH)_2.12H_2O$ -----(6.4)
(tricalcium aluminate)   (tetracalcium aluminate hydrate)

$3CaO.Al_2O_3$ + $10H_2O$ + $CaSO_4.2H_2O$ = $3CaO.Al_2O_3.Ca(OH)_2.12H_2O$ -----(6.5)
(tricalcium aluminate)   (gypsum)   (calcium monosulfoaluminate)

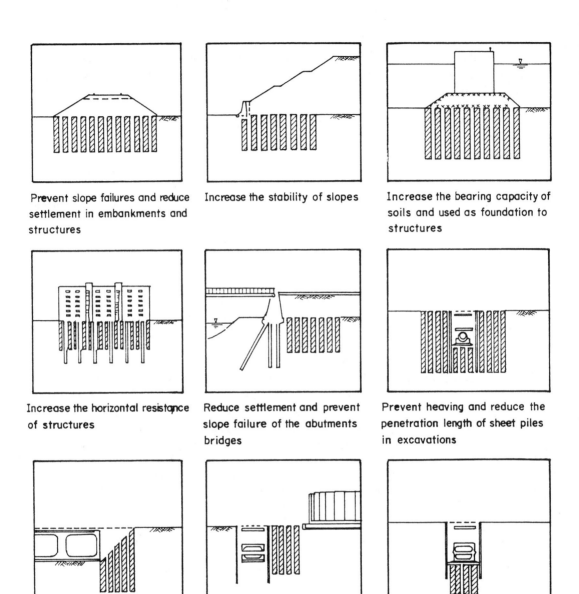

Fig. 6.1 Some Applications of Cement Columns (DJM Research Group, 1984)

The first two equations (Eqs. 1 and 2), whose materials constitute 75% of the Portland cement, show that the hydration of the two calcium silicate types produces new compounds: lime and tobermorite gel, with latter playing the leading role with regards to strength, since bondage, strength, and volume variations are mainly governed by them. The reactions which take place in soil-cement stabilization can be represented in the equation given below. The reactions given here are for tricalcium silicate (C_3S) only, because they are the most important constituents of Portland cement:

$$C_3S + H_2O \longrightarrow C_3S_2H_x \text{ (hydrated gel)} + Ca(OH)_2 \quad \text{-------------------- (6.6)}$$
$$\text{primary cementitious products}$$

$$Ca(OH)_2 \longrightarrow Ca^{++} + 2(OH)^- \quad \text{-------------------------(6.7)}$$

$$Ca^{++} + 2(OH)^- + SiO_2 \text{ (soil silica)} \longrightarrow CSH \quad \text{------------------------(6.8)}$$
$$\text{(secondary cementitious product)}$$

$$Ca^{++} + 2(OH)^- + Al_2O_3 \text{ (soil alumina)} \longrightarrow CAH \quad \text{-----------------------(6.9)}$$
$$\text{(secondary cementitious product)}$$

When pH < 12.6, then the following reaction occurs:

$$C_3S_2H_x \longrightarrow C_3S_2H_x \text{ (hydrated gel)} + Ca(OH)_2 \quad \text{------------------------ (6.10)}$$

In order to have additional bonding forces produced in the cement-clay mixture, the silicates and aluminates in the material must be soluble. The solubility of the clay minerals is equally affected by the impurities present, the crystalline degree of the materials involved, the grain size, etc. In the above equations, the cementation strength of the primary cementitous products is much stronger than that of the secondary ones. At low pH values (pH < 12.6), the reactions given by the Eq. 6.10 will occur. However, the pH drops during pozzolanic reaction and a drop in the pH tends to promote the hydrolysis of $C_3S_2H_x$, to form CSH. The formation of CSH is beneficial only if it is formed by the pozzolanic reaction of lime and soil particles, but it is detrimental when CSH is formed at the expense of the formation of the $C_3S_2H_x$, whose strength generating characteristics are superior to those of CSH. The cement hydration and the pozzolanic reaction can last for months, or even years, after the mixing, and so the strength of cement treated clay is expected to increase with time.

Thus, it means that in the soil cement containing fine clay particles, primary and secondary cementing substances are formed. The primary products harden into high-strength additives and differ from the normal cement hydrated in concrete. The secondary processes increase the strength and durability of the soil cement by producing an additional cementing substance to further enhance the bond strength between the particles.

6.3 MECHANISM OF LIME STABILIZATION

The major strength gain of lime treated clay is mainly derived from three reactions, namely: dehydration of soil, ion exchange, and pozzolanic reaction. Other mechanisms such as carbonation cause minor strength increase and can be neglected. Short term reactions include hydration (for quicklime) and flocculation (ion exchange). Longer term reactions are cementation and carbonation. The use of lime as a stabilizing additive is mainly due to its well-known effects when mixed with soils. The natural stabilizing agent for cohesive soils is calcium hydroxide, hydrated lime, or slaked lime. Calcium hydroxide is not itself a binder, but will produce a binder (consisting mainly of calcium silicate hydrates) by slow chemical reactions principally with the silicates in the clay mineral of cohesive soils (Assarson et al., 1974).

6.3.1 Hydration

A large amount of heat is released when quicklime (CaO) is mixed with clay. This is due to the hydration of quicklime with the pore water of the soil. The increase in temperature can, at times, be so high that the pore water starts to boil (Broms, 1984). An immediate reduction of natural water contents occurs when quicklime is mixed with cohesive soil, as water is consumed in the hydration process. Assarson et al. (1974) reported that at the slakening of the lime, a part of the soil water, about 0.3 kg/kg CaO, is consumed. Moreover, a considerably larger amount of the pore water evaporates because of the heavy heat release, i.e., as the hydration of the quicklime proceeds and the temperature increases, the amount of pore water is reduced. This drying action is particularly beneficial in the treatment of the moist clays. Thus, if a reduction of the natural water content in a cohesive soil is desirable, quicklime (or unslaked lime) instead of calcium hydroxide is used. It is important that the water content of the base clay must be sufficient for the complete slakening of the quicklime. Furthermore, to make the ion exchange possible between calcium ions of hydrated lime and the alkali ions of the clay minerals, there must be enough water after the evaporation caused by the heat release at the slakening of the quicklime. During the placement of lime columns and layers, the heat generation and the expansion of lime further affect the consolidation phenomena.

$$CaO + H_2O \longrightarrow Ca(OH)_2 + HEAT \quad (280 \text{ Cal/gm of CaO}) \text{----------(6.11)}$$

The calcium hydroxide, $Ca(OH)_2$, from the hydration of quicklime or when using calcium hydroxide as the stabilizer, dissociates in the water, increasing the electrolytic concentration and the pH of the pore water, and dissolving the SiO_2 and Al_2O_3 from the clay particles.

$$Ca(OH)_2 \longrightarrow Ca^{++} + 2(OH)^- \text{----------------------(6.12)}$$

These processes will result in ion exchange, flocculation, and pozzolanic reactions.

6.3.2 Ion Exchange and Flocculation

When lime is mixed with clay, sodium and other cations adsorbed to the clay mineral surfaces are exchanged with calcium. This change in cation complex affects the structural component of the clay mineral. Within a period of a couple of minutes up to some hours after mixing, the calcium hydroxide is transformed again due to the presence of carbonic acid (H_2CO_3) in the soil (Kezdi, 1979). The presence of carbonic acid in the soil is due to the reaction of carbon dioxide of the air in the soil and the free water. The reaction results in the dissociation of the lime into Ca^{++} (or Mg^{++}) and $(OH)^-$ which modifies the electrical surface forces of the clay minerals. A transformation of the soil structure begins, i.e., flocculation and coagulation of soil particles into larger sized aggregates or grains and an associated increase in the plastic limit. Lime causes the clay to coagulate, aggregate or flocculate. The clay plasticity (measured in terms of Atterberg Limits) is reduced making it more easily workable and potentially increasing its strength and stiffness. The change in the soil structure is a consequence of cation exchange caused by dissociated bivalent calcium ions in the pore water replacing such univalent alkali ions that normally are attracted to the negatively charged clay particles (Assarson et al. 1974). This results in the flocculation of the clay particles.

$$Ca^{++} + Clay \longrightarrow Ca^{++} \text{ exchanged with monovalent ions } (K^+, Na^+) \quad (6.13)$$

In general, the order of replaceability of common cations associated with soils follows the lyotropic series (or Hofmeister series): $Na^+ < K^+ < Ca^{++} < Mg^{++}$, with highly metallic ions replacing the weaker one on the surface of clay particles. The crowding of Ca^{++} ions onto the surface of the clay particles (adsorption) brings about flocculation (Herrin and Mitchell, 1961). The cation exchange capacity highly depends on the pH of the soil water and on the type of clay mineral in the soil. Among the types of clay mineral, Montmorillonites have the highest and Kaolinites have the lowest cation exchange capacities (Assarson et al. 1974). Brandl (1981) concluded that cation exchange capacity of a soil is not a criterion for its reactivity with lime, and instead suggested that the amount of semi-removable silica is a useful criterion of lime reactivity for practical purposes.

The fact that flocculation of clay occurs as a consequence of the addition of lime is a well-known phenomenon, but the achievement of flocculation is not necessarily a main mechanism by which lime stabilizes the soils. Diamond and Kinter (1965) argued that although calcium saturation is required for stabilization, many natural soils which are largely saturated still exhibit deficiencies associated with problematic soil for use as subgrades, and thus require stabilization. Furthermore, many chemical agents other than lime induce immediate flocculation when mixed with clays (Brandl, 1981), yet are valueless for stabilization.

6.3.3 Pozzolanic Reaction

The shear strength of the stabilized soil gradually increases with time mainly due to pozzolanic reactions. Calcium hydroxide in the soil water reacts with the silicates and aluminates (pozzolans) in the clay to form cementing materials or binders, consisting of calcium silicates and/or aluminate hydrates (principally dihydrates) (Diamond and Kinter, 1965). The dissolved dissociated Ca^{++} ions react with the dissolved SiO_2 and Al_2O_3 from the clay particle's surface and form hydrated gels, resulting in the combination of the soil particles (Diamond and Kinter, 1965).

$$Ca^{++} + 2(OH)^- + SiO_2 \longrightarrow CSH \quad\quad\quad\quad (6.14)$$

$$Ca^{++} + 2(OH)^- + Al_2O_3 \longrightarrow CAH \quad\quad\quad\quad (6.15)$$

The gel of calcium silicates (and/or aluminate hydrates) cements the soil particles in a manner similar to the effect produced by the hydration of Portland cement, but the lime cementing process is a much slower reaction which requires considerably longer time than the hydration of cement. The main part of the reaction does not start until a couple of days after the mixing of lime (Assarson et al. 1974). As a rule, it is not finished until one to five years later (Diamond and Kinter, 1965). The solubility of the pozzolans and thus their inclination to react with lime depends on the pH of the soil water. The rate of reaction also increases with increased soil temperature (Broms, 1984).

6.3.4 Carbonation

Lime reacts with carbon dioxide in the atmosphere or in the soil to form relatively weak cementing agent, such as calcium carbonate or magnesium carbonate (Ingles and Metcalf, 1972). The strength of calcium carbonate which is formed by this process can be discounted, and its significance on the soil lime stabilization can be dismissed (Broms, 1984). Diamond and Kinter (1965) even suggested that carbonation is probably a deleterious rather than helpful phenomenon in the soil stabilization.

6.4 SCHEMATIC ILLUSTRATIONS OF CEMENT-IMPROVED SOIL

With the schematic diagrams shown in Fig. 6.2, it was proposed to illustrate the conditions of hardening (Saitoh et al. 1985). Figure 6.2(a) shows the condition immediately after mixing a cohesive soil and a hardening agent slurry. It is considered that even if the cohesive soil and hardening agent slurry are thoroughly mixed, clay particles will form to a cluster which will be surrounded by the slurry. Figure 6.2(b) shows the condition of the cohesive soil and hardening agent slurry that have formed a hardened body. Here the hardening agent slurry [shown in Fig. 6.2(a)] produces hydrated calcium silicates, hydrated calcium aluminates, $Ca(OH)_2$, etc., and forms hardened cement bodies. The pozzolanic reaction between the clay and the $Ca(OH)_2$ obtained from the cement hydration reaction produces hardened soil bodies. It is considered that the strength of the improved soil will depend upon the strength characteristics of both types of hardened bodies.

6.5 PREDOMINANT FACTORS THAT CONTROL HARDENING CHARACTERISTICS OF CEMENT TREATED CLAY MATERIALS

The hardening characteristics of cement treated soil mixtures are developed by a number of factors. Owing to the large number of alternatives and combinations, it is impossible to tabulate the various mechanical properties as functions of these factors, so the experimental determination is indispensable in most cases. There are, nevertheless, some predominant factors presented below, but they only provide information outlining order-of-dominance value, and illustrating the effect of these factors on the strength and stiffness of the cemented clay. An outline of some superficial factors exerting an influence on the properties of cement treated soils are illustrated in Fig. 6.3.

6.5.1 Type of Cement

The differences in improvement of cement treated clays by using different types of Portland cement have been investigated. The stabilization by Type III Portland cement renders better improvement of soil than the Type I cement does. However, the Type I Portland cement is the most popular cement used in soil stabilization. This is because it is the most readily available and cheapest compared with other types of cement.

6.5.2 Cement Content

In general, it has been found that the greater the cement content, the greater is the strength of the cement treated clay (Broms, 1984). This behavior is different in the case of lime treated clay. In the case of lime, there is a maximum strength limit that can be obtained at the optimum lime content.

6.5.3 Curing Time

In a manner similar to that of concrete and lime treated soils, the shear strength of cement treated clay increases with time. The rate of increase of strength is generally rapid in the early stages of the curing period. Thereafter, the rate of increase of strength decreases with time. The rate of increase of strength for cement treated clay is greater than that of lime treated clay at the early stage.

6.5.4 Soil Type

The effectiveness of cement and lime decreases with increasing water content and organic content. The improvement decreases generally with increasing plasticity index of the clay (Broms, 1986). The strength increase of cement treated clay on organic soils is often very low. However, cement is more effective than lime in the stabilization of organic soils (Miura et al. 1986).

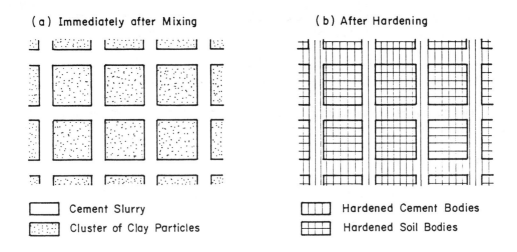

Fig. 6.2 Schematic Illustrations of Improved Soil (Saitoh et al, 1985)

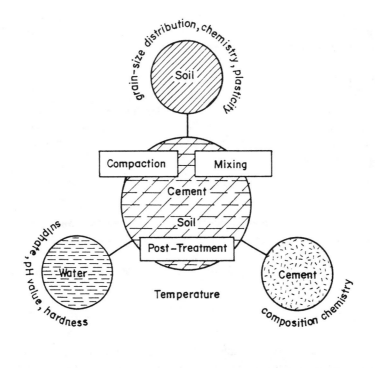

Fig. 6.3 Factors Affecting the Properties of Cement Soils (Kezdi. 1979)

The effects of cement gradually decrease with increasing clay content and increasing plasticity index (Woo, 1971). In general, when the activity of a soil is very high, the increase of the shear strength of the soil treated with cement, is low. However, these are in the reverse order in the case of lime, since the strength of the lime treated clay depends mainly on the participation of the clay particles in the pozzolanic reactions. But for cement treated clay, it depends mainly on the cementation from the cement hydration. The increase of the shear strength due to the flocculation is often relatively small for marine clays deposited in salt water, since these clays already have a flocculated structure (Broms, 1984).

6.5.5 Curing Temperature

The increase of temperature accelerates the chemical reactions and solubility of the silicates and aluminates, thus increasing the rate of strength gain of the treated soil.

6.5.6 Soil Minerals

In the case of soil with the property of higher pozzolanic reactivity, the strength characteristic of the treated soil is governed by the strength behavior of the hardened cement bodies. But in the case of soils having lower pozzolanic reactivity, the strength characteristics of the treated soils are governed by the strength characteristics of the hardened soil bodies (Saitoh et al. 1985). Therefore, if improvement conditions are equal, greater strength is obtained from the soil with higher pozzolanic reactivity. Hilt and Davidson (1960) observed that montmorillonitic and kaolinitic clayey soils were found to be effective pozzolanic agents, as compared to clays which contain illite, chlorite or vermiculite. Wissa et al. (1965) also explained that the amount of secondary cementitious materials that are produced during pozzolanic reaction of the clay particles and hydrated lime ($Ca(OH)_2$) is dependent on the amount and mineral composition of the clay fraction as well as the amorphous silica and the alumina present in the soil. The montmorillonite clay mineral will probably react more readily than the illites and kaolins because of their poorly defined crystallinity.

6.5.7 Soil pH

The long-term pozzolanic reactions are favored by high pH values, since the reactions are accelerated due to the increased solubility of the silicates and the aluminates of the clay particles. When the pH value of the treated clay is lower than 12.6, the reaction of the Eq. 6.10 occurs, where $C_3S_2H_x$ is used up to produce the CSH and the hydrated lime [$Ca(OH)_2$]. This will reduce the strength of the treated clay at the expense of stronger cementitious material, $C_3S_2H_x$, to produce the weaker cementitious material, CSH.

6.6 PREDOMINANT FACTORS THAT CONTROL HARDENING CHARACTERISTICS OF LIME TREATED CLAY

6.6.1 Type of Lime

The efficiency of lime stabilization depends in part on the type of lime material used. Quicklime is generally more effective than hydrated lime (Kezdi, 1979), but generally it needs care in handling for soils with high moisture contents. Unslaked lime or quicklime is more effective since water will be absorbed from the soil and more importantly, the hydration will cause an increase in temperature which is favorable to strength gain (Broms, 1984).

6.6.2 Lime Content

The strength of lime soil mixtures, provided they are properly cured, increases as the lime content is increased. There appears to be no optimum lime content which produces a maximum strength in a lime stabilized soil under all conditions. However, it can be stated that for a particular condition of curing time and soil type, there is a corresponding optimum lime content which causes the maximum strength increase (Herrin and Mitchell, 1961).

6.6.2.1 Lime Fixation Point

The lime fixation point is defined as the point at which the percentage of lime is such that additional increments of lime produce no appreciable increase in the plastic limit. Handy et al. (1965) referred to this point as the "lime retention point". Based on extensive investigations at Iowa State University, the concept of the lime fixation point was suggested. Lime contents equal to the lime fixation point for a soil will generally contribute to the improvement in soil workability, but may not result in sufficient strength increases (Hilt and Davidson, 1960).

6.6.2.2 Optimum Lime Content

Methods of determining the optimum lime requirement for lime stabilization have been proposed. Eades and Grim (1966) suggested that the amount of lime consumed by a soil after one hour affords a quick method of determining the percentage of lime required for stabilization, i.e., the lowest percentage of lime required to maintain a pH of 12.6 is the percentage required to stabilize the soil. However, a strength test is still necessary to show the percentage of strength increase. McDowell (1959) pointed out that short-time or quick tests probably will not identify optimum lime contents, but are essential in checking against the use of non-reactive soils for treatment of lime. On the other hand, while long-term tests would do a better job of identifying optimum lime contents, they may be impractical from the standpoint of time, and may even suggest the use of insufficient amounts of lime due to the ideal conditions under which they are run. Hilt and Davidson (1960) gave a correlation which showed that the amount of lime fixation is in proportion to the type and amount of clay present and is independent of the absorbed cation present in the clays. The relationship is given as:

$$\text{Optimum Lime Content} = \frac{\% \text{ of clay}}{35} + 1.25 \quad \text{--------------------(6.16)}$$

6.6.3 Curing Time

Broms (1984) reported that the shear strength of stabilized clays will normally be higher than that of untreated clay after mixing. Figure 6.4 shows a typical plot of the increase of shear strength with time for various types of soils. The shear strength of clay stabilized with lime will normally be higher than that of undisturbed clay about one to two hours after mixing even when the sensitivity of the clay is relatively high (Broms, 1984). The undrained final shear strength of stabilized clay can be, under favorable conditions, as high as 10 to 50 times the initial shear strength (Assarson et al. 1974). The shear strength of the stabilized soil gradually increases with time through pozzolanic reactions when the lime reacts with the silicates and aluminates in the soil (Broms, 1984). The rate of increase is generally rapid at the early stage of curing time; thereafter, the rate of increase in strength decreases with time. Lime has an initial reaction with soil taking place during the first 48-72 hours after mixing, and a secondary reaction which starts after this period and continues indefinitely (Taylor and Arman, 1960).

Several attempts have been made to express the strength of lime stabilized soils as a function of curing time. Broms (1984) found that the shear strength of stabilized soils as determined by unconfined compression tests increased linearly with time when plotted in log-log scale (log C_u, log t). Brandl (1981) and Okamura and Terashi (1975), however, found that the time-dependent increase in shear strength was approximately linear with the logarithm of time.

6.6.4 Type of Soil

For lime treatment to be successful, the clay content of the soil should not be less than 20% and the sum of the silt and clay fractions should preferably exceed 35%, which is normally the case when the plasticity index of the soil is larger than 10 (Broms, 1984). The shear strength increase of the stabilized soil is highly dependent on pozzolanic reactions, i.e., the reactions of lime with the silicates and aluminates in the soil.

6.6.4.1 Grain Size Distribution

Figure 6.5 shows the effect of grain size distribution on the lime stabilization method. The increase in strength with time is in general highest for normally consolidated silty clays, with low plasticity index and a low water content. The strength increase in lime treated organic soils is often very low; even a relatively small amount of organic material can have a large effect on the strength increase (Broms, 1984). Gypsum has often been used together with unslaked lime to stabilize organic soils when lime alone is not effective (Broms and Anttikoski, 1983). Generally, the effect of lime decreases with increasing water content (Holm et al. 1983; Miura et al. 1987).

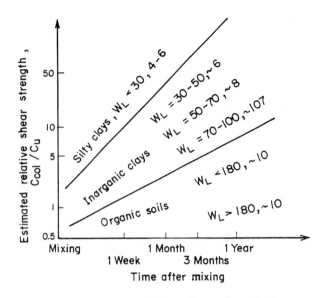

Fig. 6.4 Increase of Shear Strength with Time

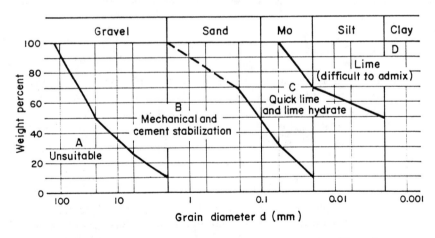

Fig. 6.5 Effect of Grain Size Distribution on the Applicability of Lime Stabilization (Kezdi, 1979)

6.6.5 Clay Minerals

Eades and Grim (1966) reported that the quantity of lime needed to effectively treat a clay is dependent on the type of clay mineral present. Eades and Grim (1966) observed that although kaolinites, illites, montmorillonites and other mixed-layered clays all react with lime to give greater strengths, the quantity of lime needed to treat a clay is dependent on the type of mineral present. Hilt and Davidson (1960) found that from the unconfined compression test results, kaolinitic and montmorillonitic clayey soils are effectively stabilized with lime alone. Whereas illitic clays require addition of fly-ash to obtain a significant strength gain. Lee et al. (1982) found that in terms of strength increase, lime treatment has a greater effect in montmorillonites than kaolinitic soils.

6.6.6 Soil pH

Lime addition will increase the pH of the water content in the soil, and give rise to increased solubility. The base exchange is low when the pH-value is less than 7. The long-term chemical reactions in lime stabilized soils are favored by a high pH-value (pH > 12) since the reactions are accelerated due to the increased solubility of the silicates and aluminates (pozzolans) present in the clays (Broms, 1984). Davidson et al. (1965) suggested that a minimum pH of approximately 10.5 is necessary for pozzolanic reaction to take place, while Eades and Grim (1966) suggested that the lowest percentage of lime required to maintain a pH of 12.40 is the percentage required to stabilize a soil. Broms (1984) pointed out that the pH of the treated soil will normally exceed 12 even when only a few percent of lime has been added to the soil.

6.6.7 Curing Temperature

The chemical reactions in the soil are favored by a high temperature (Broms, 1984). For lime-soil mixture at the same age, the effect of increasing the curing temperature is to increase strength (Ruff and Ho, 1966). The curing temperature has been found to affect the long term reactions between lime and clay. Broms (1984) attributed the favorable effects of high curing temperature to the increased solubility of the silicates and aluminates (pozzolans) in the clay at high temperatures. For lime stabilized clays, Metcalf (1964) found that the curves (UC strength versus temperature) were different for different clays, and that there was an abrupt change in the slope in the vicinity of 45°C. Ruff and Ho (1966) extended the work of Metcalf (1964), and suggested that different reaction products are formed at different curing temperatures and that the cut-off temperature is from 23°C to 40°C. Furthermore, it was found that there was increase of strength with time at all temperatures, with greater rate of increase at the higher temperature. Chaudry (1966) reported that the compacted lime stabilized Bangkok clay cured at 100°F had higher strength values than those cured at 70°F.

6.7 RELEVANT CHARACTERISTICS OF SOFT BANGKOK CLAY

The relevant soil properties for the soft Bangkok clay at Nong Ngu Hao is given in Fig. 6.6. For typical soft Bangkok clay, the percentage of clay varies from 30% to 70%, silt from 20% to 60%, and sand from 0 to 15% (Balasubramaniam and Bergado, 1984). Typical values of natural water content range from 40% to 130%. The liquid limit varies in the range of 50% to 130%, while plastic limit ranges from 20% to 60%. The plasticity index ranges from 20% to 80% and the liquidity index from 0.4 to 1.2. The organic and salt contents are also given in Fig. 6.6. The organic contents seemed to vary from 2% to 5% with occasional maximum value of 9%. Balasubramaniam et al. (1985) reported salt contents of 0 to 0.5% (0 to 5 g/liter) in the upper portions in the weathered crust and 0.5% to 2% (5 to 20 g/liter) in the lower soft clay portions. The sensitivity of Bangkok clay can sometimes be as high as 8 and averaging about 5 or 6. Regarding the clay minerals, illites seemed to dominate, ranging from 20% to 70%, while montmorrillonite ranged from 5% to 30% and kaolinites about 10%.

6.8 EFFECTS OF QUICKLIME, NATURAL WATER CONTENT, SALT, AND ORGANIC CONTENTS

Miura et al. (1986) previously reported that the improvement mixing with quicklime is more effective for clays located nearshore than onshore locations for soft Ariake clay in Japan. One reason for this difference is the amounts of natural moisture contents being larger in the latter than the former (Fig. 6.7). However, it was suggested that the salt content of clay may have influenced the degree of improvement. Ariizumi (1977) described the improvement effect of quicklime on clay that was accelerated by addition of salts. Tests were made where small amounts of salts were added to the clay sample together with quicklime and the results are shown in Fig. 6.8. In these tests, salt up to 5 percent of quicklime weight and quicklime at 5, 7.5, and 10 percent of dry soil weight were added and then mixed. As seen in Fig. 6.8, the strength of improved soil increased with the increase in salt content up to a certain limit. According to Ariizumi (1977), the addition of salt, NaCl, may act as catalyzer and the ions Cl^-, Na^+, Mg^{+2} may have accelerated the pozzolanic reaction. The effects of quicklime and natural water contents are demonstrated in Fig. 6.9. For clays with high organic contents of more than 8%, the use of cement instead of quicklime becomes advantageous (Miura et al. 1987). This is probably because high organic clay generally requires much amount of quicklime and hence there may remain non-reacted excess admixture.

6.9 STRENGTH OF CEMENT TREATED CLAY

A most comprehensive review of the strength properties of cement stabilization was presented by Mitchell et al. (1974). The unconfined compressive strength, q_u, is generally described as increasing linearly with the cement content percentage, A_w. This increase is more pronounced for coarse-grained soil than for silt and clays. Like q_u, other strength parameters such as cohesion intercept and the friction angle increase with A_w and curing time. Mitchell et al. (1974) gave the following relationships between curing time and q_u.

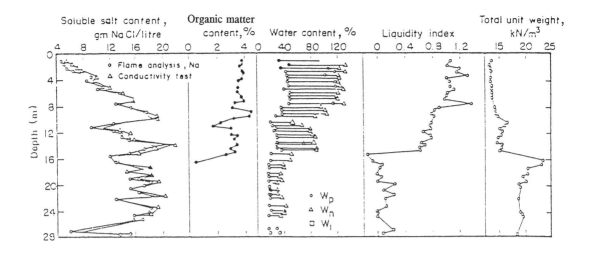

Fig. 6.6 Variation of Soluble Salt Content, Organic Matter Content, Water Content, Liquidity Index, and Total Unit Weight at Nong Ngu Hao (Bangkok Clay)

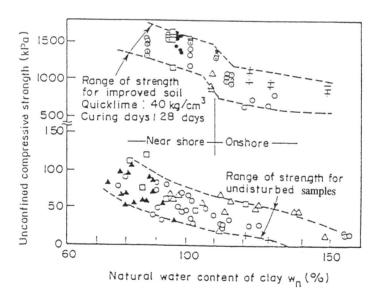

Fig. 6.7 Relationship Between Natural Water Content and Unconfined Compressive Strength of Undisturbed Sample and Improved Soil

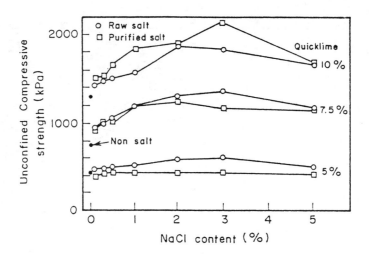

Fig. 6.8 Influences of NaCl Content on the Quicklime Improvement Effect of Ariake Clay (Hasuike Area)

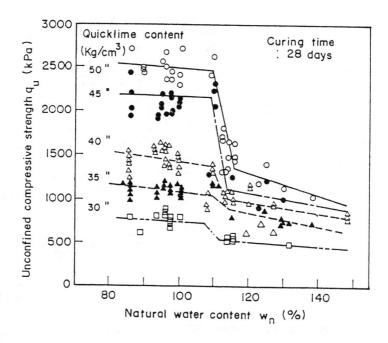

Fig. 6.9 Influences of Natural Water and Quicklime Content on the Improvement Effect

$$q_u = q_u(t_o) + K \cdot \log \frac{t}{t_o} \tag{6.17}$$

where: $q_u(t)$ = Unconfined compressive strength at t days, kPa
$q_u(t_o)$ = Unconfined compressive strength at t_o days, kPa
A_w = Cement content, % by mass
K = 480 A_w for granular soils and 70 A_w for fine grain soil
t = Curing time

6.10 LIME STABILIZATION OF SOFT BANGKOK CLAY

Bangkok clay is well known for its low strength and high compressibility. In this region, there are extensive deposits of soft marine clay which encounter problems in settlement and stability. The sub-soil conditions in the Bangkok area are typical of deltaic plains. The geotechnical problems encountered there are often associated with the presence of a relatively thick layer of soft clay (about 10 metres) with low strength and high compressibility. Various ground improvement techniques are increasingly being explored for the construction works in this area in order to achieve the desired engineering properties of the soil and to minimize the settlement to within the tolerable values.

Extensive laboratory studies on the use of lime stabilized columns have been carried out at the Asian Institute of Technology (Balasubramaniam et al. 1988, 1989). Addition of 5 to 10% quicklime is the optimum mix proportions for the soft Bangkok clay. The addition of quicklime increased the unconfined compressive strength to about 5 times and increased the preconsolidation by as much as 3 times. The vertical coefficient of consolidation also increased by 10 to 40 times and the effective strength parameters also increased, especially the angle of internal friction from 24° to 40°.

6.11 CEMENT TREATMENT OF SOFT BANGKOK CLAY

Broms (1984) has suggested that in Southeast Asia, it is preferable to use cement rather than lime because of the following reasons: low cost of cement compared to lime; difficulty of storing unslaked lime in a humid and hot climate; higher strength can be obtained while there is a limit for maximum strength for lime. At the Asian Institute of Technology, Kamaluddin (1995) conducted studies on the strength and deformation characteristics of cement treated soft Bangkok clay under unconfined compression, oedometer, constant stress ratio tests, and triaxial drained and undrained tests. The study addressed the effect of the variability in terms of quantity of the hardening agent, the pre-shear consolidation pressure, the stress conditions imposed during testing, the drainage condition, and the time dimension. The Oedometer Tests revealed that the cement treatment caused substantial improvement of consolidation properties. Normalized intrinsic compression curves offered confirmation of the hardening effect in the void index plane. The prevalent role of pre-shear consolidation pressure was manifested to annihilate the cementation effects attributing ductility to the treated matrix. Kamaluddin (1995) observed that the main effect of cement treatment is to modify the behavior of the soft clay from normally consolidated to overconsolidated state. The existence of a small strain domain denoted as 'Initial Small Strain Phase' was ascertained based on the characteristics of undrained stress paths and from the (q, ε_v) relationships of drained tests, establishing a family of loci in the (**q**, p) stress space which governs the onset phase transformation of the treated clays. Beyond the small strain domain and up to the curved failure envelope, a work hardening and elasto-plastic type of behavior with large strain is observed. The stress-dilatancy relations on the Spatial Mobilized Plane (SMP) exhibit a remarkable phenomenon such that the stress ratio, τ/σ_N, and the normal and shear strain increment vector, $-d\varepsilon_N/d\gamma$, constitute a specific correlation between treated and untreated clays. In Constant Stress Ratio (CSR) Tests, strain paths produced sets of stress-dilatancy relationships. Heavily overconsolidated (rigid) behavior was observed inside the loci of the transition points of bilinear strain paths obtained from CSR tests. A composite conceptual model to describe the mechanical behavior of cement treated clay is proposed in which three zones and four subzones have been envisaged. The treated clays have been found to be strain-softened after failure in an 'Unstable Phase' with the residual stress state lying on a failure envelope identified as 'Destructured Envelope', which is close to the critical state line of the untreated clay. At this envelope, the cohesion is destroyed and the treated clay then behaves as purely frictional material.

6.11.1 Fundamental Behavior of Cement Treated Clay

The base soil used in the study was soft Bangkok clay taken from a site within the AIT campus, obtained from depths of 3-4 m. The properties of this clay are shown in Table 6.1. The sample preparation was done by mixing the cement slurry and the base clay thoroughly. Organic matter content of the base clay was 5.6%; whereas organic carbon was 2.87. A considerable amount of cement was required to neutralize the large reserve of potential acidity or buffering capacity of the clay. The properties of the treated samples after curing are listed in Table 6.2.

Table 6.1 Physical Properties of the Soft Bangkok Clay

Liquid Limit	103%	Sand	3%
Plastic Limit	43%	Total Unit Wt., (kN/m^3)	14.3
Plasticity Index	60%	Dry Unit Wt., (kN/m^3)	7.73
Water Content	76-84%	Initial Void Ratio, e	2.2
Liquidity Index	0.62	Activity	0.87
Silt	28%	Sensitivity	7.3
Clay	69%		

Table 6.2 Properties of Cement Treated Soft Bangkok Clays

Cement Content (%)	Curing Time (month)	Total Unit Weight (kN/m^3)	Dry Unit Weight (kN/m^3)	Void Ratio	Water Content (%)	Degree of Saturation (%)
5	1	14.85-15.12	8.16-8.67	2.03-2.22	73.35-83.77	95.77-100.00
7.5	1	14.90-15.25	8.31-8.76	1.99-2.15	772.92-81.4	96.75-100.00
10	1	14.95-15.43	8.44-8.90	1.93-2.09	69.33-78.53	95.56-99.95
12.5	1	15.03-15.52	8.49-9.21	1.90-2.06	68.51-77.70	95.96-99.96
15	1	15.04-15.40	8.60-9.05	1.87-2.02	66.79-75.37	94.65-98.88
5	2	14.90-15.51	8.28-8.98	1.99-2.16	71.66-80.89	96.15-100.00
7.5	2	14.97-15.55	8.32-9.10	1.92-2.12	70.92-79.98	97.89-100.00
10	2	15.00-15.48	8.62-9.15	1.91-2.05	69.19-77.46	98.78-99.87
12.5	2	15.05-15.60	8.68-9.22	1.85-1.98	67.67-74.22	96.57-98.97
15	2	15.28-15.61	8.81-9.31	1.82-1.95	67.06-73.49	97.67-99.87
Untreated	0, 1 and 3	14.1-14.96	7.74-8.24	2.20-2.44	81.60-86.00	99.81-100.00

The effect of cement content and curing time on the liquid limit was insignificant. It was the plastic limit which was increased with cement content (A_w) and curing time (t). The reduction of plasticity index was due to an increase of plastic limit. The immediate decrease in the water content after mixing cement with clay was from 5% to 10%. Sharp reduction of water content occurred up to the vicinity of 12.5% cement content (reduced by 8.3% to 17.3%), and then the effect of the reduction to a cement content of 40% was small (4.8% to 9.1%). Water content reduces with curing time rapidly up to 10 weeks and then slows down, but continues for a longer time. Cement causes significant reduction in specific gravity and this change is dependent on cement content and curing time. Variation of specific gravity with cement content follows a pattern; the curvilinear variation of lowest curing time gradually turns into a linear variation at the longest curing time. The influence of curing time is larger at the initial stage of curing. After 3 months, the influence becomes minor. A comparative study with untreated base clay revealed that cement treated samples have lower void ratio, higher degree of saturation, and higher unit weight. An investigation showed that the slopes of the lines in the unit weight vs. cement content (γ_t, A_w) plane were found to be almost constant, but the intercepts increased sharply up to 12 weeks curing period and thereafter decreased. In the relationship of unit weight versus curing time (γ_t, t), curvilinear variation was observed up to 8 weeks, then the variation was of linear pattern.

6.11.2 Unconfined Compression Tests

Stress-strain curves for treated samples were found to increase abruptly until the peak compressive strengths, then suddenly decreased to very low residual values upon further straining. Zonal demarcation showing the effectiveness of cement treatment has been illustrated in Fig. 6.10. The q_u (unconfined compressive strength), A_w (percentage of cement by weight) plane was sub-divided into 3 regions (A, B and C) on the basis of gradient development. The properties of the treated soil exhibited significant increase in strength and modulus of deformation, but simultaneously the clay material was changed mainly to brittle and quasi-brittle materials. In general, stress-strain curves of the treated samples were found to increase abruptly to peak values, then suddenly decreased to low residual values. From the aspect of stress-strain relationships, the overall behavior was categorized into brittle, quasi-brittle and ductile. Higher strain, low strength and mild peak were found to be associated with ductile behavior, whereas low strain, higher strength, and sharp peak exhibited brittle behavior. Since the reduction of water content is a function of A_w in the clay matrix, it was concluded that strength gain due to decrease in water content is smaller for low cement content such as 5%; whereas, for higher cement content, a part of the strength increase is due to reduction of water content. From the aspect of unconfined compression condition, 15% to 20% cement content and 1 to 2 months curing period were regarded as optimum. In (ε_f, q_u) relationships, the samples with higher strength produce lower values of failure strain which corresponds to brittle behavior. Ductile samples were found to be more scattered than the brittle samples. For the case of ductile samples, significant reduction of failure strain was observed with curing time without substantial development of strength. Failure strain reduces significantly as the cement content and curing time increase. The (ε_f, A_w) relationships showed that substantial reduction of failure strain occurred up to $A_w=20\%$. On the other hand, failure strain reduced tremendously with a curing period up to 8 weeks, and beyond 16 weeks, the failure strain reaches an asymptotic value. The initial tangent modulus vs. unconfined compressive strength (E_u, q_u) relationships produced a straight band of narrow scatter. In the (E_u, A_w) relationship, sharp increase of the modulus can be noticed up to cement content of 20%. An asymptotic value is reached at about 35% to 40%. A similar type of trend can be noticed in (E_u, t) relationship. Shear type of failure mode was noticed in the case of treated samples. The higher the values of A_w and t, the greater is the degree of stiffness and brittleness of the treated clay. Though the effect of curing time is similar to that of cement content, the result shows that the curing hardening has less influence on failure mode and brittleness than that of cement.

6.11.3 Consolidation Behavior

6.11.3.1 Oedometer Tests

The void ratio-axial stress plot (Fig. 6.11) shows the similarity in the shapes of the void ratio-axial stress curves for 1 to 6 month curing period. The (e, log σ_v) relationships show that the treated curve crosses the untreated curve much before its preconsolidation pressure and then is displaced from untreated curve with increasing values of σ_v, indicating instinct characteristics of lower compressibility than the untreated one. The higher the value of cement content, the greater is the enhancement of the preconsolidation pressure and the decrease of compression index accompanied with gradual reduction of compressibility. The gradual reduction process of compressibility of treated clay with time is quite obvious from the test results. But it was observed that at low value of A_w such as 5%, the increase of curing time could not help in developing significant hardening effect in the clay matrix. The results show that the hardening potential gained by increasing cement content (e.g., 5% to 7.5%) is much greater than that of enhancing curing time (e.g., 1 month to 2 months). With low cement content, the curing time parameter becomes latent. It can be postulated that the value of A_w in

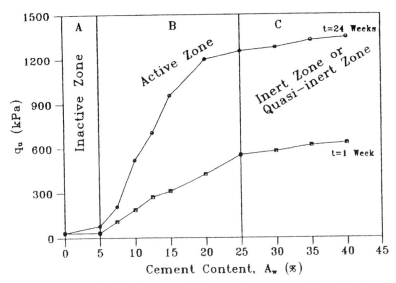

Fig. 6.10 Influence of Cement Content on Unconfined Compressive Strength

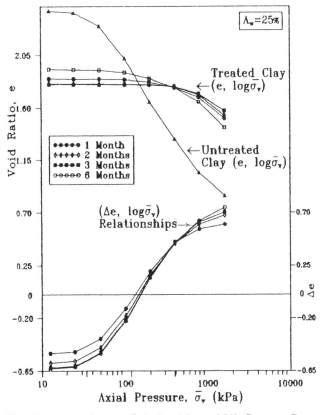

Fig. 6.11 (Δe, log σ_v) Relationship at 25% Cement Content

a sample is a more active parameter than curing time which is a secondary and auxiliary parameter of the former. The swelling ratio (SR) is reduced greatly due to cement treatment, implying a rigid mechanism that is employed during the swelling process. The high pressure tests [σ_v (max)=6000 kPa] showed that the (e, log σ_v) relationships at the highest range of stress tend to proceed nearly parallel with the normally consolidated line of untreated base clay. At large strains, the cohesion of the clay is destroyed and the treated material then behaves as purely frictional material. At a higher stress level, a breakage in (e, log σ_v) relationship was discernible. At this point, a (e, log σ_v) relationship possessing lower values of compression index (C_c) evolved. At sufficiently higher stress level, the relation of (e, log σ_v) treated clay tends to run parallel to that of untreated clay. This stage can be referred to as normally compressed response of the treated clay. Thus, a postulation can be made that the (Δe, log σ_v) relationship (Fig. 6.11) reaches an asymptotic value when the treated clay will reach the state of normally compressed response. The initial part of the compression plane (Initial Compression Line) deviates upward for low cement content and curing time. On the other hand, the later part of the compression plane displaces with increasing value of σ_v gradually as the cement content and curing time increase. A way of finding the value of σ_v at which treated clay reaches the normally compressed response is to extrapolate the (Δe, log σ_v) plot to asymptotic value where the axial pressure is the required σ_v.

The void index, I_v, versus log σ_v plot (Fig. 6.12) renders excellent confirmation of the overconsolidation effect of the treated clays. The untreated clay proceeds with ICL (Intrinsic Compression Line). As the cement content increases, the curve is displaced with an increasing value of σ_v. A method can be explored to find the equivalent consolidation pressure, p_e, from the shifted ICL. The intersecting point of the shifted ICL and the horizontal line drawn at the void index of the sample is the value of σ_e for the said sample. One important effect of cement treatment is to increase the values of coefficient of consolidation. The Cv value generally decreases approximately linearly with increasing consolidation pressure. Results show that the higher the cement content, the greater the value of Cv. The highest enhancement of Cv value with cement content occurs in the vicinity of 15% cement content. Curing period of 1 to 2 months is found to be the most effective length of time for enhancement of Cv values and can be designated as optimum curing time. The compression index (C_c) value was found to decrease with cement content. After a certain reduction of C_c value with 5% cement content, gradual reduction was noticed up to 15% cement content and, thereafter, changes were very minor. Substantial reduction of C_c value occurred during the period of 1 to 2 months, and thereafter the C_c values were almost the same. High pressure oedometer tests render similar compressibility characteristics as those of other tests [σ_v (max)=1600 kPa], having approximately identical C_c values.

6.11.3.2 Constant Stress Ratio (CSR) Tests

Application of higher stress ratio exhibited greater compressibility characteristics. The effect of variation of stress ratio on the (e, ln p) relationships (Fig. 6.13) is small for higher cement content and curing time. Higher volumetric strains are associated with low cement

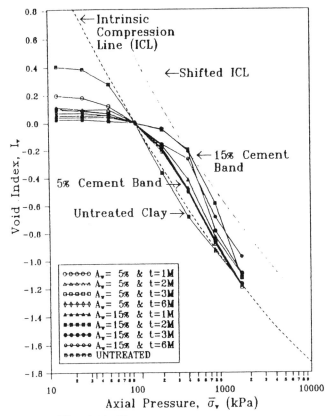

Fig. 6.12 Normalized Intrinsic Compression Curves of the Treated Clays

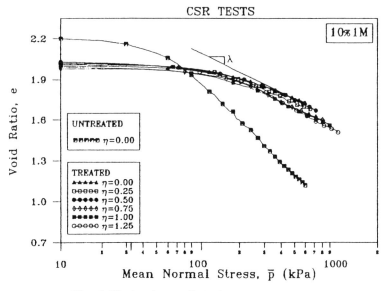

Fig. 6.13 (e - log p) Relationships from CSR Tests (10% Cement and 1 Month Curing)

content and curing time and higher stress ratio. The locus of the transition point between small strain and large strain is influenced by the cement content and curing time. It was found that the higher the stress ratio, the greater is the shear strain incurred by the sample. The variation of shear strain with stress ratio is much larger at low cement content and curing time. The strain path showed (Fig. 6.14) bilinear characteristics. The loci of the transition points of small strain and large strain of the bilinear relationships (Fig. 6.15) are affected by the cement content and curing time; higher cement content and curing time expand the loci considerably.

The strain increment ratio was considered in two parts, inside and outside the transition point. It was found that higher cement content and curing time render greater values of dilatancy at both inside and outside the transition points. The stress-dilatancy relationship (Fig. 6.16) of Modified Cam Clay provides the relationship at lower strain than the experimental curves.

6.11.4 CIU Triaxial Compression Tests

6.11.4.1 Deviator Stress-Shear Strain Relationships

The effect of equivalent pressure, p_e, is very gradual for low range of A_w such as 5% (Fig. 6.17) since pre-shear consolidated volumes are affected largely by the p_o, which in turn affect the (q, ε_s) relationship. On the other hand, for higher cement content such as 15%, the (q, ε_s) relationship (Fig. 6.18) is largely affected by the p_o when it is at the higher range. The ε_s is found to be linearly dependent on deviator stress for the values of q up to 60% to 70% of q_{max}. Beyond this linear elastic portion, the deviation takes place at varying rates according to p_o. Generally, the shear strain at maximum deviator stress is reduced as the curing time increases. With higher curing time, q value rises sharply toward well-defined peak followed by greater amount of strain softening. The (q, ε_s) relationships were found to plot under varying rates depending on the parameter A_w; maximum deviator stress increases with increasing value of A_w. The ε_s at q_{max} is reduced when cement content is increased. Low A_w (5% and 7.5%) results in mild peak and low strain softening beyond peak; whereas sharper peak and larger strain softening can be observed from higher A_w.

6.11.4.2 Excess Pore Pressure-Shear Strain Relationships

Cement treatment modifies pore pressure response behavior by reducing the strain at peak pore pressure (Figs. 6.19 and 6.20). For low p_o values, the maximum pore pressure occurs prior to the failure of the sample, whereas for higher values of p_o, the maximum pore pressure occurs after the failure of the sample. The strain at peak pore pressure keeps on increasing as p_o increases. Lower p_o render greater value of negative pore pressure. Reduction of strains at peak pore pressure occurs when samples are cured for a longer curing time. Moreover, in (Δu, ε_s) relation, a distinct peak and significant amount of pore pressure dissipation after peak are common criteria for longer curing time. Secondly, long term cured specimens tend to develop greater negative pore pressure than that of short term cured specimens. The higher A_w mobilizes lower values of strain at Δu_{max}. As A_w increases, the

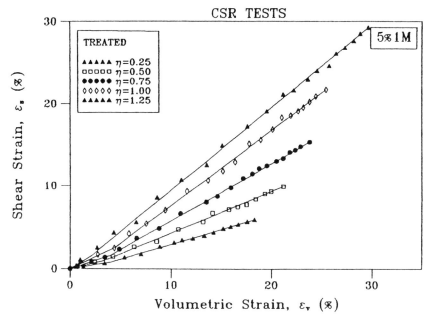

Fig. 6.14 Bilinear Relationship of (ε_s, ε_v) Plot
(5% Cement and 1 Month Curing)

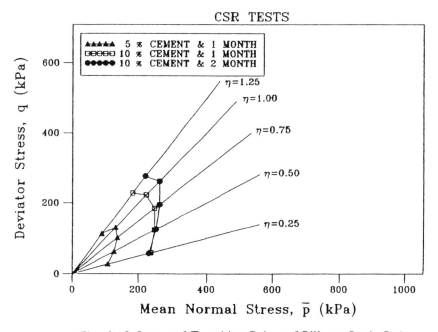

Fig. 6.15 Locus of Transition Points of Bilinear Strain Paths

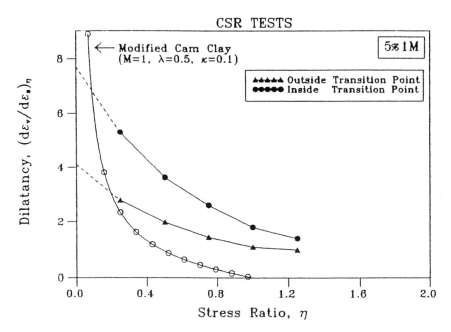

Fig. 6.16 Variation of Strain Increment Ratio with Stress Ratio (5% Cement and 1 Month Curing)

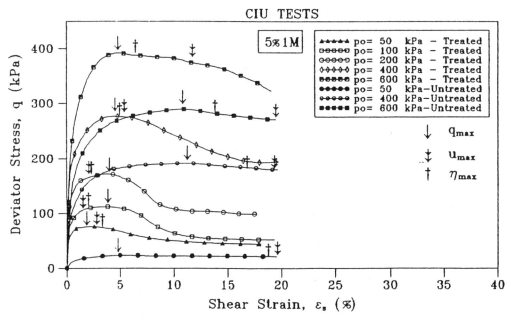

Fig. 6.17 q vs. ϑ_s Plot for Cement Treated Samples (5% Cement Content and 1 Month Curing)

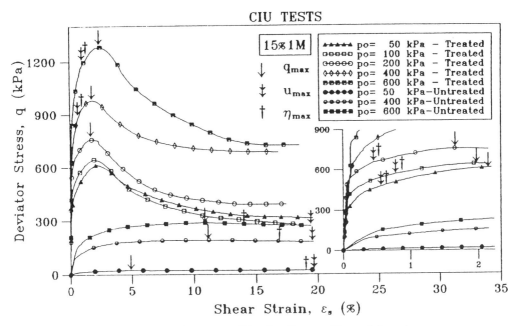

Fig. 6.18 q vs. ε_s Plot for Cement Treated Samples
(15% Cement Content and 1 Month Curing)

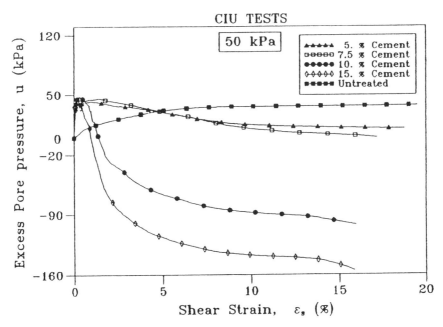

Fig. 6.19 u vs. ε_s Plot for Cement Treated Clay
(p_o=50 kPa and 2 Months Curing)

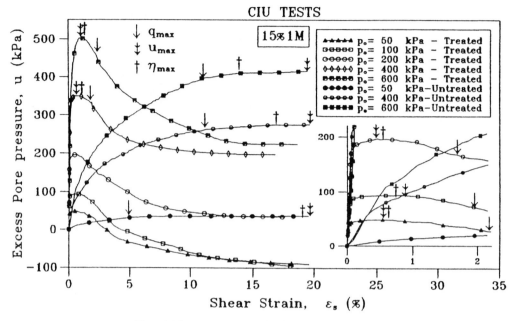

Fig. 6.20 u vs. ε_s Plot for Cement Treated Clay
(15% Cement Content and 1 Month Curing)

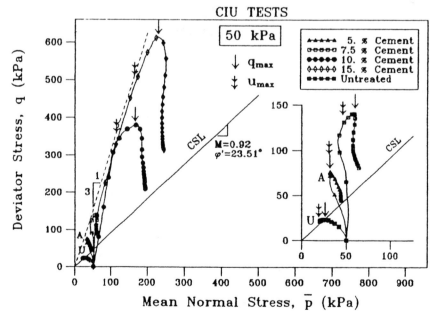

Fig. 6.21 Undrained Stress Paths of Cement Treated Clay
(p_o = 50 kPa and 1 Month Curing Time)

peak pore pressure tends to develop at lower strain. For higher values of A_w, the maximum pore pressure mobilization occurs much before failure. On the other hand, sample with low cement content, Δu_{max} occurs after failure state. The higher the A_w, the greater the negativity of pore pressure.

6.11.4.3 Undrained Stress Paths

The effective stress paths (Figs. 6.21 to 6.23) of treated clays raise several interesting aspects on the undrained behavior of the cement treated clays. It was observed that cement treatment changes the initial normally consolidated characteristics of the clay to that of overconsolidated, rendering more stiffness and rigidity of the clay. A lower A_w value such as 5% and lower does not make a greater alteration of the physico-chemical aspect of the base clays. An 'Initial Small Strain Phase' with constant-p path was found to be followed by large strain phase and generation of greater pore pressure. A conforming picture of phase transformation emerged. Outside the small strain domain and up to the curved failure envelope, a work-hardening, elasto-plastic type of behavior was observed. The phase transformation occurs below the angle of contraflexure. The angle of contraflexure has been defined as an arc tangent of the stress ratio at which the stress path in the (**q**, p) space change its direction from decreasing **p**-value to increasing **p**-value. A very high A_w (e.g., 15%) accompanied by higher t (e.g., 2 months) exhibited rigid behavior and did not show any phase transformation all through. The A_w has been found as the principal hardening parameter that controls the stiffness and rigidity of the treated clay as a prime factor. The hardening effects are influenced mainly by the parameters A_w and t. At the curing period of 1 month, the values of A_w=5% to 15% were found to have completely different effects; the cement content as low as 5% renders less beneficial effects. The curing time also plays a role in increasing the rigidity of the treated samples; but for samples of A_w=5% and lower, the influence of curing time is nominal. The curing time has been marked as secondary hardening parameter in the soft clay solidification process by cementation, whereas A_w has been found as primary parameter. The third variable p_o was also found to influence markedly the behavior of the treated clays. It was seen that higher p_o tends to reduce the beneficial effects of cement treatment by reducing the overconsolidation ratio of the soil. So, it is possible to transform a heavily overconsolidated clay to a normally consolidated clay by employing a sufficiently higher p_o, whose magnitude is a dependent function of A_w and t mainly, among others. It was observed that at low levels of pre-shear consolidation pressure, cement treatment resulted in the samples having stress paths that rise parallel to q-axis, indicating that the mean normal stress p, does not vary much during shear. This behavior can be considered to be a manifestation of the elastic wall concept used extensively in the critical state concepts (Calladine,1963). The stress paths of heavily overconsolidated samples that resulted from cement treatment indicate that samples remaining on the dry side, first move on a constant-p line, then approach Hvorslev strength envelope and tend to seek failure state either moving on the Hvorslev envelope or moving parallel along the tension cut-off line at 1:3 line until they reach the failure envelope of the corresponding treated clay. After reaching the q_{max} level, the stress path incurs strain softening by falling in a line sub-parallel to the q-axis; at this level the sample possesses destructured state. The stress paths of the treated samples follow paths that

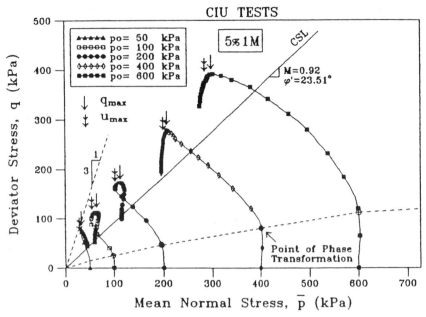

Fig. 6.22   Undrained Stress Paths of Cement Treated Clay
(5% Cement Content and 1 Month Curing Time)

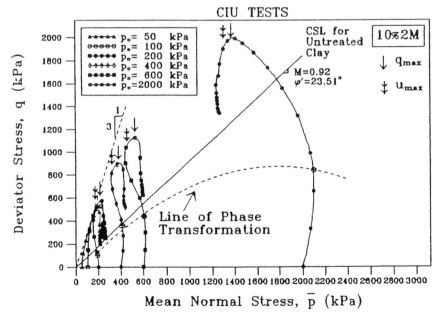

Fig. 6.23 Undrained Stress Paths of Cement Treated Clay
(10% Cement Content and 2 Months Curing)

rise to maximum value of the deviator stress, and then return to a lower value of q at the end of the tests. This falling characteristic of the undrained stress paths beyond the peak value of the deviator stress is a remarkable aspect. The treated clay strain-softened with the residual states lying close to the CSL of the untreated clay.

6.11.4.4 Some Salient Aspects of Undrained Behavior

The overall aspects of undrained behavior discussed so far can be put into a framework as illustrated in Fig. 6.24. The undrained behavior of cement treated clay can be explained by sub-dividing (**q, p**) space into 3 zones. The behavior after the q_{max} condition can be regarded as unstable; strain softening occurs in constant-p route. The residual points of treated clay do not fall exactly on CSL, rather they plot above the CSL in a manner that they constitute an envelope referred as Destructured Envelope. It was postulated that if complete destructuration of the treated clay were achieved, and neglecting any permanent changes in the soil structure due to cement treatment, a destructured envelope would be achieved. At this envelope, cohesion is destroyed and the treated clay then behaves as purely frictional material. The failure envelopes were found to be curved. The overall curvature depends on range of p_o applied. The envelopes shift upward as the values of A_w and t go higher. The normalized pore pressure exhibited bilinear trend with stress ratio, η propounding the possibility of predicting undrained stress path and pore pressure development. The value of η_{tp} increases and slopes of first and second linear portions decrease as A_w and t go higher. A linear type of variation of the relationship (u_f, p_o) exists where all the data fall within a narrow band of scatter. The influence of A_w and t on u_f is not large at low level of p_o. Higher A_w and t shows lower values of shear strain at failure (ε_{sf}); but it reaches asymptotic value at about $A_w=15\%$. Larger consolidation stress results in greater values of ε_{sf}. An increase of A_w reduces the strain at which η_{max} occurs. With a higher value of p_o, the (η, ε_s) relationship tends to be close to untreated clay. A transitional phase (from N.C. to O.C. state) can be clearly discerned in this relationship. Shear strain is found to increase with stress ratio. Increase of cement content attributes rigidity in the sample which manifests lower range of shear strain. Generally, the greater the pre-shear consolidation stress, the larger is the shear strain and ductility. During strain softening, the shear strain increases with the reduction of stress ratio. Higher cement content exhibits lower pore pressure generation. Enhancement of consolidation pressure results in the generation of higher pore pressure. Reduction of pore pressure can be noticed when the stress ratio decreases during the strain softening process resulting in even negative pore pressure. c and ϕ were found to be increased with cement content and curing time but seems to reach an asymptotic value at the vicinity of $A_w=15\%$. The curved nature of the failure envelopes obtained from CIU tests indicated that increase of p_o tends to reduce the positive effect of cement treatment. It is observed that failure envelopes are affected by the A_w, t, and p_o. The initial, peak and residual stress states have curved projections on the (q, p) and (e, log p) planes (Fig. 6.25), which refer to the characteristics generally associated with overconsolidated behavior. The stress path of treated clay was found to be initially sub-parallel to q-axis. Phase transformation was manifested by the abrupt change in curvature of the stress paths (Figs. 6.22 and 6.23). A family of curved loci was obtained from these transformation points.

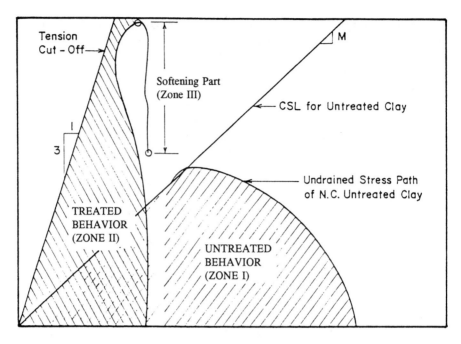

Fig. 6.24 Undrained Behavior of Treated and Untreated Clays in (q, p) Stress Space

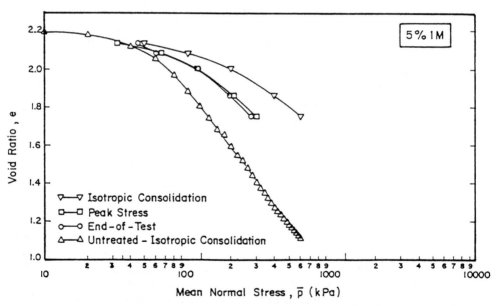

Fig. 6.25 (e, ln p) Relationships from CIU Tests Showing Consolidation Curves ($A_w = 5\%$ and $t = 1$ Month)

6.11.5 CID Triaxial Compression Tests

6.11.5.1 Stress - Strain Relationships

Higher values of A_w and t render potential in developing q values and modulus with substantial low strains showing brittle behavior especially at low p_o (Fig. 6.26). The samples with higher cement content showed dilative nature at the residual states. The (q, ε_s) relationships show that they contain an initial linear relationship where the stress states possess low strains followed by a break point and, thereafter, the sample is subjected to higher strain. The (ε_v, p) relationships (Fig. 6.27) during strain hardening showed bilinear characteristics with a transition point of small strain and large strain. (q, ε_v) relationships showed that lower p_o produces smaller ε_v of the treated samples. A transition stress state exists which make demarcation between initial small strain phase and the remaining large strain phase. For higher A_w and low p_o, sample's dilation occurred with negative volumetric strains. With the increase of pre-shear consolidation pressure, the dilatant nature of the samples diminishes. Thus, the role of p_o is to eradicate the cementation effect. At low p_o, the indication of phase transformation is not observed.

6.11.5.2 Stress Ratio - Shear Strain Relationship

From experimental results, it was found that higher p_o reduces the magnitude of η_{max}, increases the shear strain at which $(\sigma_1/\sigma_3)_{max}$ occurs and tends to lessen the hardening potential of cement treatment. The stress ratio for specimens with low p_o increased at a greater rate with increasing strain. The samples at higher curing period reach higher values of stress ratios accompanied by lower values of strain at which η_{max} occur.

6.11.5.3 Some Salient Aspects of Drained Behavior

The treated soil initially shows a stiff and low strained response on drained compression until it reaches the transition point of small strain and large strain. The curved nature of the failure envelopes reveals that pre-shear consolidation stress tends to reduce the cementation effects. During strain hardening, maximum volumetric deformation is attained with stress ratio at the level of the failure envelope. Enhancement of consolidation stress attributes ductility to the sample. Higher cement content produces greater value of η_{max}. Minor reduction of volumetric strain was noticed during strain softening as the η decreased from η_{max}. Higher values of A_w render greater potentiality in the increase of q_{max} with curing time and consolidation stress. For low A_w values, the (q_{max}, p_o) relationships are almost linear but become curved at higher A_w values. Further, the rate of increase of q_{max} is greater in the region of higher p_o. Drained failure envelopes were found to be curved and are influenced by A_w and t. In general, the higher the value of A_w, the farther is the envelope above the CSL of untreated clay. A majority of the end-of-test stress states did not reach the failure envelope of the untreated clay even at much larger strains; rather they were found to cluster on an envelope referred to as the destructured envelope. It is found that an increase of A_w and t lessens the

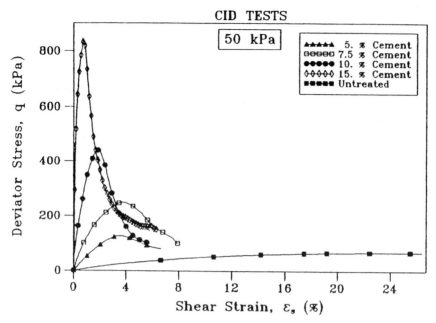

Fig. 6.26 q vs. ε_s Plot for Cement Treated Clay
(p_o = 50 kPa and 1 Month Curing)

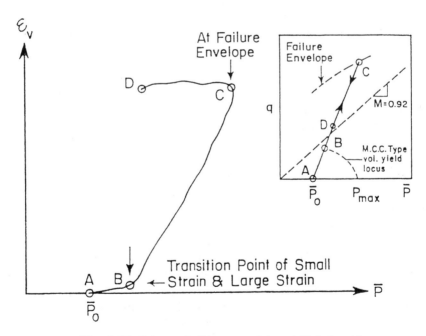

Fig. 6.27 Schematic Diagram of (ε_v, p) Relationship Showing Transition Point

volumetric failure strain. Furthermore, due to enhancement of p_o, the values of ε_{vf} increase at higher rate at the low level of p_o. The values of ε_{vf} are somewhat unaffected by the value of cement content and curing time at higher values of p_o. The (e, ln p) and (e, ln q) plots indicate that the states at the peak and end-of-test conditions have projections which are curved. It is found that the curves for CID tests are different than those of CIU tests. It is noticed that the relationships (q, ε_s) and (q, ε_v) from drained test exhibit the existence of a small strain region referred to as 'Initial Small Strain Phase' at the outset of the drained shearing. This phase can be clearly discerned in (q, ε_v) relationships. The stress state at transition point has been found to be affected mainly by cement content and curing time. The stress-dilatancy relationships (Figs. 6.28 and 6.29) for the treated and untreated clay show some specific pattern in the plane of (τ/σ_N, $d\varepsilon_N/d\gamma$) relationship. The variation of the relationship between stress ratio (τ/σ_N) and strain increment ratio ($-d\varepsilon_N/d\gamma$) can be categorized with respect to cement content and curing time. The stress-dilatancy for untreated normally consolidated clay is linear and that for treated clay is expressed by the dotted line at the initial stage of shearing and then the relation coincides with the untreated line XY after the dotted line intersects with the solid line. The larger the cement content and curing time are, the higher the stress ratio at the intersection is, as shown in Fig. 6.29 (lines A, B, C)..

6.12 DRY JET MIXING METHOD

Deep mixing method may be classified into two categories, namely: a) mechanical mixing method and b) slurry jet mixing method. Chida (1982) proposed a method that uses cement powder or quicklime instead of slurry called the "Dry Jet Mixing Method (DJM method)." In this method, the cement or quicklime powder is injected into the deep ground through a nozzle pipe with the aid of compressed air and then the powder is mixed mechanically by rotating wings. The detail of this equipment is given in Fig. 6.30. In the DJM method, no water is added to the ground, and hence, much higher improvement is expected than using slurry. When quicklime is used, the hydration process generates some amounts of heat resulting in additional drying effects to the surrounding clay and the improvement can be done more effectively (Yamanouchi et al. 1982).

For the design of dike foundations on the Ariake clay in Japan improved by DJM method, a tentative code has been proposed (Miura et al. 1986). The code states that the unconfined compression strength of improved soil in-situ should be larger than 400 kPa at 28 days curing. It is necessary to carry out a series of laboratory tests in order to estimate the appropriate percentage of quicklime to be mixed in practice that satisfies the design strength. Considering the difficulties of quality control in the deep mixing method, the strength of improved ground may be significantly lower than the laboratory test. Miura et al. (1986) showed that the laboratory strength should not be less than 4 times the design strength required in the field for a certain content of quicklime.

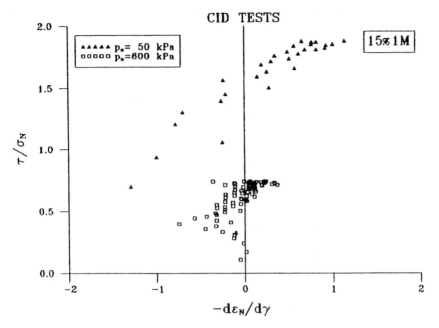

Fig. 6.28 Stress-Dilatancy Relationships on Spatial Mobilized Plane ($A_w = 15\%$ and $t = 1$ Month)

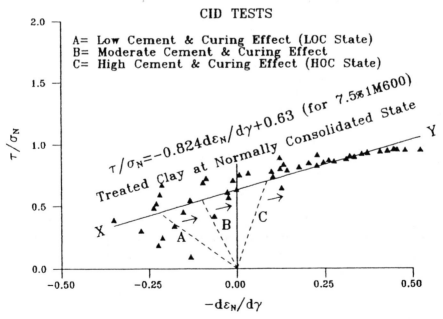

Fig. 6.29 Generalized Stress-Dilatancy Relationships on Spatial Mobilized Plane

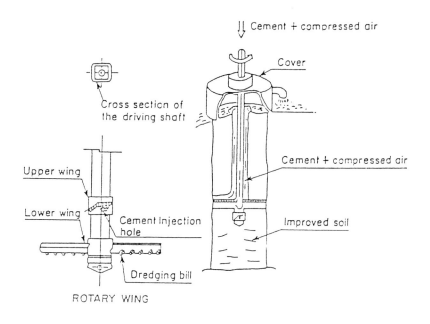
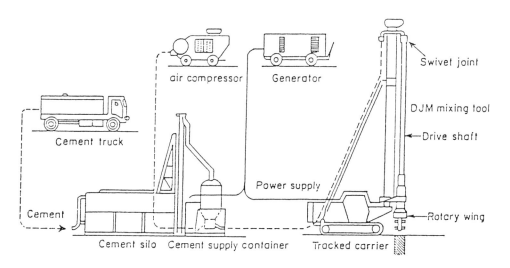

Fig. 6.30 Installation of Cement Column by DJM (DJM Research Group, 1984)

6.13 WET JET MIXING METHOD

Another improvement method is the Wet Jet Mixing (WJM) method or slurry jet grouting method in which slurry lime/cement is jetted into the clay by a pressure of 20 MPa from a rotating nozzle (Chida, 1982). In this method, the machine is relatively light and easy to carry to the project site. A jet grouting system is shown in Fig. 6.31. The main disadvantage is that the diameter of improved column tends to vary with depth according to the variations of the subsoil shear strengths.

A case study is presented by Miura et al. (1987) concerning a site in Ariake clay, Japan. Table 6.3 presents the basic properties of the clay and Fig. 6.32 shows the relevant soil properties with depth. It was found that the field to laboratory strength ratio averaged to about 70 percent. In this case, it was decided to add 190 kg of cement per cubic meter of clay in slurry of water/cement ratio of 2.3.

Table 6.3   Physical Properties of Ariake Clay

Depth (m)	Water Content (%)	Liquid Limit (%)	Plastic Limit (%)	Plastic Index (%)	Liquidity Index
2.5	162	125	55	70	1.57
6.5	122	97	43	54	1.46
10.5	108	79	39	40	1.73

A case of jet grouting work on soft Bangkok clay to stabilize an excavation project is shown in Fig. 6.33. An schematic diagram of the equipment for jet grouting is shown in Fig. 6.34. In this project, 150 kg of cement per cubic meter of clay was added for a water/cement ratio of 1.11. The average diameter of the resulting cement columns was 1.6 m with center to center spacing of 1.4 m. Up to 40 m of cement stabilized columns can be completed in one day.

6.14 TEST EMBANKMENT ON DMM IMPROVED SOFT BANGKOK CLAY

A full scale test embankment, 5.0 m high, was constructed on soft Bangkok clay improved with deep mixing method (dry method) using cement powder by Honjo et al. (1991). The plan and sections of the test embankment is shown in Fig. 6.35. The cement piles were constructed in two different patterns such as the wall type (south side) and the pile type (north side). The lateral movements and settlements of the wall type improved ground are less than that of pile type improved ground when subjected to the same loading conditions as shown in Figs. 6.36 and 6.37, respectively. Furthermore, the deformation pattern of the pile type is tilting, i.e. simple shear deformation. On the other hand, the deformation pattern of the wall type was sliding. Thus, the wall type is more effective in reducing the lateral and vertical deformation. The unconfined compressive strength in the laboratory-mixed stabilized soil was up to 20 times the original value for 28 days curing for cement content of 10 percent (100 kg of Portland cement to 1.0 cubic meter of clay). The in-situ strength was found to be one-half

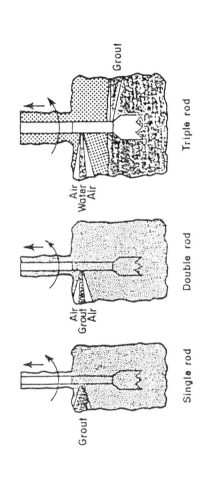

Fig. 6.31 Jet Grouting Systems

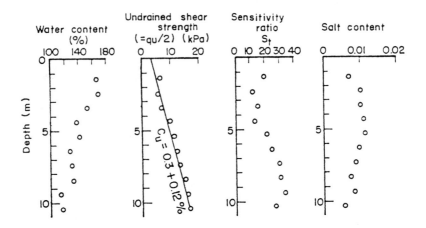

Fig. 6.32 The Properties of the Clay in Ariake District

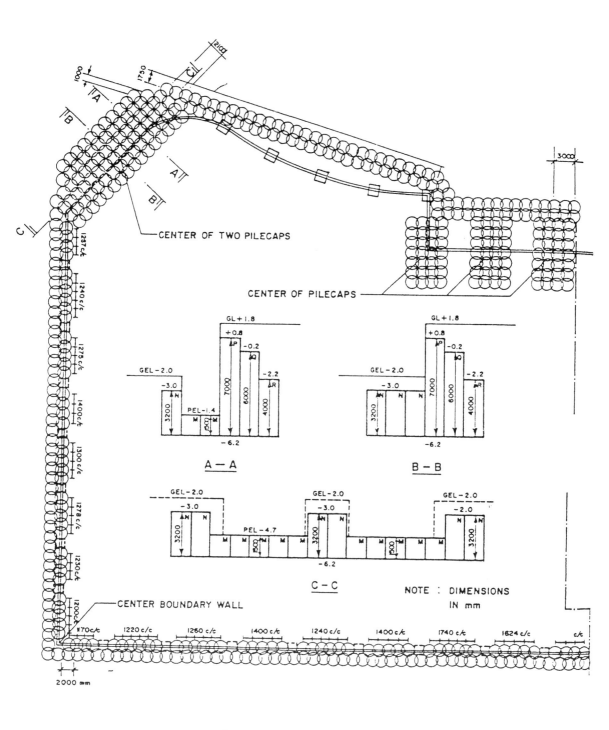

Fig. 6.33 Jet Grouting Layout

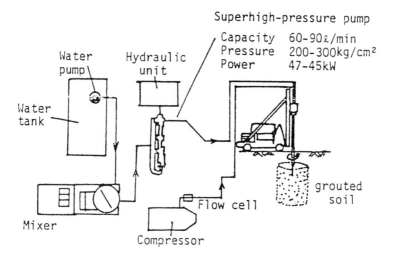

Fig. 6.34 Equipment Used for Jet Grouting

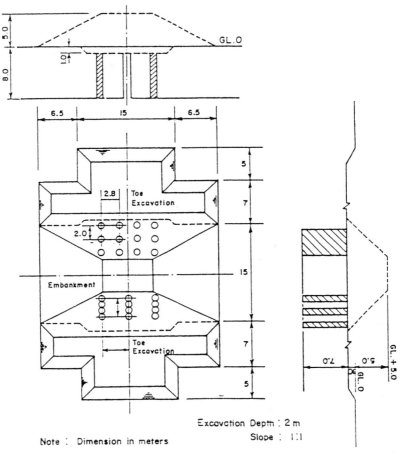

Fig. 6.35 Plan and Section of the Test Embankment with Soil-Cement Piles

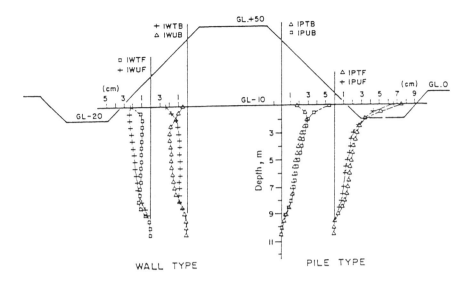

Fig. 6.36 Comparison of Lateral Movement on Improved and Unimproved Ground (71 days)

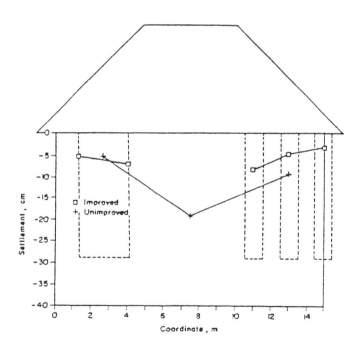

Fig. 6.37 Comparison of Settlement of Improved and Unimproved Parts at 1 m Depth (71 days)

of the corresponding laboratory test results. Figure 6.38 shows a photograph of the equipment to install the cement piles using DMM method.

6.15 CALCULATION METHODS FOR LIME COLUMNS

6.15.1 Ultimate Bearing Capacity of Single Lime Column

The bearing capacity of a single lime column is either governed by the shear strength of the surrounding soft clay (soil failure) or by the shear strength of the column material (column failure). The former mode of failure depends on both the skin friction resistance along the surface of the column and on the point resistance, while the later depends on the shear strength of the column material. The short-term ultimate bearing capacity of a single column in soft clay at soil failure can be calculated from the given expression:

$$Q_{ult, soil} = (\pi d H_{col} + 2.25 \pi d^2) C_u \tag{6.18}$$

where d is the diameter of the column, H_{col} is the column length, and C_u is the average undrained shear strength of the surrounding soft clay as determined by field tests, e.g., fall-cone and vane shear tests. It has been assumed that the skin resistance is equal to the undrained strength of the clay (C_u) and that the point resistance corresponds to 9 C_u. Experience with driven piles indicated that the skin friction resistance of the single column corresponds to at least the undrained shear strength of the surrounding soft clay when C_u is less than 30 kPa. When the undrained shear strength exceeds 30 kPa, a reduced shear strength, e.g. 0.5 C_u, should be used in the calculation of skin resistance. The point resistance of floating columns which do not penetrate the compressible strata is generally low compared with the skin resistance. A high point resistance is expected when the column extends through the compressible strata down to the underlying firm layer with high bearing capacity. A large part of the applied load will be transferred to the underlying soil through the bottom of the column. However, the point resistance cannot exceed the compressive strength of the column.

For the case where bearing capacity is governed by column failure, the behavior of the column material is considered to be similar to that of a stiff fissured clay. This was observed based from load tests on single excavated columns which indicated that failure takes place along the joint planes in unconfined columns and that the shear strength along the joints or fissures will govern the compressive strength rather than the shear strength of the clay matrix in the lumps or aggregates. The shear strength of the clay matrix in the lumps or aggregates represents an upper limit. This limit as determined by fall-cone or vane shear tests is about two or four times the shear strength along the joints as determined by unconfined compression tests. The corresponding failure envelope is shown in Fig. 6.39. The short term ultimate bearing capacity at column failure at depth z can be estimated from the relationship:

$$Q_{ult, col} = A_{col}(3.5 C_{col} + 3\sigma_h) \tag{6.19}$$

Fig. 6.38 Installation of Cement Piles by DMM Method

where C_{col} is the cohesion of the column material and σ_h is the total lateral pressure acting on the column at critical section. It has been assumed that the angle of internal friction of the soil is 30°. The factor 3 corresponds to the coefficient of passive pressure, K_p, at $\phi_{u,col} = 30°$. It has also been assumed that $\sigma_h = \sigma_v + 5 C_u$, where σ_v is the total overburden pressure and C_u is the undrained shear strength of the surrounding unstabilized clay. It is proposed for the time being that only the total overburden pressure should be used in design because of the large lateral displacement required to mobilize the passive earth pressure.

The long-term ultimate strength of the columns may be lower than the short-term strength due to creep. The creep strength of the columns $Q_{creep,col}$ is 65% to 85% of $Q_{ult,col}$. The load-deformation relationship shown in Fig. 6.40 is normally used to calculate the load distribution assuming that the load-deformation relationship is linear up to the creep, $Q_{creep,col}$ of the columns and that the slope corresponds to the compression modulus of the column material. When the creep strength is exceeded, the load in the columns assumed to be constant.

6.15.2 Ultimate Bearing Capacity of Lime Column Groups

The ultimate bearing capacity of lime column groups depends on both the shear strength of the untreated soil between the columns and on the shear strength of the column material. Failure is either governed by the bearing capacity of the block with lime columns (Fig. 6.41a) or by the local bearing capacity of the block along the edge (Fig. 6.41b) when the spacing of the lime column is large. In certain cases, shear resistance along a failure surface which cuts through a whole block may govern the bearing capacity. The ultimate bearing capacity of the lime column group at block failure is given as:

$$Q_{ult, group} = 2C_u H[B+L] + (6 \text{ to } 9)C_u BL \tag{6.20}$$

where B, L, and H is the width, length and height of the column group, respectively. The factor 6 corresponds for a rectangular foundation where the length is large compared with the width (i.e. L >> B), while the factor 9 corresponds for a square foundation. It is suggested that end bearing resistance should not be utilized in the design because a relatively large deformation of about 5-10% of the width of the loaded area is required to mobilize the maximum bearing resistance. The ultimate bearing capacity with respect to a local failure along the edge of the lime column block depends on the average shear strength of the soil along the approximately circular failure surface as shown in Fig. 6.41b. This average shear strength can be calculated in the same way as for the slope stability (see Eq. 6.29). The ultimate bearing capacity with respect to local failure can be estimated from the expression:

$$q_{ult} = 5.5 C_{av}(1 + 0.2 \frac{b}{l}) \tag{6.21}$$

where b and l are the width and the length of the locally loaded area, respectively, and C_{av} is the average shear strength along the assumed failure surface. The average shear strength of the

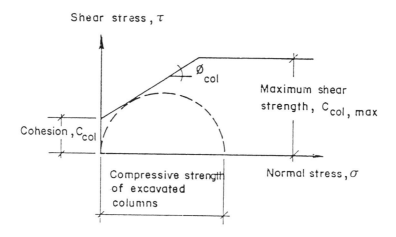

Fig. 6.39 Assumed Failure Diagram of Lime Stabilized Soil

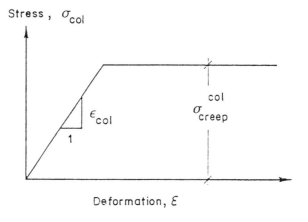

Fig. 6.40 Assumed Stress-Strain Relationship of Lime Stabilized Soil

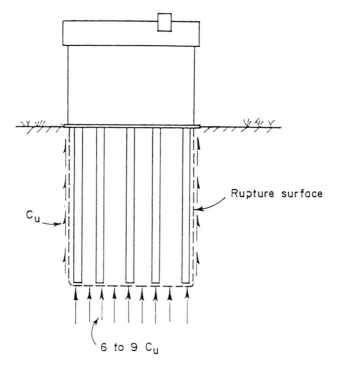

a) Block failure

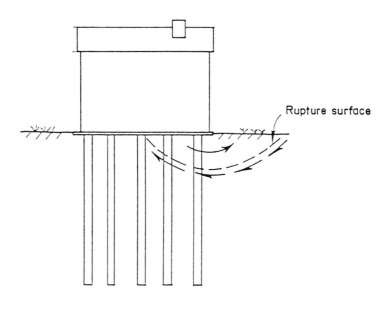

b) Local shear failure

Fig. 6.41 Failure Modes of Lime Column Foundations

stabilized area is affected by the relative column area, a, (b x 1) and by the shear strength of the column material. A factor of safety equal to 2.5 is proposed to be used in the design calculations.

6.15.3 Total Settlements

The total settlement of a structure supported on lime columns can be calculated as illustrated in Fig. 6.42. The maximum total settlement is taken equal to the sum of the local settlement of the reinforced block (Δh_1) and the local settlement of the unstabilized soil below the block (Δh_2). Two cases have to be investigated for the calculation of total settlement. In the first case (case A), the applied load is relatively low and the creep of the columns will not be exceeded. In the second case (case B), the applied load is relatively high and the axial load in the columns will correspond to the creep limit.

Previous studies indicate that the columns and the untreated soil between the columns deform as a unit and that the axial shortening of the columns corresponds to the settlement of the surrounding soil. However, it was also reported that the vertical deformation of the surrounding soil closed to the ground surface has been somewhat larger than the axial deformation of the columns. The maximum difference, though, is only a few percent of the total settlement even when the spacing of the columns is 3 to 4 diameters (1.5 to 2 m). When the axial shortening of the column corresponds to the settlement of the surrounding soil, the load distribution will depend on the relative stiffness of the column material with respect to the surrounding untreated soil as long as the axial stress in the columns (σ_{col}) is less than the creep limit of the material of the lime column ($\sigma_{col,creep}$). The load distribution will then depend on the compression modulus of the column material and of the stabilized soil. At the same relative deformation, the axial stress in the columns is expressed as:

$$\sigma_{col} = \frac{Q_{col}}{A_{col}} = \frac{q}{a + (\frac{M_{soil}}{M_{col}})(1-a)} \quad (6.22)$$

where q is the average contact pressure, a is the relative column area, (NA_{col}/BL) which is the ratio of the total area of the columns (NA_{col}) and the stabilized area (BL), M_{soil} and M_{col} are the compression moduli of the surrounding soil and of the column material, respectively.

Case A: In this case the relative stiffness of the columns with respect to the unstabilized soil will govern the load distribution between the columns and the enclosed unstabilized soil. The relationship M_{col} = 50 to 100 C_{col} is often used to estimate the settlements where M_{col} is the compression modulus and C_{col} is the cohesion of the column material. It is preferable, however, to evaluate M_{col} from oedometer tests. Normally, the modulus is 15 to 25 MPa. For rough calculations, a value of 20 MPa may be used. The average axial stress in the columns is governed by the compression modulus ($M = \Delta\sigma/\Delta\varepsilon$) of the surrounding unstabilized clay. When the clay is overconsolidated and the preconsolidation pressure will not be exceeded, the compression modulus can be estimated from the empirical relationship, $M_{soil} = 250C_u$, where C_u is the undrained shear strength of the surrounding clay as determined by field vane tests. When the soil is normally consolidated or slightly overconsolidated, the compression modulus should be estimated from oedometer tests. For soft Bangkok clay, the empirical relationship based from back-analyzed data, $M_{soil} = 150C_u$, may be used.

The stress increase, q, caused by a structure or a fill, is carried partly by the columns (q_1) and partly by the surrounding soil (q_2). At the same relative displacement, the following relationship can be derived:

$$\frac{q_1(BL)}{NA_{col}(M_{col})} = \frac{q_2(BL)}{(BL - NA_{col})M_{soil}} \tag{6.23}$$

which is simplified to:

$$\frac{q_1}{aM_{col}} = \frac{q_2}{(1-a)M_{soil}} \tag{6.24}$$

The local settlement, Δh_1 (see Fig. 6.42), can then be calculated by the expression:

$$\Delta h_1 = \frac{qH}{aM_{col} + (1-a)M_{soil}} \tag{6.25}$$

The stress increase, q, is assumed constant throughout the height of the block and that the load in the block is not reduced by the perimeter shear stress. This is a conservative assumption.

The local settlement, Δh_2, below the block can be calculated in the same way as at a standard settlement calculation. The increase in stress at any point below the block can be estimated by the 2:1 method as illustrated in Fig. 6.42. The settlement reduction ratio (β) which is the ratio of the total settlements down to the bottom of the reinforced block with and without lime columns can be estimated from the relationship:

$$\beta = \frac{M_{soil}}{aM_{col} + (1-a)M_{soil}} \tag{6.26}$$

Case B: In this case the applied load is so high that the axial load in the columns corresponds to the creep limit. The settlement, Δh_1, can be calculated as illustrated in Fig. 6.43. The applied load is divided into one part, q_1, which is carried by the columns and a second part, q_2, which is carried by the surrounding soil, exactly the same as Case A. The part q_1, is governed by the creep load of the columns given as:

$$q_1 = \frac{NQ_{creep\ col}}{BL} \tag{6.27}$$

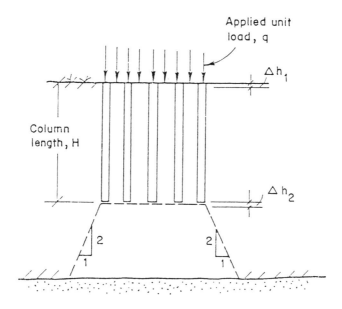

Fig. 6.42 Calculation of Settlement When the Creep Strength of Lime Columns is Not Exceeded

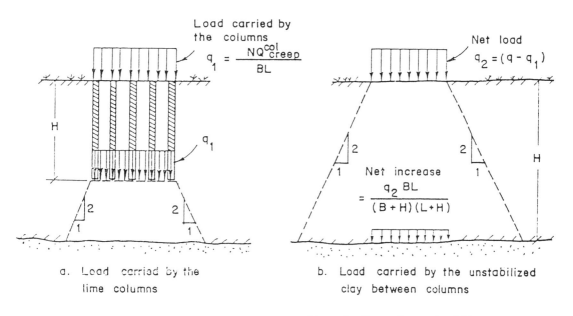

a. Load carried by the lime columns

b. Load carried by the unstabilized clay between columns

Fig. 6.43 Calculation of Settlement When the Creep Strength of Lime Columns is Exceeded

where N is the total number of columns. The part $q_2 = q-q_1$, can be used to calculate the local settlement, Δh_1, down to the bottom of the reinforced block. This settlement can be determined by dividing the block into layers. The settlement of each layer is calculated separately.

The settlement, Δh_2, below the reinforced block can be calculated from the assumption that load q_1 is transferred down to the bottom of the reinforced block, while load q_2 is acting at the ground surface.

6.15.4 Differential Settlements

The differential settlements will be small as long as the average shear stress along the perimeter of the reinforced block is less than the average shear strength of the surrounding clay. The angle change (α) between two column rows (refer Fig. 6.44) will be proportional to the average shear stress (τ_{ave}) along the perimeter of the column reinforced block and an average shear modulus (G_{soil}) as expressed by the relationship:

$$\alpha = \frac{\tau_{ave}}{G_{soil}} \tag{6.28}$$

The above expression has been based on the assumption that the columns are stiff in comparison with the surrounding soil and that the contribution of the axial shortening of the columns is small and negligible. The soil between two adjacent column rows will behave in a similar way as in direct shear test. The two rows correspond to the top and the bottom parts of a direct shear box. The average shear stress in the soil corresponds to the average shear stress along the perimeter of the reinforced block.

6.15.5 Settlement Rate and Slope Stability

The rate of settlement can be estimated similar to that for vertical drains. The stability of a slope in soft clay can be improved with lime columns as illustrated in Fig. 6.45. The techniques used for stability analysis of granular piles-improved ground can be used for lime columns-stabilized ground. The average shear strength along a potential failure or rupture surface in the soil can be estimated using the following equation:

$$C_{ave} = C_u(1-a) + S_{col}\, a \tag{6.29}$$

where S_{col} and C_u are the average strength of the column and the initial shear strength of the surrounding clay, respectively.

6.15.6 Stability of Excavation and Trenches

Lime columns can be used instead of sheet piles in deep trenches and excavations to increase the stability as illustrated in Fig. 6.46. Failure by overturning or by bottom heave

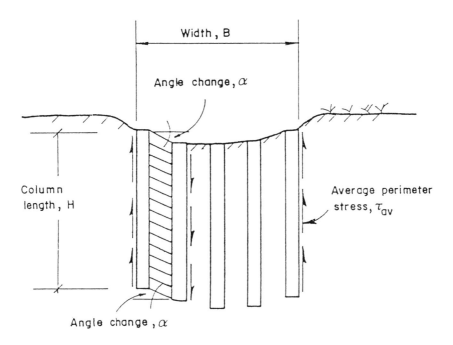

Fig. 6.44 Calculation of Differential Settlement

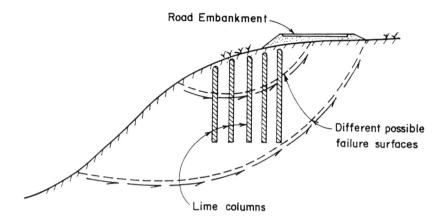

Fig. 6.45 Stabilization of a Slope with Lime Columns

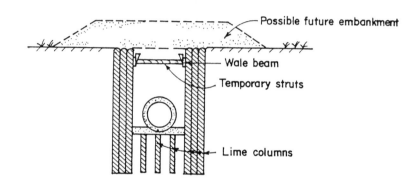

Fig. 6.46 Trench Stabilized with Lime Columns

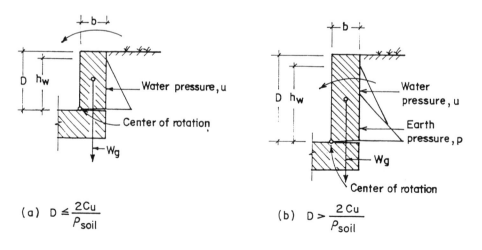

(a) $D \leq \dfrac{2C_u}{P_{soil}}$ 　　(b) $D > \dfrac{2C_u}{P_{soil}}$

Fig. 6.47 Failure by Overturning

should be considered as well as shear failure of the lime column wall with respect to a circular or plain failure surface through the columns.

6.15.6.1 Overturning

The water and earth pressures causing failure of the column wall by overturning are resisted by the weight of the wall in the same way as gravity retaining wall. The moment preventing failure with respect to the centre of rotation should be larger than the earth pressure behind the wall as illustrated in Fig. 6.47. The centre of rotation is located at the bottom of the excavation close to the surface of the wall. When the wall is loaded by water pressure only, the following relationship had to be satisfied (Fig. 6.47a)

$$0.5 \, W_g \, b > 1/6 \, h_w^3 \rho_w \qquad (6.30)$$

where W_g and b are the weight and width of the wall, respectively, h_w is the water level in the crack behind the wall with respect to the bottom of the excavation and ρ_w is the unit weight of water. When h_w is equal to the depth D of the excavation Eq. 6.30 is simplified to

$$b/D \geq 0.58 \sqrt{\rho_w / \rho_{col}} \qquad (6.31)$$

where ρ_{col} is the unit weight of the column material.

6.15.6.2 Shear Failure

The stability of a lime column wall with respect to shear failure must also be considered in the design. The stability can be calculated by assuming that failure takes place along an inclined rupture surface as illustrated in Fig. 6.48 or along a horizontal plane through the bottom of the excavation. If the cracks behind the wall are completely filled with water, the following relationship can be derived

$$h_w / b \leq \sqrt{3} + \sqrt{\frac{4c_{col}}{b\rho_w} + \frac{\rho_{col}}{\rho_w} - \frac{2D\rho_{col}}{3b\rho_w}} \qquad (6.32)$$

where c_{col} is the cohesion of the column material. When (D-1.7b) is larger than ($2c_u/\rho_{soil}$), the earth pressure behind the wall will also affect the stability as illustrated in Fig. 6.48b.

6.15.6.3 Bottom Heave

Failure by bottom heave can occur when the depth of the excavation is larger than the value of $6c_u/\rho_{soil}$ and the length of the excavation is larger compared with the depth. The bottom can be stabilized with lime columns as illustrated in Fig. 6.49. When the columns overlap the undrained shear strength c_{col} of the stabilized soils in the columns will govern.

The length of the columns (H_d) which is required to prevent failure by bottom heave can be calculated from the relationship

$$H_d > \frac{BD\rho_{soil}}{2s_{col}} - \frac{3c_u B}{s_{col}} \qquad (6.33)$$

The shear strength s_{col} is affected by the normal pressure along failure surface.

6.15.7 Transfer Length and Perimeter Stress for Lime Column

The length which is required to transfer the load in a single column to the surrounding soil is relatively small due to the large surface area and high skin friction resistance in comparison with the axial load in the columns (Fig. 6.50). The lime columns and the soil enclosed by the columns act as a rigid block (Fig. 6.51) when the spacing between the individual lime columns does not exceed 1.5 -2.0 m. The columns will then function as vertical reinforcement in the soil. The interaction of the 'reinforced block' and the surrounding soil is primarily governed by the relative dimension of the block, the magnitude of the applied load and the shear strength of the surrounding untreated soil. At low load levels, the load on the block is transferred to the surrounding soil mainly through the periphery of the block (Fig. 6.51) since a very small relative displacement is required to mobilize the shear strength of the soil. The differential settlements will be small if the columns are long and the average shear stress along the perimeter of the block is less than the average shear strength (C_u) of the surrounding soil. The average shear stress along the perimeter of the reinforced block (τ_{per}) can be calculated from the expression

$$\tau_{per} = \frac{0.8W_g}{2(B+L)H} < \frac{c_u}{f_c} \qquad (6.34)$$

where B, L, and H are the width, length and height, respectively, of the block, W_g is the total weight of the structure, and f_c is a partial safety factor which should be at least 1.5.

6.16 CALCULATION METHODS FOR DEEP CEMENT MIXING (DCM)

The following methods of calculation are employed on the design procedure of wall-shaped cement pile-stabilized ground for bulkhead foundation as shown in Fig. 6.52.

6.16.1 Sliding

The factor of safety against sliding (Fig. 6.53) should exceed 1.20 in static condition and 1.0 in seismic condition.

$$FS_s = \frac{Rf + F_u + P_p(E)}{H} \qquad (6.35)$$

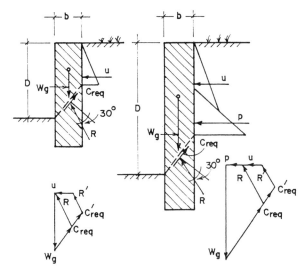

(a) $D \leq \dfrac{2C_u}{\rho_{soil}} + b\sqrt{3}$ (b) $D > \dfrac{2C_u}{\rho_{soil}} + b\sqrt{3}$

Fig. 6.48 Shear Failure

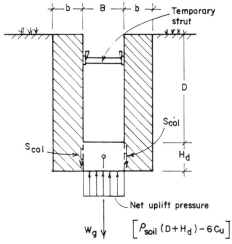

Fig. 6.49 Bottom Heave

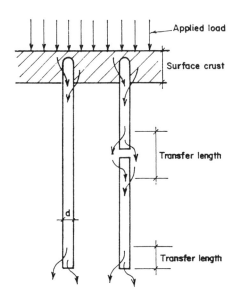

Fig. 6.50 Transfer Length for Lime Columns with Imperfections

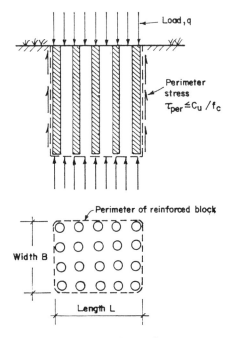

Fig. 6.51 Perimeter Shear Stress

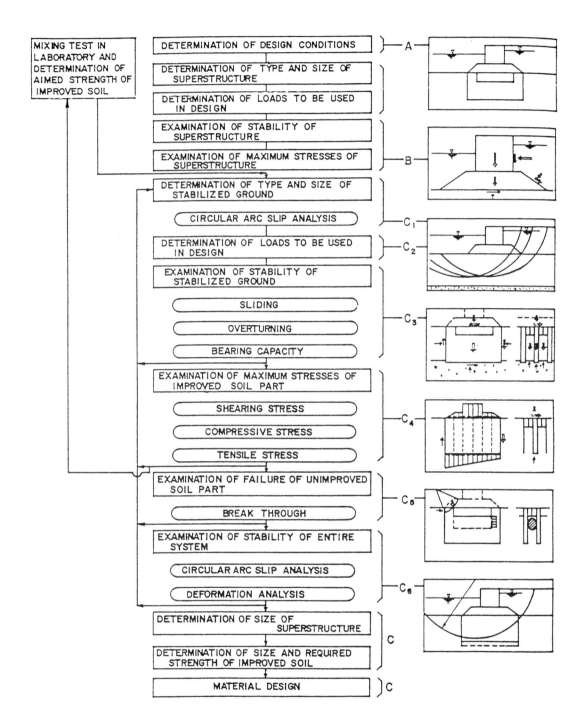

Fig. 6.52 Outline of Simplified Design Procedure

where:

N_{wt}	=	summation of vertical load on bottom surface of long wall
	=	$W_{S(E)} + W_T + P_{AV(E)}$
H	=	summation of horizontal loads
	=	$P_{AS(E)} + P_{AH(E)} [+H_{WT} + H_{WS} + H_{WU}]$
		( ): only in seismic condition
Rf	=	$N_{WT} \tan \phi$
F_u	=	minimum value ($W_u \tan \phi$ or $C_{(Z=D)}$BLS)
ϕ	=	angle of shear resistance of stiff soil layer
$C_{(Z=D)}$	=	undrained shear strength of soft clayey soil at the depth of D

6.16.2 Overturning

The factor of safety against overturning should exceed 1.2 in static condition and 1.1 in seismic condition. The loads are shown in Fig. 6.54.

$$FS_0 = \frac{M_{RWT} + M_{WU}}{M_D} \qquad (6.36)$$

where:

M_{RWT}	=	summation of resistance moments against overturning
	=	$M_{WT} + M_{WS(E)} + M_{pp(E)}$
M_D	=	summation of driving/overturning moments
	=	$M_{PAH(E)} (M_{HWS} + M_{HWT} + M_{HWV})$
		( ): only in seismic condition
M_{WT}	=	moment of load W_T around point o
$M_{WS(E)}$	=	moment of load $W_{S(E)}$ around point o
$M_{PAV(E)}$	=	moment of load $P_{AV(E)}$ around point o
$M_{PAS(E)}$	=	moment of load $P_{AS(E)}$ around point o
$M_{PAH(E)}$	=	moment of load $P_{AH(E)}$ around point o
M_{HWS}	=	moment of load H_{WS} around point o
M_{WHT}	=	moment of load H_{WT} around point o
M_{HWU}	=	moment of load H_{WU} around point o
M_{WU}	=	moment of load W_U around point o

6.16.3 Allowable Bearing Capacity

The allowable bearing capacity should exceed the maximum contact pressure. The contact pressures, q_1, and q_2 (see Fig. 6.55) are given as:

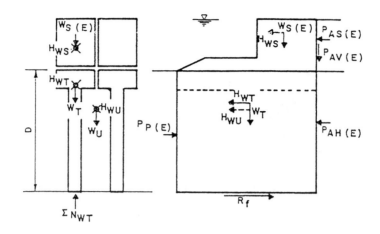

Fig. 6.53 Sliding of Stabilized Ground

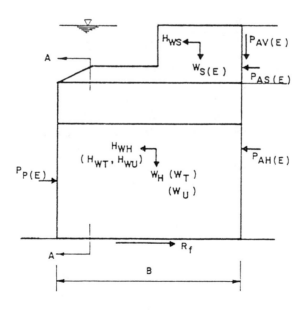

Fig. 6.54 Overturning and Resisting Loads

a) if $e_1 \leq B/6$:

$$q_1 = -\frac{N_{WT}}{BL_1}(1 \pm \frac{6e_1}{B}) \qquad (6.37)$$

b) if $e_1 > B/6$:

$$q_1 = \frac{2N_{WT}}{3d_1 L_1} \qquad (6.38)$$

where:

d_1 = distance of resultant load of contact pressure = $\dfrac{N_{RWT} - M_D}{N_{WT} + W_U}$

e_1 = eccentric distance of resultant load of contact pressure = $B/2 - d_1$

L_1 = thickness of long wall

The allowable bearing capacity is expressed as:

$$Q_A = \frac{1}{FS}(\beta \gamma_1 L_1 N_\gamma + \gamma_2 D N_q) + \gamma_2 D \qquad (6.39)$$

where the geometric dimensions are shown in Fig. 6.56 and the parameters are defined as follows:

FS = factor of safety = 2.5 in static condition and 1.5 in seismic condition

γ_1 = total unit weight of original soil above the plane of contact pressure (bouyant unit weight below the underground water level)

γ_2 = total unit weight of original soil below the plane of contact pressure (bouyant unit weight below the underground water level)

N_γ, N_q = bearing capacity factors

ß = shape factor (0.50)

6.16.4 Maximum Vertical and Horizontal Shear Stresses

Both maximum vertical and horizontal shear stresses should not exceed the strength of the improved soil. The maximum vertical shear stress (Fig. 6.57) is given as:

$$\tau_x = \frac{Q - W}{A} \qquad (6.40)$$

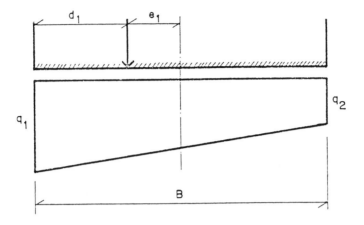

Fig. 6.55 Contact Pressure on Bottom Surface of Stabilized Ground

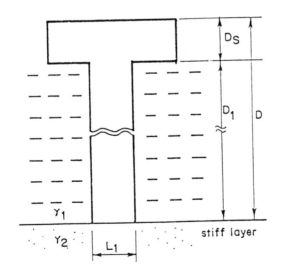

Fig. 6.56 Bearing Capacity of Stiff Layer

where:

Q = resultant load of pressure in the offshore-side vertical surface = $\dfrac{B_x(q_1 + q_x)L_1}{2}$

W = total weights of improved parts and the superstructure which is included in distance B_x

A = vertical cross-sectional area
= $LD - L_s D_1$

The maximum horizontal shear stress (Fig. 6.58) is expressed as:

$$\tau_y = \dfrac{P_{AS} + P_A - P_P}{A} \qquad (6.41)$$

where:

P_{AS} = horizontal load acting on superstructure
P_A = resultant load of active earth pressure and additional water pressure
$= D_y \dfrac{(P_{A1} + P_{Ay})}{2} L$
A = horizontal cross-sectional area = $L_1 B$

6.16.5 Breakthrough of Unimproved Soil

The examination of the stability of breadthrough in the unimproved soil is shown in Fig. 6.59. It is assumed that the breakthrough of the unimproved soil between piles forms a tunnel-shaped with oval section in direction to the offshore-side. The factor of safety against this type of failure must exceed 2.5 in static condition and 1.5 in seismic condition. The factor of safety is expressed as:

$$FS_b = \dfrac{P_r}{P_d} \qquad (6.42)$$

where:

P_r = resistant force = $B_s (\pi a + 2ma) [C_0 + C_1 (D_s + 2(m + 1)/2]$
P_d = breakthrough force = $\Delta P (\pi a^2/4 + ma^2)$
ΔP = lateral force increment on backfill side of stabilized ground
B_s = width of short wall
D_s = height of short wall
a = thickness of unimproved part (soft original soil between improved soil parts)
m = shape factor = $\dfrac{\pi}{4}\sqrt{1 + \dfrac{8}{a}(\dfrac{C_0}{C_1} + D_s - \dfrac{\pi - 2}{4}a)} - 1$

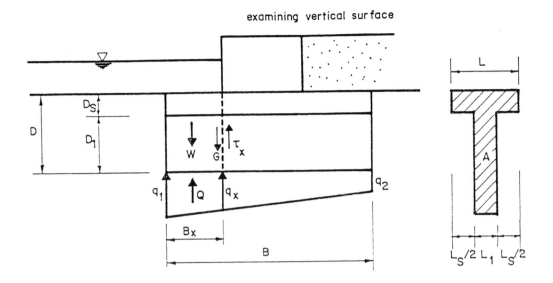

Fig. 6.57 Examination of Vertical Shear Stress

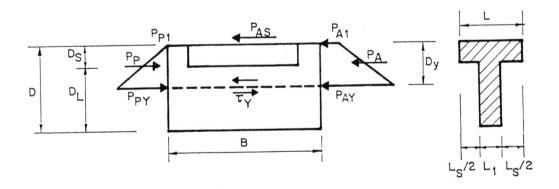

Fig. 6.58 Examination of Horizontal Shear Stress

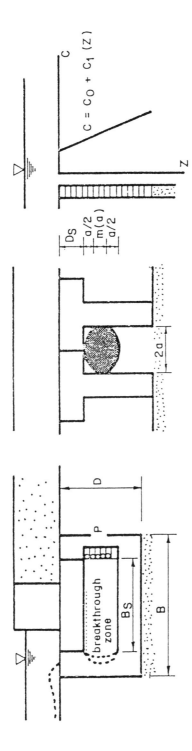

Fig. 6.59 Examination of Breakthrough of Unimproved Part

6.17 CONCLUSIONS

The use of lime and/or cement deep mixing method proved to be effective methods of improving the properties of the soft Bangkok clay. With the addition of 5 to 10% quicklime, which is the optimum mix proportions, the unconfined compressive strength increased to about 5 times. Moreover, the preconsolidation pressure improved to as much as 3 times while the coefficient of consolidation increased by 10 to 40 times. The effective strength parameters also increased wherein the angle of internal friction increased by about 67%. Mixing 10% of cement with soft Bangkok clay improved the unconfined compression strength and the preconsolidation pressure up to 10 to 20 times and by 2 to 4 times, respectively. The coefficient of consolidation increased by 10 to 40 times. From UC tests, overall consideration suggested that the percentage of cement (by weight) of 10% to 15% and and curing time of 1 to 2 months can be treated as optimum. Oedometer test results showed that cement treatment caused substantial improvement of consolidation properties, enhancement of preconsolidation pressure, reduction of compression index and simultaneously raised the value of coefficient of consolidation (Cv). The normalized intrinsic compression curves of the treated clays and the plot of (I_v, log σ_v) in the void index plane confirmed of the hardening effect brought about by cement treatment.

The strain paths from constant stress ratio (CSR) tests showed bilinear characteristics. The locus of transition point of strain paths are similar to the limit state curves for natural clays. The configuration and orientation of these loci of transition points are mainly controlled by the cement content and the curing time. Heavily overconsolidated behavior was observed inside the locus of transition points. The stress-dilatancy relationships showed that Modified Cam Clay provides a relationship at lower stress ratio than the experimental curves. From triaxial tests, it was observed that cement treatment causes transformation of the initial normally consolidated characteristics of the base clay to lightly overconsolidated and then to heavily overconsolidated rendering more stiffness and rigidity of the clay with increasing cement content, Aw, and time, t. The degree of overconsolidation was noticed to be affected by the A_W, t, p_o and applied stress path. In the initial small strain phase, the generation of pore pressure remains very small and concurrently shear strain is small and recoverable. The phase is followed by large strain phase with greater pore pressure, which is extended up to the curved failure envelope. A consistent picture of the phase transformation emerged evolving a family of loci. Outside the small strain domain, a work-hardening, elasto-plastic type of behaviour was observed. The prevalent role of p_o is to annihilate the cementation effects attributing ductility to the treated matrix. The behavior after the maximum deviator stress condition i.e. during strain softening can be regarded as an unstable phase where the stress paths fall along approximately in constant-p path up to residual states and the shear strain increases with the reduction of stress ratio. It was postulated that if complete destructuration of the treated clay could be achieved, and neglecting any permanent changes in the soil structure due to cement treatment, a destructured envelope would be achieved. At this envelope, cohesion is destroyed and the treated clay then behaves as purely frictional material. The pre-shear failure and end-of-test stress states for cement treated clays have curved projection on the (**q**, p) and (e, ln p) planes, showing characteristics which are typical of overconsolidated clays.

A consistent set of transition points of small strain and large strain was derived from relationships of (ε_v, p) of drained tests. The existence of the initial small strain phase is also observed at the outset part of the drained tests during strain hardening, as was discerned from the transition stress state in the relationship of (q, ε_v) which governs onset propagation of large strain of treated clays. The transition stress state was found to be affected mainly by cement content and curing time. The stress-dilatancy relationships on the spatial mobilized plane (SMP) exhibit a remarkable phenomenon such that the stress ratio, τ/σ_N, and the normal and shear strain increment vector, $-d\varepsilon_N/d\gamma$, constitute a specific correlation between treated and untreated clays. The pattern manifests gradual transformation behavior in the (τ/σ_N, $-d\varepsilon_N/d\gamma$) plane as the degree of overconsolidation of the treated mass changes. The enhancement of effective strength parameters (ϕ, c) was discerned to occur substantially after cement treatment for both the drained and undrained conditions.

The in-situ strength of the cement stabilized soft Bangkok clay was found to be one-half of the corresponding laboratory test results. Based on full-scale test embankment on soil-cement piles, the deformation of the improved ground depends on the type of construction pattern. The lateral movements and settlements of the wall type construction were lower than that of the pile type. Thus, the wall type is more effective in reducing the lateral and vertical deformation.

The method of calculating the bearing capacity of lime columns either for single column and column groups, total and differential settlements, and slope stability are presented. Methods of calculation for the design procedure of bulkhead foundation stabilized with cement piles using the Deep Cement Mixing (DCM) method is also outlined.

6.18 REFERENCES

Ariizumi, M. (1977), Mechanism of lime stabilization, J. of Japan Soc. Soil Mech. and Found. Eng'g., Tsuchi-to-Kiso, Vol. 25, pp. 9-16.

Assarson, K., Broms, B., Granholm, S., and Paus, K. (1974), Deep stabilization of soft cohesive soils, Linden Alimark, Sweden.

Balasubramaniam, A.S., and Bergado, D.T. (1984), Geotechnical problems related to construction activities in soft Bangkok clays, Proc. Symp. on Soil Improvement and Construction Techniques in Soft Ground, pp. 174-185.

Balasubramaniam, A.S., Bergado, D.T., and Sivandran, C. (1985), Engineering behavior of soils in Southeast Asia, In Geotechnical Engineering in Southeast Asia - A Commemorative Volume of the Southeast Asian Geotechnical Society, pp. 25-96, A.A. Balkema Printers.

Balasubramaniam, A.S., Phien-wej, N., and Kuhananda, M. (1988), Coastal development in soft clay deposits, Proc. Kozai Club Seminar 1988, Bangkok, Thailand.

Balasubramaniam, A.S., Honjo, Y, Law, K.H., Phien-wej, N., and Bergado, D.T. (1989), Ground improvement techniques in Bangkok subsoils, Proc. Kozai Club Seminar 1989, Bangkok, Thailand.

Brandl, H. (1981), Alteration of soil parameters by stabilization with lime, Proc. 10th Intl. Conf. Soil Mech. and Found. Eng'g., Stockholm, pp. 587-594.

Broms, B.B. (1984), Stabilization of soft clay with lime columns, Proc. Seminar on Soil Improvement and Construction Techniques in Soft Ground, Nanyang Technological Institute, Singapore.

Broms, B.B., and Anttikoski, U. (1983), Soil stabilization, Specialty Session No. 9, General Report, Proc. 8th Europ. Conf. Soil Mech. and Found. Eng'g., Helsinki, pp. 1289-1315.

Broms, B.B. (1986), Stabilization of soft clay with lime and cement columns in Southeast Asia, Applied Research Project RP10/83, Nanyang Technical Institute, Singapore.

Calladine, C. R. (1963), Correspondence, Geotechnique, Vol. 13, pp. 250-255.

Chaudry, H. (1966), Stabilization of two tropical clays, M. Eng. Thesis, SEATO Graduate School of Eng., Bangkok.

Chida, S. (1982), Dry jet mixing method, State-of-the-Art on Improvement Methods for Soft Ground, Japan Soc. Soil Mech. and Found. Eng'g., pp. 69-76.

Davidson, L.K., Demirel, T., and Handy, R.L. (1965), Soil pulverization and lime migration in soil lime stabilization, Highway Research Record No. 92, Highway Research Board, Washington, D.C., pp. 103-126.

Diamond, S., and Kinter, E. B. (1965), Mechanisms of soil-lime stabilization - An interpretative review, Highway Research Record No. 92, Highway Research Board, Washington, D.C., pp. 83-102.

DJM Research Group (1984), The manual for the dry Jet mixing method, pp. 20-31.

Eades, J. L., and Grim, R. E. (1966), A quick test to determine lime requirements for lime stabilization, Highway Research Record No. 139, Behavior Characteristics of Lime-Soil Mixtures, Highway Research Board, Washington, D.C., pp. 61-75.

Handy, R. L., Demirel, T., Ho, C., Nady, R. M., Ruff, C. G., et al (1965), Discussion, Mechanism of soil lime stabilization - An interpretative review, by Diamond, S. and Kinter, E. B., pp. 96-99.

Henkel, D. J. (1958), Correspondence, Geotechnique, Vol. 8, No. 3, pp. 134-136.

Herrin, M. and Mitchell, H. (1961), Lime soil mixture, Bulletin No. 304, Highway Research Board, Washington, D.C., pp. 99-138.

Hilt, G. H., and Davidson, D. T. (1960), Lime fixation in clayey soils, Bulletin No. 262, Highway Research Board, Washington, D.C., pp. 20-32.

Holm, G., Trank, R., and Ekstrom, A. (1983), Lime columns under embankments - A full-scale test, Proc. 8th Europ. Conf. Soil Mech. and Found. Eng'g., Helsinki, pp. 909-912.

Honjo, Y., Chen, C.H., Lin, D.G., Bergado, D.T., Balasubramaniam, A.S., and Okumura, R. (1991), Behavior of the improved ground by the deep mixing method, Proc. Kozai Club Seminar, Bangkok, Thailand.

Ingles, O. G., and Metcalf, J. B. (1972), Soil stabilization, principles and practice, Butterworths Pty. Ltd., Melbourne.

Kamaluddin, M. (1995) Strength and deformation characteristics of cement treated Bangkok clay, D. Eng. Dissertation No. GT-94-1, Asian Institute of Technology, Bangkok, Thailand.

Kawasaki, T., Niina, A., Saitoh, S., Suzuki, Y., and Honjo, Y. (1981), Deep mixing method using cement hardening agent, Proc. 10th. Int'l. Conf. Soil Mech. Found. Eng'g., Stockholm, pp. 721-724.

Kezdi, A. (1979), Stabilization with lime, Development in Geotechnical Eng., Vol. 19, Elsevier Scientific Publ. Co., Amsterdam, pp. 163-174.

Lea, F.M. (1956), The chemistry of cement and concrete, Edward Arnod (Publishers) Ltd., London.

Lee, H.J. (1983), Lime columns stabilization of Bangkok clay, M. Eng. Thesis, Asian Institute of Technology, Bangkok, Thailand.

Lee, S. L., Ramaswamy, S. D., and Aziz, M. A. (1982), A study of the soil type suitable for stabilization with lime, Proc. 7th Southeast Asian Geotech. Conf., Hongkong, pp. 615-629.

Law, K.H. (1989), Strength and deformation characteristics of cement treated clay, M. Eng. Thesis, Asian Institute of Technology, Bangkok, Thailand.

McDowell, C. (1959), Stabilization of soils with lime, lime flyash and other lime flyash materials, Bulletin 231, Highway Research Board, Washington, D.C., pp. 60-66.

Metcalf, J. B. (1964), The effect of high curing temperatures on the unconfined compressive strength of a heavy clay stabilized with lime and with cement, Proc. Australia-New Zealand Conf. Soil Mech. and Found. Eng'g., Vol. 4, pp. 126-130.

Mitchell, J. K., Veng, T. S., and Monismith, C. L (1974), Behavior of stabilized soils under repeated loading, Dept. of Civil Eng'g., Univ. of California, Berkeley, Calif.

Miura, N., Koga, Y., and Nishida, K. (1986), Application of a deep mixing method with quicklime for the Ariake clay ground, J. of Japan Soc. Soil Mech. and Found. Eng'g., Vol. 34, no. 4, pp. 5-11.

Miura, N., Bergado, D.T., Sakai, A., and Nakamura, R. (1987), Improvements of soft marine clays by special admixtures using dry and wet jet mixing methods Proc. 9th Southeast Asian Geotech. Conf., Bangkok, Thailand, pp. 8-35 to 8-46.

Okamura, T., and Terashi, M. (1975), Deep lime mixing method of stabilization for marine clays, Proc. 5th Asian Reg. Conf. Soil Mech. and Found. Eng'g., Bangalore, India, Vol. 1, pp. 69-75.

Roscoe, K. H., Schofield, A. N., and Thurairajah, A. (1963), Yielding of clays in states wetter than critical, Geotechnique, Vol. 13, pp. 211-240.

Ruff, C. G., and Ho, C. (1966), Time-temperature strength-reaction product relationships in lime-bentonite-water mixtures, Highway Research Record No. 139, Behavior Characteristics of Lime-Soil Mixtures, Highway Research Board, Washington, D.C., pp. 42.60.

Saitoh, S., Suzuki, Y., and Shirai, K. (1985), Hardening of soil improvement by deep mixing method, Proc. 11th Intl. Conf. Soil Mech. and Found. Eng'g., Helsinki, Finland, pp. 947-950.

Suzuki, Y. (1982), Deep chemical mixing method using cement as hardening agent, Proc. Symp. on Soil and Rock Improvement Techniques Including Geotextiles, Reinforced Earth and Modern Piling Method, AIT, Bangkok, Thailand.

Taylor, W. H., and Arman, A. (1960), Lime stabilization using preconditioned soils, Bulletin No. 262, Highway Research Board, Washington, D.C., pp. 1-11.

Terashi, M., Tanaka, H., and Okumura, T. (1979), Engineering properties of lime-treated marine soils and deep mixing method, Proc. 6th Asian Regional Conf. on Soil Mech. and Found. Eng'g. Vol. 1, pp. 191-194.

Wissa, A.E.Z., Ladd, C.C., and Lambe, T. W. (1965), Effective stress strength parameters of stabilized soils, Proc. 6th Intl. Conf. Soil Mech. and Found. Eng'g., Montreal, Vol. 1, pp. 412-416.

Woo, S. M. (1971), Cement and lime stabilization of selected lateritic soils, Thesis No. 409, Asian Institute of Technology, Bangkok, Thailand.

Yamanouchi, T., Miura, N., Matsubayashi, N., and Fukuda, N. (1982), Soil improvement with quicklime and filter fabric, Proc. ASCE, Vol. 108, pp. 935-965.

Chapter 7

MECHANICALLY STABILIZED EARTH (MSE)

7.1 GENERAL

Mechanical stabilization of soils by reinforcement with foreign materials is not a new idea, but has been used since time immemorial. The quality of adobe bricks has been improved by adding straw. Mechanically stabilized earth (MSE) consists of reinforcing the soil using polymer, steel, or natural materials. The reinforcement which is strong in tension effectively combines with the soil which is strong in compression, forming a semi-rigid composite material. A French engineer, Henry Vidal, was the first to formalize the rational design of modern reinforced earth with his patented "Reinforced Earth" technique (Vidal, 1969). In this technique, a strip metal reinforcement usually made of galvanized steel is inserted into high quality backfill soil materials consisting of sand and gravel (Fig. 7.1). The strip reinforcements are usually associated with the use of relatively high quality but expensive clean sands and gravel backfill to be able to generate the required frictional resistance between backfill soil and reinforcements. The use of grid type of reinforcements has become necessary because grids have more pullout resistance than strip reinforcements (Fig. 7.2). In so doing, cheaper low-quality, locally available, and cohesive-frictional backfill soils can be used. Steel grids were started to be used in California (Chang et al. 1977). Another type of grid reinforcements is the welded wire mats made of galvanized welded steel wires or welded steel bars patented by an American engineer, Bill Hilfiker, in the late 1970s (Peterson and Anderson, 1980). Subsequently, production of strong polymer grids called Tensar and also Tenax, with high extension stiffness and resistance to corrosion, had increased the use of grid reinforcement with cohesive-frictional backfill soil. Later on, the utilization of polymer geotextiles, both woven or non-woven, as reinforcing materials has further popularized the construction of MSE embankments.

7.2 RELEVANCE TO DEVELOPMENT

MSE construction is quite relevant to developed and developing countries. Easy to construct with no need of specialized equipment, the reinforced earth wall can lead to savings up to 50% in comparison to concrete gravity retaining wall or anchored sheet pile wall construction (Cheney, 1990). Construction of MSE embankments/walls is simple, quick, and labor intensive. It can be built in several stages if necessary and does not demand skilled labor. It involves erecting a layer of facing elements connected to the reinforcement, placing the fill, and compacting adequately. The process is repeated until the entire structure reaches the required height.

Clean granular soils with not more than 15% smaller than 0.074 mm in particle diameter, are considered as ideal backfill materials. However, in the coastal plains including Chao Phraya Plain of Bangkok, Thailand, clean sands and gravel are not readily available, and thus expensive due to large transportation costs from far away sources. Savings in construction costs can be

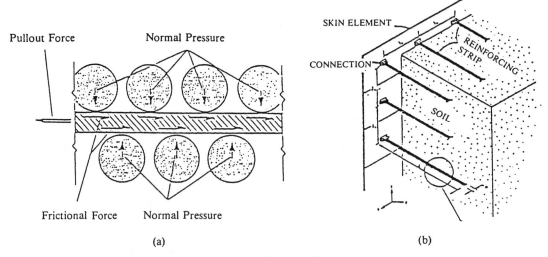

Fig. 7.1　Frictional Transfer Between Soil and Reinforcement and Schematic Diagram of Strip Reinforced Wall

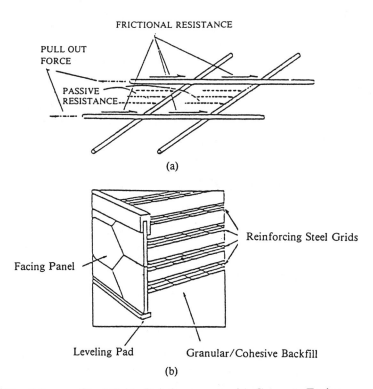

Fig. 7.2　Steel Grids Reinforcement with Concrete Facing

as much as 6 to 10 times if locally-available, low-quality and cohesive-frictional backfill soil material are utilized. Furthermore, the pullout resistance of cohesive-frictional backfill soil materials are utilized. Furthermore, the pullout resistance of cohesive-frictional backfill soil against grid reinforcement is high due to the passive resistance generated by the transverse member. For instance, the pullout capacity of grid reinforcements can be as high as 6 times the pullout capacity of strip reinforcements (Chang et al. 1977). Thus, the use of locally-available, lower quality, and cohesive frictional backfill soil materials reinforced by grid reinforcements is imperative for economic reasons.

7.3 MSE EMBANKMENT/WALL ON SOFT GROUND

The important coastal areas where the major cities are located in Southeast Asia is shown in Fig. 7.3. As stated previously, the main foundation problem in coastal plain areas is the presence of thick and soft clay deposits which is very weak and compressible material. Associated with the low strength of the subsoil is the problem of low bearing capacity and slope instability (Fig. 7.4). To solve these problems, the conventional approach is to use very gentle slope of 3H to 1V and construct low embankments. Due to subsequent settlements (Fig. 7.4) costly reconstruction is needed to raise the embankments above maximum flood levels after 5 to 10 years. MSE construction will allow construction of steeper slopes as well as higher embankments. Steeper slopes mean savings in construction materials. Increased embankment height extend the design life of the embankment structure by compensating for its sinking on weak ground.

Another problem occurring on earth structures on soft ground is the phenomenon of lateral spreading (Fig. 7.4) that will also contribute to total and differential settlements. For MSE embankments/walls on soft ground, the reinforcement tend to hold the outward thrust of the embankment fill in equilibrium with its tensile forces (Fig. 7.5). The reinforcement also restrain the surface of the foundation soil against lateral displacement (Jewell, 1980). Thus, not only is the lateral spreading minimized, but also the slope stability and bearing capacity are increased.

Furthermore, MSE construction can tolerate differential settlements common in areas of subsiding ground. MSE structure will function as a stiff raft floating over compressible stratum and will redistribute and load uniformly, thereby, reducing differential settlements. Thus, MSE structures can be used as foundation for residential houses, industrial buildings, oil storage tanks, and other lightly-loaded structures.

7.4 DESIGN CRITERIA FOR MSE STRUCTURES

The basic design criteria for MSE structures involves satisfying: (i) external stability and (ii) internal stability (Lee et al. 1973; Anderson et al. 1985; Mitchell and Villet, 1987). External stability is evaluated by considering the entire reinforced soil mass as semi-rigid structure which is checked for the conventional criteria as shown in Fig. 7.6, such as: (a) overturning (b) sliding (c) bearing capacity and (d) deep stability (conventional slope stability

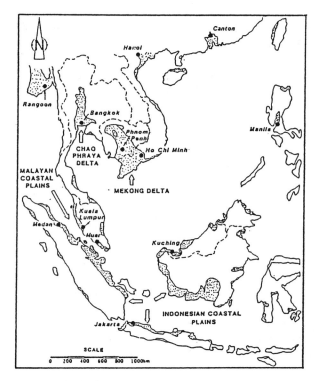

Fig. 7.3　　Soft Clay Deposits of Southeast Asia (Brand and Premchitt, 1989)

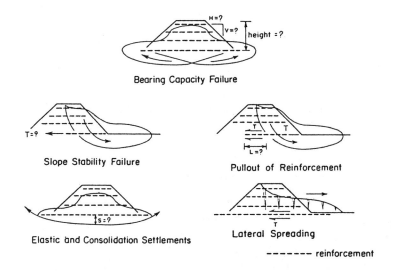

Fig. 7.4　　Design of Mechanically Stabilized Embankment on Soft Ground

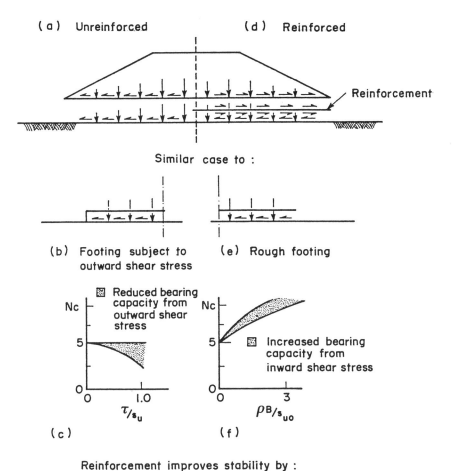

Fig. 7.5 Mechanics of Reinforced Embankments on Soft Ground

with failure surface below the reinforced mass). The internal stability of reinforced structures requires the evaluation of: (a) tension in reinforcing elements and (b) pullout resistance of reinforcing elements (Fig. 7.7). As pointed out by Ingold (1983a), the pullout test is the most realistic model to study the soil-reinforcement interaction.

7.5 THEORETICAL BACKGROUND FOR DESIGN OF MSE STRUCTURES

As described in the design criteria section, for design of MSE structures, the calculations of external and internal stabilities are needed. For external stability, the theory used is exactly the same as that for design the conventional earth structures. However, for internal stability, the calculations shall be done to check the required and available resistance force from the reinforcement. Both of these calculations need the knowledge about the mechanical properties of reinforced earth and soil/reinforcement interaction behavior.

7.5.1 The Basic Principles of Reinforced Earth

To understand the mechanisms of reinforced earth, several experimental and theoretical investigations have been done. The comprehensive triaxial tests using the aluminum disks reinforced sand samples had been carried out. Their results indicated that the reinforced samples have higher shear strength than unreinforced samples. The results were interpreted using two different assumptions: the anisotropic cohesion assumption and the enhanced confining pressure assumption (Ingold, 1982).

The anisotropic cohesion concept is based on the assumption that when the reinforced soil sample is at failure state and if the major principal stress is kept the same as the unreinforced soil sample, the minor principal stress is reduced. Therefore, the failure envelope of the reinforced soil sample will lie above that of the unreinforced sample (Schlosser and Long, 1973). Hausmann (1976) pointed out that at low normal stress levels the reinforced sample fails by slippage, and there is no apparent anisotropic cohesion intercept but only the internal friction angle is increased. At high normal stress, however, the reinforced earth fails by breakage of reinforcements and has an anisotropic cohesion intercept but the angle of internal friction is the same for both the reinforced and unreinforced soil samples, as shown in Fig. 7.8a.

The enhanced confining pressure concept is based on the assumption that the horizontal and vertical planes are no longer the principal stress planes, due to the shear stresses induced between the soil and the reinforcement. The minor principal stress within the reinforced soil sample increases when the major principal stress is increased, resulting in the shifting of the Mohr's circle of stress. The additional strength for the reinforced soil can be attributed to the enhanced confining pressure effect. The failure envelope is the same for both reinforced and unreinforced samples as shown in Fig. 7.8b (Yang, 1972). In Fig. 7.8b, the dashed line is showing the anisotropic cohesion concept for comparison.

Thus, under low confining stresses in a given reinforcement, the MSE system tends to fail by slippage or pullout between soil backfill and reinforcement while under high confining

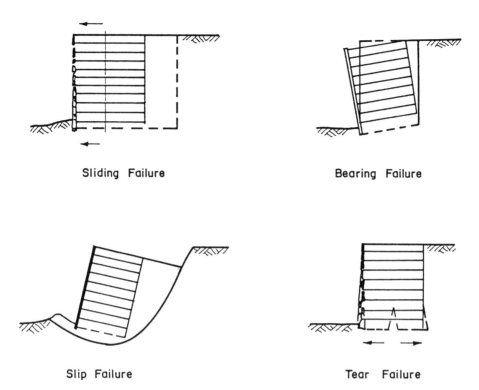

Fig. 7.6   External Stability Criteria for Reinforced Soils Walls

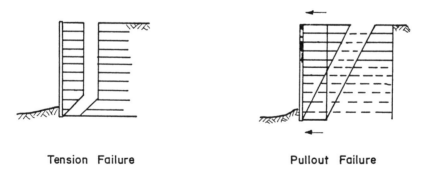

Fig. 7.7   Internal Stability Criteria for Reinforced Soil Walls

pressures the same systems fail by breakage of reinforcement (Mitchell and Villet, 1987). As shown in Fig. 7.9, the zones of reinforcement breakage or slippage are indicated. Both anisotropic cohesion and enhanced confining pressure concepts explain the same phenomenon that due to interaction between soil and reinforcement, the reinforced soil has a higher strength than unreinforced soil. The interaction mechanism developed in a reinforced soil is characterized by the mobilization of shear stress along the soil/reinforcement interface. This process, consequently, results in the generation of tension forces in the reinforcement.

7.5.2 Soil/Reinforcement Interaction Mechanisms

The mechanism governing soil/reinforcement interaction in a reinforced earth structure is concerned with the mobilization of soil/reinforcement friction resistance, soil passive bearing resistance on reinforcement bearing members, and the bending movement in the reinforcement. Realistically, the influence of bending movement on the behavior of the reinforced earth structure is small, and it can be ignored under working stress condition (Schlosser and De Buhan, 1990). Therefore, the soil/reinforcement interaction mechanism can be simplified into two types, namely: soil sliding over the reinforcement or direct shear mechanism and pullout of the reinforcement from the soil or pullout mechanism. Direct shear and pullout tests are used to simulate these two different mechanisms, respectively. Direct shear test provides a local shear stress/shear displacement along the reinforcement. Figure 7.10 shows a typical reinforced earth structure in a reinforced soil slope. Assuming that the dashed line in the figure is a potential failure surface, the reinforcement behind the potential failure surface, position A, will be subjected to pullout interaction mechanism. At position B, the direct shear mechanism is likely to occur. The interaction between the soil and grid reinforcement is more complicated and more general than that of the strip and sheet type reinforcements.

7.5.2.1 Direct Shear Resistance

Generally, the direct shear resistance between the grid reinforcement and the soil has three components: (a) the shear resistance between the soil and the reinforcement plane surface area; (b) the soil to soil shear resistance at the grid opening; and (c) the resistance from soil bearing on reinforcement apertures (Jewell et al. 1984). Since the last part is difficult to assess, usually, the influence of the reinforcement apertures on direct shear resistance is treated as to increase the skin friction resistance between the soil and the reinforcement plane surface. Therefore, the direct shear resistance can be expressed in terms of the two contributions from shear between soil and plane surface area of reinforcement and shear between soil and soil. The general expression for direct sliding resistance proposed by Jewell et al. (1984) is given as follows:

$$f_{ds}.\tan\phi_{ds} = \alpha_{ds}.\tan\delta + (1-\alpha_{ds}).\tan\phi_{ds} \tag{7.1}$$

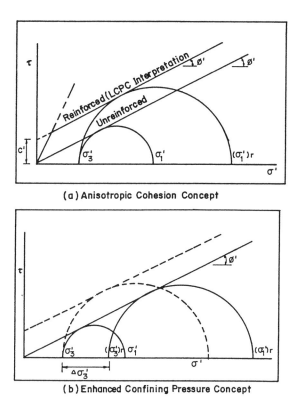

Fig. 7.8 Mechanics of Reinforced Soil (Ingold, 1982)

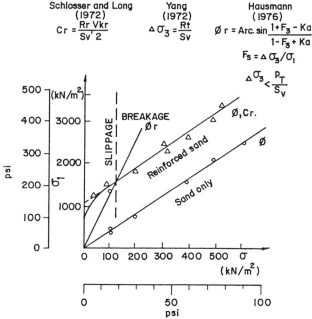

Fig. 7.9 Strength Envelopes for Sand and Reinforced Sand

where: f_{ds} = coefficient of resistance to direct shear
ϕ_{ds} = angle of friction for soil from direct shear test
δ = angle of skin friction
α = fraction of grid surface area providing direct shear resistance

When α_{ds} equals zero, it is the case of soil shear over soil and f_{ds} is 1.0. When α_{ds} equals 1.0, it is the case of soil shearing over the reinforcement plane surface and f_{ds} is $\tan\delta/\tan\phi_{ds}$.

7.5.2.2 Pullout Resistance

The pullout resistance of grid reinforcement consists of two parts. One is frictional resistance, P_f, which is developed between the soil and the frictional surface of the grid reinforcement. The magnitude of friction resistance depends on the angle of skin friction and the normal effective stress between the soil and the reinforcement surface, as shown in Eq. 7.2.

$$P_f = A_s \cdot \sigma'_s \cdot \tan\delta \tag{7.2}$$

where: A_s = friction area
σ'_s = average normal stress, equals $0.75\sigma_v$ (Anderson and Nielsen, 1984)
δ = skin friction angle between reinforcement and soil

The other part of the pullout resistance is passive resistance of soil bearing on the grid bearing members. Three different failure mechanisms have been proposed to estimate the maximum pullout bearing resistance of an isolated bearing member. One is general shear failure mode (Peterson and Anderson, 1980), one is punching failure mode (Jewell et al. 1984), and another is modified punching failure mode (Chai, 1992).

The general shear failure mode assumes a characteristic field as shown in Fig. 7.11a. However, the bearing capacity equation given is the Prandtl's solution (Prandtl, 1921) of the failure stress for a shallow smooth strip foundation, and with assumption that the horizontal effective stress is equal to applied vertical pressure σ_n. The maximum bearing stress σ'_{bm} us as follows:

$$\sigma'_{bm} = c' \cdot N_c + \sigma'_n \cdot N_q \tag{7.3}$$

$$N_q = \exp(\pi \cdot \tan\phi') \cdot \tan^2\left[\frac{\pi}{4} + \frac{\phi'}{2}\right] \tag{7.4}$$

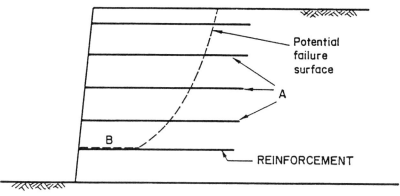

Fig. 7.10 Typical Reinforced Slope Showing the Soil Reinforcement Interaction Models: A - Pullout; B - Direct Shear

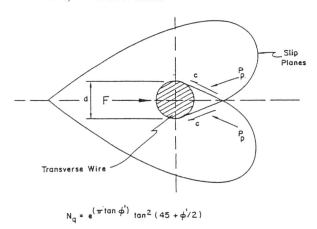

$$N_q = e^{(\pi \tan \phi')} \tan^2(45 + \phi'/2)$$

(a) Bearing Capacity Failure (After Peterson & Anderson, 1980)

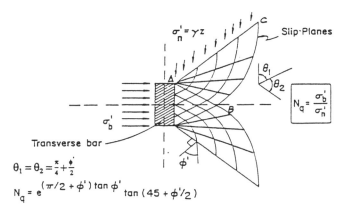

$\theta_1 = \theta_2 = \frac{\pi}{4} + \frac{\phi'}{2}$

$N_q = e^{(\pi/2 + \phi') \tan \phi'} \tan(45 + \phi'/2)$

(b) Punching Shear Failure (After Jewell et al, 1984)

Fig. 7.11 Pullout Bearing Failure Mechanisms

$$N_c = \cot\phi'.(N_q-1) \tag{7.5}$$

where: c' = effective intercept based on effective stresses
 ϕ' = friction angle based on effective stress

If the problem satisfies the above assumptions and the soil behaves as perfectly plastic material, this solution is the exact solution. However, during the pullout test, the bearing member is more or less like deep strip foundation, so the characteristic field may not be the same as that for shallow foundation. Vesic (1963) did a large-scale model test of foundation embedded in sand. The results showed that when the ratio between foundation depth and foundation width is over 6, the failure model was dominated by punching failure.

Punching failure mode is based on the stress characteristic field as shown in Fig. 7.11b. The relationship between maximum bearing stress σ'_{bm} and applied normal stress σ_n for cohesionless soil used by Jewell et al. (1984) is as follows:

$$\sigma'_{bm} = \sigma'_n.N_q \tag{7.6}$$

$$N_q = \exp\left\{(\frac{\pi}{2}+\phi').\tan\phi'\right\}.\left[\frac{\pi}{4}+\frac{\phi'}{2}\right] \tag{7.7}$$

This relationship assumes that: (a) only two failure zones: active failure zone, ABD, and rotational failure zone, BDC (there is no passive failure zone); (b) the stress acting on inclined line AC with a density of σ'_n, that is, the normal stress acting on the boundary stress characteristic line AC is $\sigma'_n.\cos\phi'$; (c) the strength is fully mobilized on AC.

Several test results showed that the general shear and punching shear failure mechanisms provided apparent upper and lower bounds for actual pullout test results as shown in Fig. 7.12 (Palmeira and Milligan, 1989; Jewell, 1990). The bearing capacity factors for modified punching shear failure mode are derived from the stress characteristic field as shown in Fig. 7.13. The bearing capacity factors, N_c and N_q, can be expressed as follows:

$$N_q = \left[\frac{1+k}{2}+\frac{1-k}{2}.\sin(2.\beta-\phi)\right].\frac{1}{\cos\phi}.\exp^{[2.\beta.\tan\phi]}.\tan\left(\frac{\pi}{4}+\frac{\phi}{2}\right) \tag{7.8}$$

$$N_c = \frac{1}{\sin\phi}.\exp^{[2.\beta.\tan\phi]}.\tan\left(\frac{\pi}{4}+\frac{\phi}{2}\right)-\cot\phi \tag{7.9}$$

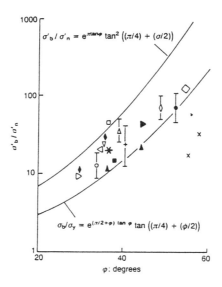

Fig. 7.12   Comparison of Measured and Predicted Pullout Bearing Stress (Jewell, 1990)

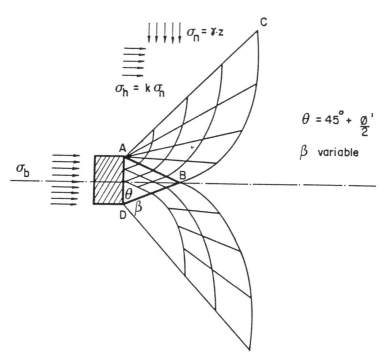

Fig. 7.13   Proposed Stress Characteristic Field for Pullout Bearing Failure

where k is the horizontal earth pressure coefficient; β is the angle of rotational failure zone; c is the cohesion; and ϕ is the friction angle of backfill soil. The β angle is variable. If the soil is more compressible, the angle is smaller (Vesic, 1963). If β angle equals $\pi/2$ and k equals 1.0, the proposed formula predicts the laboratory pullout test data well (Chai, 1992).

7.5.2.3 Pullout Resistance/Pullout Displacement Relationship

For the design of the MSE structure, the important item is to calculate available pullout resistance from the reinforcement under certain pullout displacement. Therefore, besides the theory for maximum pullout resistance of single bearing member, there are two points that also need to be considered. Firstly, for actual grid reinforcement, several transverse members are placed within regular intervals. During the pullout of the grid from soil, the bearing member may influence each other. Secondly, the MSE structures are not always at limit equilibrium state, and understanding the pullout resistance mobilization process is essential at working stress levels. The pullout bearing resistance of an individual bearing member of grid reinforcement is simulated by a hyperbolic function as follows:

$$\sigma_b = \frac{d_n}{\frac{1}{E_{ip}} + \frac{d_n}{\sigma_{bult}}} \tag{7.10}$$

where σ_b is the pullout bearing resistance (stress); d_n is the normalized pullout displacement which is defined as pullout displacement divided by grid bearing member thickness; E_{ip} is the initial slope of bearing resistance/normalized displacement curve; and σ_{bult} is the ultimate value of bearing resistance.

The factors controlling the initial slope of normalized pullout bearing resistance curve mainly consist of the backfill soil stiffness and the bearing member (deflection) rigidity. The dimensionless bearing member deflection rigidity index, I_d, is defined as follows:

$$I_d = \frac{E.I.d}{L^4.D.P_a} \tag{7.11}$$

where L is the spacing between two neighboring longitudinal members; E is the elastic modulus of reinforcement; D is the bearing member thickness; I is the moment of inertia of the bearing member cross-sectional area; d is the unit length to make I_d dimensionless; and P_a is the atmospheric pressure. The ratio between the bearing member index, I_d, and soil stiffness index, I_r, is defined as the stiffness ratio, R_r, given as follows:

$$R_r = \frac{I_d}{I_r}.(100\%) \tag{7.12}$$

where I_r has been defined by Vesic (1972) as the shear modulus divided by the shear strength of soil. The initial slope of the normalized pullout bearing resistance curve is expressed by an empirical equation as follows:

$$E_{ip} = \frac{\ln R_r}{\ln R_{rc}} \cdot R_{io} \cdot E_i \qquad (7.13)$$

where E_i is the initial slope of triaxial compression test stress/strain curve of the backfill soil; R_{io} is the initial slope ratio between normalized pullout bearing resistance curve and triaxial stress/strain curve of the backfill soil for the case of rigid bearing member; and R_{rc} is the limit stiffness ratio. From test results (Chai, 1992), it is determined that R_{io} is equal to 0.1 and R_{rc} is equal to 250%. When R_r is larger than R_{rc}, take it as R_{rc}.

The ultimate value of pullout bearing resistance, σ_{bult}, is related to the maximum value by pullout bearing failure ratio, R_{fp} ($\sigma_{bult} = \sigma_{bm}/R_{fp}$). R_{fp} is approximately the same as that of triaxial test failure ratio of backfill material (Chai, 1992). The maximum pullout bearing resistance of an individual bearing member of grid is expressed as the maximum pullout bearing resistance of isolated single member (Eqs. 7.8 and 7.9) multiplied by the bearing resistance ratio, R, which is defined as follows:

$$R = \frac{P_n}{n \cdot P_o} \qquad (7.14)$$

where P_o is the pullout bearing force of a single bearing member and P_n is the total bearing force of grid reinforcement with n bearing members. A dimensionless parameter of bearing member spacing ratio, S/D, is used to express the influence of grid geometry on pullout resistance in which S is the center to center spacing between two neighboring bearing members, and D is the thickness of the bearing member. The bearing resistance ratio, R, is the function of S/D ratio as follows:

$$R = a + b \cdot \left(\frac{S}{D}\right)^{nr} \qquad (7.15)$$

in which a, b, and nr are constants, and nr is between 0.5 and 1.0. The S/D ratio is varied from 1 to 45. When S/D is larger than 45, the value of R is 1.0.

The superscript, nr, is the function of backfill soil friction angle. It is suggested that when the backfill soil friction angle, ϕ, is larger than 45°, take nr as 0.5; when ϕ is between 35° to 45°, take nr as 2/3; when ϕ is in the range of 25° to 35°, take nr as 3/4; and when ϕ is smaller than 25°, take nr as 1.0. The pullout bond coefficient, f_b, has been defined as the total pullout force of the grid reinforcement divided by the shear strength of backfill soil multiplied by the upper and lower soil/reinforcement interface areas (Jewell et al. 1984). Then the constants a and b in Eq. 7.15 can be determined by using two conditions: (a) when S/D equals 1, the pullout bond coefficient, f_b, is 1.0, and (b) when S/D equals 45, the pullout bearing resistance ratio, R, is 1.0. When S/D is larger than 45, R equals 1.0 (Chai, 1992).

For a grid reinforcement, the mobilized resistance along the reinforcement may not be uniform because the elongation of the grid longitudinal members. Take this factor into account, the pullout resistance of a grid reinforcement can be analytically determined by above proposed hyperbolic model. Figures 7.14 and 7.15 show the comparison between calculated and measured pullout resistance/displacement relationship for steel grid and polymer grid, respectively (Chai, 1992).

7.5.2.4 Pullout versus Direct Shear Tests to Obtain Interaction Parameters

The pullout and the direct shear tests are two widely used test methods for determining the soil/reinforcement interface parameters. These two fundamentally different testing methods are associated with different boundary conditions and loading paths. The direct shear test provides shear stress/shear displacement relationship on a pre-determined shear surface, while the pullout resistance of grid reinforcement integrates the variation of friction resistance from the grid longitudinal members as well as passive resistance from the grid transverse members along the reinforcement. Therefore, these two test methods yield different soil/reinforcement interface parameters, including both interface shear stiffness and maximum interface resistance.

The soil/grid reinforcement interface shear resistance consists of the friction resistance between reinforcement plane surface and soil and the shear resistance between soil and soil. These kind of resistance can be mobilized with a relatively small displacements, so that it will result in a higher interface stiffness. The friction component of the pullout resistance can also be passive bearing resistance of the order of 10 times larger than that required to generate the friction resistance. As a result of these various mechanisms, when considering the pullout response of a grid under small displacements, the friction resistance is the most important component of the pullout force, and, under large displacements, the passive bearing resistance will be the most significant factor in the pullout resistance (Schlosser and De Buhan, 1990). Another factor occurs in extensible reinforcements, wherein during the pullout test, the resistance mobilization along the reinforcement is non-uniform and the peak resistance at different positions of the reinforcement is mobilized at different times. These factors result in a lower interface stiffness from pullout test. However, at present, for analysis of the reinforced earth structure, the interface parameters are determined by either direct shear test or pullout test.

The difference of maximum soil/reinforcement interface resistance determined from direct shear and pullout tests can be partially expressed by comparing the band coefficient, $\tan\delta/\tan\phi$. Figure 7.16 shows the test results on bond coefficient determined by both direct shear and pullout tests (Juran et al. 1988). Although the reinforcements tested have different geometry, generally speaking, for most cases, the direct shear test yielded higher bond coefficient than that of pullout test, except for the case of a grid placed in coarse dense sands under low confining pressure. In this case, the pullout test yielded higher bond coefficient, owing to dilatancy effect and interlock of the soil particles into the grid openings.

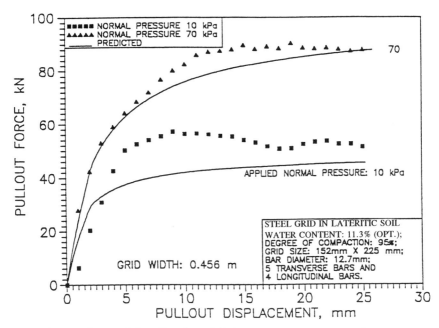

Fig. 7.14 Comparison of Predicted and Measured Pullout Force/Pullout Displacement Curves for Steel Grid in Lateritic Soil Backfill

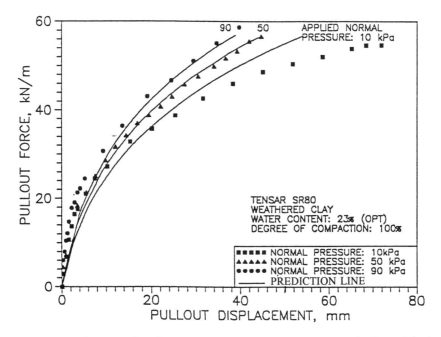

Fig. 7.15 Comparison of Predicted and Measured Pullout Force/Pullout Displacement Curves for SR80 in Weathered Bangkok Clay Backfill

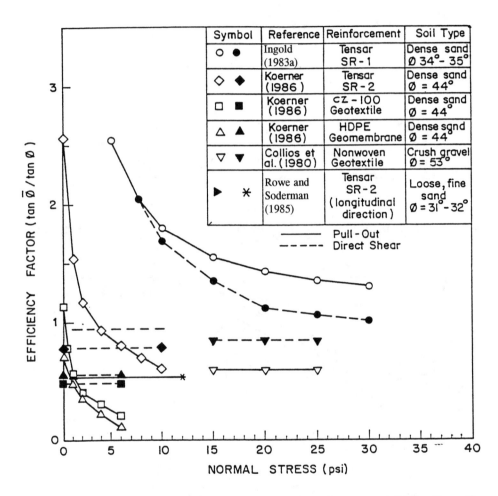

Fig. 7.16 Comparison of Bond Coefficients from Pullout and Direct Shear Tests (Juran et al. 1988)

7.5.3 Location and Magnitude of the Reinforcement Tension Force

Most of the MSE structures are related to wall or embankment construction. The reinforcement used in a reinforced wall has two effects. One is to restrain or reduce the lateral displacement of the wall by mobilized tensile force in the reinforcement. The other is modification of the strain pattern in the reinforced soil mass due to the influence of the interface shear stresses. The horizontal and vertical planes are no longer the principal stress directions. For inextensible reinforcements, one of the stress characteristic lines in the reinforced soil wall is vertical rather than inclined at $45_o + \phi'/2$, as in the case of a vertical retaining wall supporting an unreinforced horizontal backfill (Schlosser and De Buhan, 1990). Therefore, the failure planes in a reinforced wall with inextensible reinforcements are quite different from the classical Mohr-Coulomb failure surface.

7.5.3.1 Location of Maximum Tensile Force

In a reinforced soil retaining wall, the soil mass can be subdivided into two zones, the active zone and the resistance zone. In the active zone, the soil tries to move away from the structure, but is restrained by frictions developed along the reinforcements. The mobilized shear forces on the reinforcements are directed toward the front of the wall which results in an increase of reinforcement tensile force with distance from the facing. In the resistant zone, the interface shear forces on reinforcement are oriented away from the facing and provide the resistance force against the pullout failure. In this sense, the maximum tensile force is related to the boundary of the active and resistant zone, which is the potential failure surface. For the test results (e.g. Anderson et al. 1987), it is observed that the maximum tensile force line is significantly different from that predicted by the Mohr-Coulomb failure wedge for classical earth retaining wall. The presence of the horizontal and practically inextensible reinforcement restrains lateral deformations and consequently changes the stress and strain patterns in the soil. The potential failure surface for a reinforced wall with inextensible reinforcement at the top part of the wall is vertical. The approximate location of maximum tensile force is as shown in Fig. 7.17 (Anderson et al. 1987). On the other hand, extensible reinforcements induce large lateral displacements, and the maximum tensile force line is close to classical Mohr-Coulomb failure plane (Juran and Christopher, 1989), shown in Fig. 7.17 as a dashed line.

7.5.3.2 Magnitude of Maximum Tensile Force

The horizontal earth pressure developed in the reinforced soil mass is balanced by the tensile force in the reinforcement, i.e. the maximum tensile force in reinforcement is related to the horizontal earth pressure in the soil mass. The stress state in the reinforced soil mass therefore depends on the stiffness of the reinforcement. In the case of extensible reinforcements, lateral displacements will take place, especially at the top of the structure, which lead to the mobilization of active earth pressure. On the other hand, for inextensible reinforcements, horizontal earth pressure is approximately in at-rest state. Christopher et al. (1990) proposed a method to relate the horizontal earth pressure coefficient in reinforced wall with reinforcement stiffness factor, S_r, which is defined by the following formula:

$$S_r = \frac{E \cdot A}{S_v \cdot S_h} \tag{7.16}$$

where:
- E = Young's modulus of the reinforcement
- A = cross-sectional area of the reinforcement
- S_v = vertical reinforcement spacing
- S_h = horizontal reinforcement spacing

Figure 7.18 shows the variation of horizontal earth pressure coefficient with depth for various types of reinforcements.

Detailed finite element analysis shows that at the top of the wall, the tension force in the reinforcement is greater than k_o horizontal earth pressure due to the effect of compaction, and at the base of the wall is less than k_o state and approach to active (k_a) state due to the influence of the friction between the reinforced mass and rigid foundation. However, when the wall rest on soft foundation, the bending strain resulting from the foundation settlement may cause large tension the reinforcement on the base of the wall (Adib et al. 1990). It is probable that at the failure limit state, a certain homogeneity of the stress will be established, causing the lateral earth pressure coefficient to return to a value approximately equal to active earth pressure value along the entire height of the wall (Schlosser and De Buhan, 1990).

7.6 PULLOUT RESISTANCE OF STEEL GRID ON POOR QUALITY BACKFILL

The available resistance force from reinforcement is controlled either by pullout resistance of the reinforcement from soil or the tension strength of the reinforcement, whichever is smaller. As mentioned previously, due to economic consideration, the locally available, low-quality, cohesive-frictional soils are most appropriate and cheaper MSE backfill materials. The pullout resistance of steel grid in cohesive-frictional backfill materials have been comprehensively investigated in the Asian Institute of Technology (Bergado et al. 1992) through laboratory and field pullout tests by using three types of backfill soils, namely: weathered clay, lateritic soil, and clayey sand. Their properties are summarized in Table 7.1. It has been found that the cohesive-frictional soils can be used effectively as MSE backfill when compacted at about the optimum water content to over 90% of compaction of standard Proctor compaction.

7.6.1 Laboratory Pullout Test

7.6.1.1 Pullout Test Apparatus

In the laboratory, pullout tests were conducted using a 1.30 m x 0.080 m x 0.50 m test cell made up of 13.0 mm thick steel plates. A schematic diagram of the laboratory pullout testing box is shown in Fig. 7.19 and a typical test set-up is shown in Fig. 7.20. The vertical stress was supplied by an air bag fitted inside the pullout box between 6.5 mm flexible metal plates. The pullout force, measured by means of an electronic load cell, was applied to the test specimen using an electrically-controlled hydraulic cylinder mounted against the supporting frame of the pullout cell. The horizontal displacement of the mat was monitored using the linear

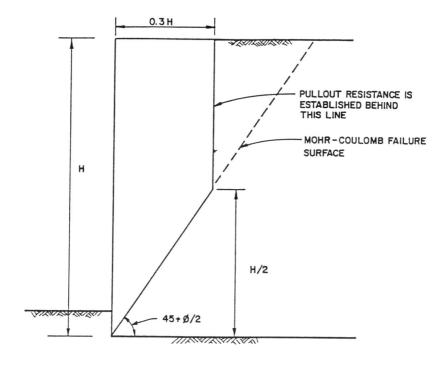

Fig. 7.17　　Location of Potential Failure Surfaces of Reinforced Wall

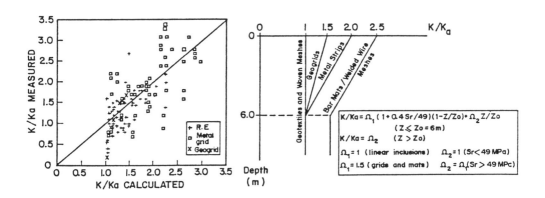

Fig. 7.18　　Variation in the Stiffness Factor as a Function of Depth and a Comparison of the Measured and Calculated Horizontal Earth Pressure Coefficients, K, for Various Types of Reinforcements (Christopher et al. 1990)

Table 7.1 Summary of Backfill Soil Parameters

Soil	G_s	ω_p (%)	ω_l (%)	I_p (%)	Passing sieve no. 200 (%)	ω_{opt} (%)	δ_{dmax} (kN/m^2)	Direct Shear		Unconsolidated Undrained		Isotropically Consolidated Undrained			
								C (kN/m^2)	ϕ (deg)	C (kN/m^2)	ϕ (deg)	C (kN/m^2)	ϕ (deg)	C (kN/m^2)	ϕ (deg)
Clayey sand	2.55	-	-	-	44.28	14.0	17.0	38.0	26.8	42.0	23.5	25.0	24.0	32.0	18.2
Lateritic residual soil	2.61	23.2	39.2	16.00	17.91	11.5	19.3	88.0	40.2	80.0	32.5	27.0	25.2	30.0	21.0
Weathered clay	2.67	21.0	45.0	24.00	82.94	22.0	16.3	129.0	30.7	118.0	31.5	19.0	25.8	24.0	15.5

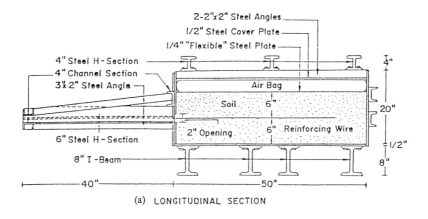

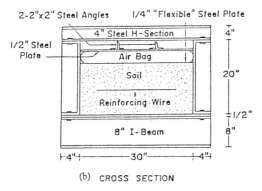

Fig. 7.19 Schematic Diagram of the Laboratory Test Apparatus

variable differential transformer (LVDT) and was pulled out at a rate of 1 mm/min. The data acquisition system (21 x micrologger) recorded the mat displacement, pullout force, and the axial strains in the longitudinal and transverse members by means of strain gauges. Steel grid reinforcements with different opening size were used. The backfill materials were compacted at 95% standard Proctor (ASTM D698) densities at dry side, optimum, and wet side of optimum water contents. A typical schematic diagram of welded steel grid reinforcement used in the laboratory tests is shown in Fig. 7.21.

7.6.1.2 Test Results

<u>Pullout force/pullout displacement curves</u>. Typical pullout force-pullout displacement curves from laboratory pullout tests with weathered clay as backfill are shown in Fig. 7.22. Generally, laboratory pullout test curves show a quicker resistance mobilization rate and maximum pullout resistance was mobilized for a pullout displacement of about 20 mm which was defined as maximum pullout resistance for laboratory test. The maximum pullout resistance increased with the same increase in normal pressure as that of high quality backfills. The curves more or less show a hyperbolic shape. The plots of strain with distance from the facing are typically shown in Fig. 7.23. The results indicate linear variation of strains in the order of 0.01% to 0.20% only. This means a maximum of 2 mm elongation of a 1.0 m longitudinal bar which is very small compared to about 25 mm pullout displacement. It is, therefore, reasonable to consider that the reinforcement moved nearly as much as the rigid body and that the pullout resistance along the reinforcement is uniformly mobilized.

<u>Maximum pullout resistance</u>. The maximum pullout resistance of steel grid reinforcement is mainly influenced by grid geometry and backfill soil strength properties. As described in the theoretical background section, the grid geometry influence is expressed by using the bearing member spacing member. Figure 7.24 shows the typical influence of S/D ratio on maximum pullout resistance. The pullout bearing resistance ratio in the figure means the pullout resistance of a bearing member of grid divided by the pullout resistance of isolated single bearing member in same test condition. It shows that pullout resistance is increased with the increase in pullout S/D ratio. When S/D is larger than 40, the bearing member interference becomes negligible. Similar results were found by Palmeira and Milligan (1989).

Over 90% of the pullout resistance of grid reinforcement is derived from bearing resistance of transverse members (Chang et al. 1977). Therefore, the backfill soil strength parameters strongly influence the grid pullout resistance. For cohesive-frictional backfill materials, the strength parameters are in a large extent controlled by the compaction water content and the degree of compaction. Figure 7.25 indicates the effect of compaction water content on the maximum pullout resistance. It can be seen that backfill soil compacted at the dry side of optimum with higher strength parameters yielded much higher pullout resistance.

<u>Effect of stage loading</u>. A multistage pullout testing procedure was followed in laboratory tests. Three pullout tests in three corresponding loading stages were conducted at each set-up. At each stage the reinforcement was pulled out by 25 mm. By so doing, more

Fig. 7.20　　Typical Set-Up of Laboratory Pullout Test

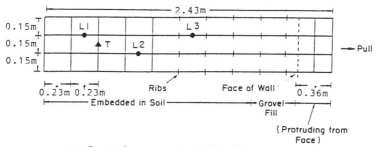

Fig. 7.21　　Schematic Diagram of Typical Dummy Steel Geogrid Reinforcement Used in the Field Pullout Test

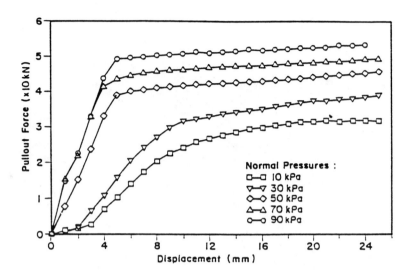

Fig. 7.22 Typical Load-Displacement curves from Laboratory Pullout Tests for a 6.5 mm Diameter Bar with 150 by 230 mm Mesh Size (Dry Side Compaction)

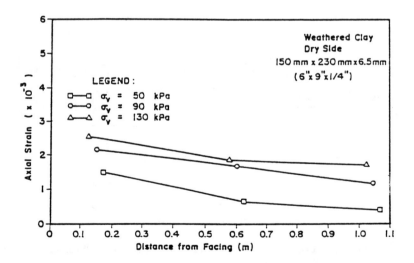

Fig. 7.23 Variation of Axial Strain with Distance from Facing (Dry Side Compaction)

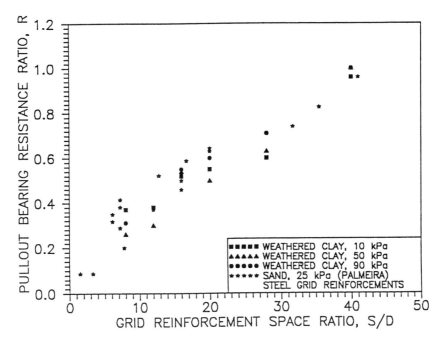

Fig. 7.24 Measured Pullout Bearing Resistance Ratio from Pullout Tests

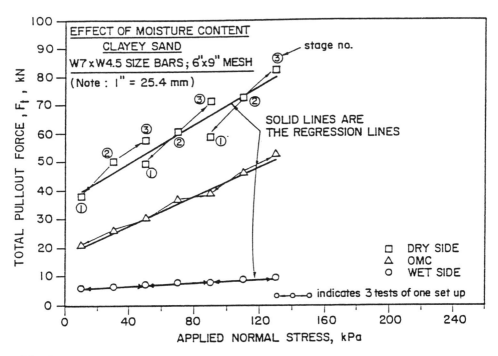

Fig. 7.25 Pullout Force with Different Confining Pressures (Clayey Sand Backfill)

information can be obtained under certain pullout set-up. Comparing the maximum pullout resistances from different loading stages, it has been found that the maximum pullout resistance at the end of 25 mm pull for a given normal stress and moisture condition were more or less the same. The only difference was in the manner in which these peak values were obtained. As shown in Fig. 7.26. the first stage load-displacement curves were found to be flatter and smoother depicting a lower modulus and corresponding to the virgin loading stage of the soil in front of the bearing member. The latter stage loadings yield a quicker resistance mobilization rate corresponding to the reloading stress history of the soil in front of the bearing member.

7.6.1.3 Prediction of the Maximum Pullout Resistance

The pullout test is labor-intensive and time-consuming. Furthermore, it is difficult to test the pullout resistance for every backfill soil and reinforcement condition. For design purposes, predicting the maximum pullout resistance becomes very important. As discussed in theoretical background section, there are three models for pullout bearing failure mechanism of isolated single bearing member. Comparing the test results with predicted values, it has been found that general shear failure mechanism (Peterson and Anderson, 1980) provided the apparent upper bound of the test results, the punching shear failure model (Jewell et al. 1984) yielded the apparent lower bound of the test results, and the modified punching shear failure model (Chai, 1992) predicted the maximum pullout resistance of isolated single bearing member reasonable well. Figure 7.27 is an example of this kind of comparison. For actual grid reinforcement, the maximum pullout resistance is strongly influence by the bearing member spacing ratio, S/D. Considering the S/D influence by Eq. 7.15, the modified punching shear failure model also yielded the good prediction for inextensible grid reinforcement as typically shown in Fig. 7.28.

7.6.1.4 Numerical Modelling of Pullout Test

A computer program, "NONLIN 1," was modified and used to model the soil-reinforcement interaction in a pullout test. "NONLIN 1" is a two-dimensional program, and it uses the initial stress method (Zienkiewicz et al. 1970) to treat the nonlinear behavior of the soil. The program is based on an analytical method suggested by Ochiai and Sakai (1987). The soil is represented by triangular and quadrilateral elements with a nonlinear elastic model criterion (Duncan and Chang, 1970). The soil parameters used in the numerical model are given in Table 7.2.

The interface properties between the soil and the reinforcement are modelled by one-dimensional joint elements, and the relative displacement between the soil and the reinforcement is allowed if the mobilized shear stress at the interface equals or exceeds the shear strength at the interface which was obtained from Mohr-Coulomb strength theory. The material parameters c and ϕ of the interface were set to be the same as those of the surrounding soil.

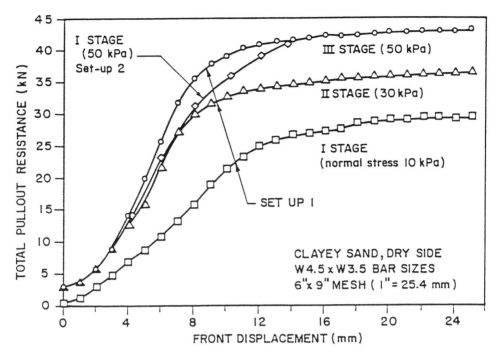

Fig. 7.26   Pullout Force-Displacement Curves (Clayey Sand Backfill)

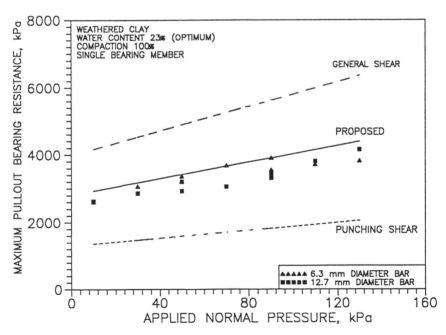

Fig. 7.27   Predicted and Measured Maximum Pullout Resistance of Single Bearing Bar in Weathered Bangkok Clay Compacted at Optimum Water Content of 23% and 100% Degree of Compaction

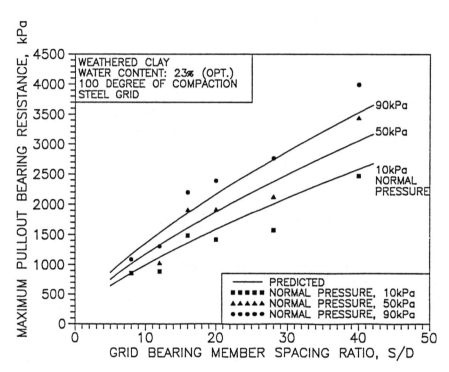

Fig. 7.28 Comparison of Predicted and Measured Maximum Pullout Bearing Resistance of Steel Grids in Weathered Bangkok Clay

Table 7.2 Soil Parameters Used for "NONLIN 1" FEM Program

Soil	δ (kN/m^3)	k_o	C (Kn/m^2)	φ (deg)	Modulus number (k)	Modulus exponent (n)	Failure (ratio) (R_f)	Poisson's ratio (ν)
Clayey sand	19.2	0.59	42.0	23.5	360	0.31	0.96	0.36
Lateritic soil	20.3	0.57	80.0	32.5	1400	0.34	0.96	0.36
Weathered clay	18.8	0.56	118.0	31.5	630	0.15	0.84	0.36

The reinforcement was represented by one-dimensional bar elements. The three-dimensional discrete bar mats were converted into two-dimensional representation (Schmertmann et al. 1989). The modulus of elasticity and yield stress of the steel grids were set equal to the known values for steel. Figure 7.29 shows the finite elements for reinforcement and reinforcement-soil interaction. For modelling the laboratory pullout test, the pullout box was represented by a fixed boundary. A graphic representation of the typical finite element mesh used for analyzing a laboratory pullout test is given in Fig. 7.30.

In order to define completely the load-displacement response in the pullout test, five displacement increments were applied to the nodal point at the free end of the reinforcement just in front of the pullout box. The comparison between the predicted load-displacement curves and experimental results for 0.15 m x 0.23 m steel mesh with 6.5 mm diameter steel bars on weathered clay backfill are presented in Fig. 7.31. The predictions from the finite element analyses are comparable with the experimental results for all three types of backfill soil with a maximum difference of about 1.5% in terms of the pullout force.

7.6.2 Field Pullout Tests

A full-scale test MSE wall/embankment was constructed inside the AIT campus (42 km north of Bangkok) and dummy reinforcement mats for field pullout tests were embedded at different levels along the face of the wall and instrumented with strain gauges. The results of the AIT test MSE wall/embankment will be used in the modelling of the MSE wall/embankment. The reinforced wall/embankment system with welded wire reinforcement was constructed in the AIT campus as part of the USAID sponsored research project. The subsoil profile at the site consists of the topmost 2.0 m thick layer of dark-brown weathered clay overlying a blackish-grey soft layer which extends to a depth of about 8 m below the existing ground. The soft clay layer is underlain by a stiff clay layer. The groundwater table fluctuated between 0.5 to 2.0 m below the ground surface. A typical subsoil profile together with the general soil properties at this site is depicted in Fig. 7.32. The welded wire mats used in the test wall/embankment system consisted of W4.5 (6.1 mm) x W3.5 (5.4 mm) galvanized welded steel wire mesh with 152 mm x 228 mm grid openings in the longitudinal and transverse directions, respectively. Each

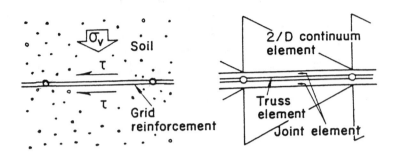

Fig. 7.29 Modelling of Reinforcement and Soil-Reinforcement Interface

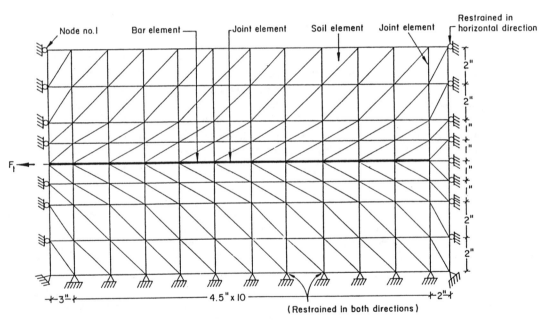

No. of nodes : 152
No. of elements : Soil elements = 194 ; Joint elements = 31 ; Bar elements = 11.

NOTE : 1" = 25.4 mm

Fig. 7.30 Typical Finite Element Mesh for Analysis of Laboratory Pullout Test

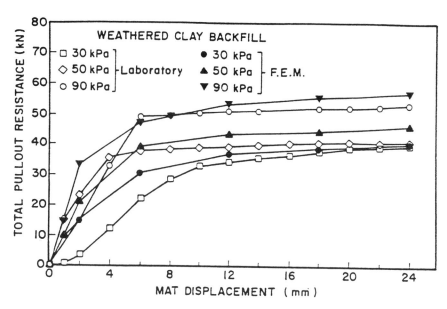

Fig. 7.31 Comparison of Experimental and Predicted FEM Load-Displacement Curves for Weathered Caly Backfill

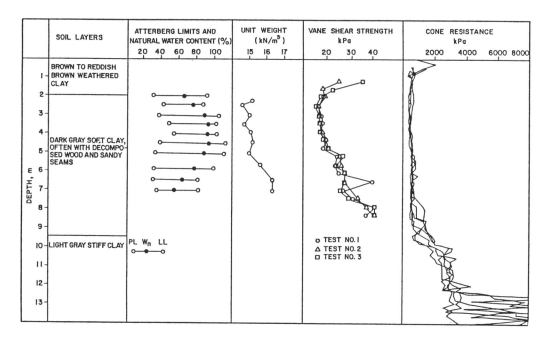

Fig. 7.32 Soil Profile and Soil Properties at MSE Test Embankment

reinforcement unit had dimensions of 2.44 m wide and 5.72 m long including the facing. The embankment was divided into three sections along its length and three different backfill materials, namely: clayey sand, lateritic soil, and weathered clay were used in each section. The embankment was constructed during a one month period from April 24 to May 24, 1989. The whole embankment is 5.8 m above the existing ground surface with about 26.0 m base length. It has three sloping faces with 1:1 slope and one vertical face (wall), as shown in Fig. 7.33a,b. The wall/embankment system was instrumented extensively for monitoring the performance of the structure. The instrumentation program included the measurement of strains, and therefore, the tension forces in the longitudinal wires. The surface and subsurface settlements, pore pressures, vertical pressures at the base of the wall, and lateral movements of the wall face were also monitored. There were 21 instrumented reinforcement mats, 19 settlement plates, 3 earth pressure cells, 10 piezometers, and 6 inclinometer casings. Figure 7.34 shows the schematic view of the field instrumentation (Bergado et al. 1991).

Field pullout tests were conducted to investigate the pullout resistance of reinforcements embedded at representative overburden, field moisture, and density conditions. A total of 15 constant strain field pullout tests were conducted on dummy mats of varying overburden pressures and bar sizes. Three of the mats had no transverse bars. These mats had only four longitudinal bars of size W4.5 (diameter of 6.1 mm) with short transverse ribs. The rest of the mats had 0.15 m x 0.23 m grid openings, with 5 to 6 transverse bars. The average length of embedment of all the mats was around 2.0 m from the face of the wall. Two dummy mats in each of three backfill materials were instrumented with strain gauges at selected points. The geometry of the instrumented mats together with the locations of strain gauges are shown in Fig. 7.35.

The pullout force was supplied by an electrohydraulic control pullout jack through a steel reaction frame butting against the wall face. The horizontal displacements of the reinforcement were monitored by using linear variable differential transformers (LVDT's). A data acquisition system, consisting of a 21X micrologger was employed to record the mat displacement, pullout forces, and the axial strains in the bars. The set-up of field pullout test equipment is shown in Fig. 7.36. The rate of pulling was electronically controlled and a pullout rate of 1.0 mm/min was adopted. All the mats were pulled out to a total displacement of about 130 mm. The loads were measured with a 222.5 kN capacity load cell.

The typical pullout force-mat displacement curves for the weathered clay backfill are shown in Fig. 7.37. All of the pullout force-displacements increased with only a small corresponding increase in pullout force. There seems to be no well-defined peak in the load-displacement curve. The pullout force was still increasing at the end of 130 mm displacement. The pullout force in such cases, however, was already near the tension capacity of the reinforcement. The comparison of the results for the same size of bars, similar configuration of mat, and overburden pressures for the three different types of backfill soil are shown in Fig. 7.38. It can be seen that the pullout resistance of the dummy bar mat in clayey sand is considerably higher. However, the pullout resistances were influenced by arching effects, which will be discussed later.

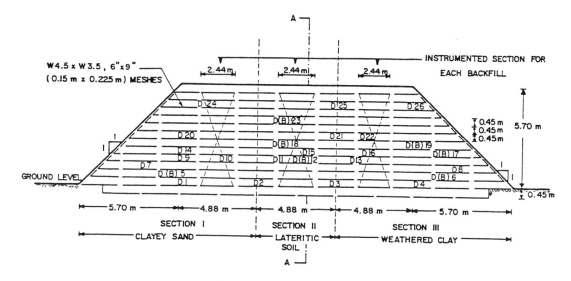

(a) Longitudinal-Section View

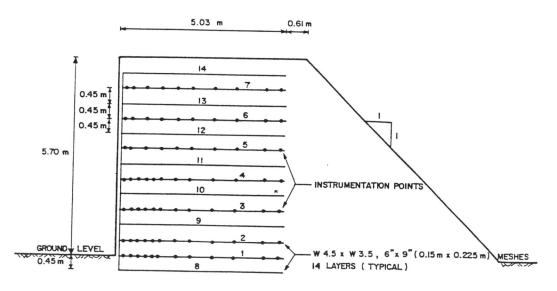

(b) Cross-Section View

Fig. 7.33 Section Views of MSE Test Embankment

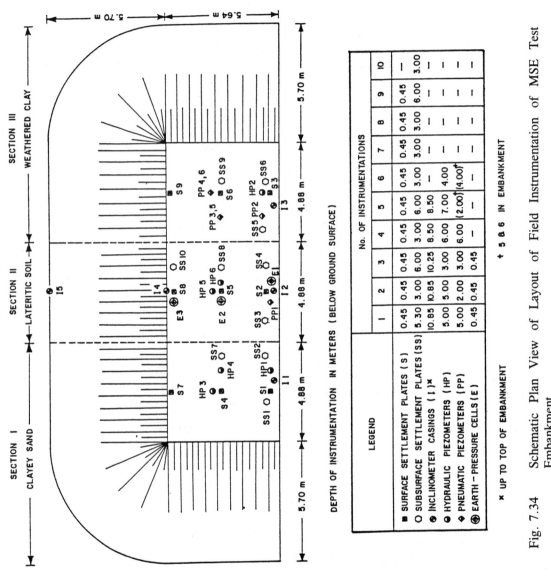

Fig. 7.34 Schematic Plan View of Layout of Field Instrumentation of MSE Test Embankment

Fig. 7.35 Typical Set-Up of Field Pullout Test

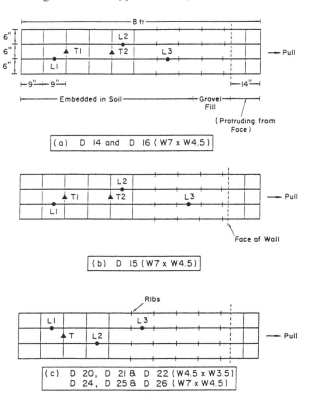

● Strain Gauge on Longitudinal Bar
▲ Strain Gauge on Transverse Bar

Note : 1" = 25.4 mm ; 1' = 0.3048 m ; 1 m = 1000 mm

Fig. 7.36 Different Configurations of Dummy Steel Grids for Field Pullout Tests

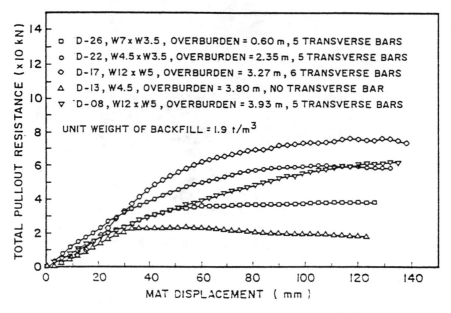

Fig. 7.37   Pullout Force-Displacement Curves from Field Pullout Tests

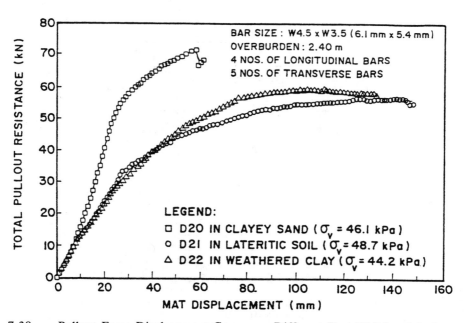

Fig. 7.38   Pullout Force-Displacement Curves on Different Backfill Materials from Field Pullout Tests

The pullout force increased almost linearly with the increase in overburden pressures for weathered clay and clayey sand backfill sections. On the contrary, in the lateritic soil backfill there was a decrease in maximum pullout resistances with the increase in the height of overburden. This phenomenon might be due to arching effects, since the lateritic soil was located in the middle section of the wall/embankment system, where it was interconnected by the facing mat to the full height of the wall which prevented the immediate response to the non-uniform settlement and caused an arching effect near the wall face. The dummy mats were embedded up to 2.0 m behind the wall face, so the arching effect influenced the pullout resistance very much (Bergado et al. 1991). Other factors, such as particle crushing of lateritic soil and the non-uniform moisture distribution within the reinforced mass, also influenced the pullout resistance although to a lesser extent.

The field pullout tests of ribbed longitudinal bars (transverse bars being cut off) were conducted in each of the three types of backfill, having the same overburden height of 3.8 m. The resistances obtained in weathered clay and clayey sand were slightly higher than that in lateritic soil. It is possible that the resistance in lateritic soil will be higher, but as mentioned above, the arching effect reduced the overburden pressure from actual overburden height of 3.8 m. Hence, the pullout resistance is slightly lower than that of the other two backfills. The total friction coefficients, f, (assuming that the ribs contributed an increase in frictional resistance of the plain longitudinal bar) in each of the three types of backfill were 2.77, 2.65, and 2.84 for clayey sand, lateritic soil, and weathered clay, respectively. For calculating the friction coefficient, f, the following equation was used (Anderson and Nielsen, 1984):

$$f = \frac{P_f}{A_s(\overline{\sigma_a})} \tag{7.17}$$

where: P_f = the peak frictional force of ribbed longitudinal bars
A_s = the surface area of longitudinal bars
σ_a = the average overburden pressure, ($\sigma_a = 0.75\sigma_v$, where σ_v is the vertical overburden pressure)

7.6.3 Comparison of Laboratory and Field Pullout Tests

The reinforcement mats used in laboratory and field pullout tests had 0.15 m x 0.23 m openings and most of them had 5 transverse bars. The length of the longitudinal bars in the laboratory tests was about 1.0 m, while in the field tests it was about 2.0 m. In the field, some of the transverse bars were cut off and only the ribbed longitudinal bars remained. The diameters of the wire used in the field and in the laboratory tests also varied. In order to compare the field and laboratory peak pullout resistance, two assumptions have been made: (1) the frictional and passive resistances are linearly proportional to the surface area of the longitudinal member and passive bearing area of the transverse member, and (2) constant values of the apparent friction coefficient between the ribbed longitudinal bar and the soils were taken as 2.77, 2.65, and 2.84 for clayey sand, lateritic soil, and weathered clay, respectively.

Based on the assumptions, the field pullout resistances were corrected to be comparable to the laboratory results. Figures 7.39 through 7.41 show the comparison between field and laboratory pullout test results in terms of maximum pullout resistance plotted against normal pressure. For weathered clay and clayey sand, it was observed that the field tests in all cases provided higher pullout resistances than the laboratory tests. The arching effects of the embankment must have increased the vertical stresses on the two end sections comprising clayey sand and the weathered clay backfills. This phenomenon resulted in increased pullout capacities at the end sections. For the lateritic backfill soil in the middle section, however, the field pullout resistance was very much affected by the arching effects. At the 2.4 m overburden condition, the field test had nearly the same pullout resistance as the corresponding laboratory tests. However, at the 3.33 m overburden condition, the field test provided much lower pullout resistance than that of laboratory tests. These results indicated that the arching effects were dominant at the lower portions of the wall. It must be noted that the envelope of laboratory pullout resistance is bilinear for lateritic soil. This is attributed to the particle-crushing phenomenon of the lateritic soils at higher stresses.

During the laboratory tests, the interaction between the soil/reinforcement system an the rigid boundaries of the laboratory pullout box (especially the front face) in the small-scale tests can affect the measured pullout resistance. As the reinforcement was pulled from the box, lateral pressure developed against the rigid front face, leading to arching of the soil over the inclusion which reduced the local vertical stress on the reinforcement, and consequently, decreased the pullout resistances (Juran et al. 1988; Palmeira and Milligan 1989). Dilatancy of soil during pullout tends to cause a localized increase in normal stresses acting on the reinforcements. Variation of the moisture content can also influence the friction between the soil and the reinforcement.

7.7 BEHAVIOR OF MSE STRUCTURES ON SOFT GROUND

In coastal areas, the MSE structures need to be constructed on soft ground. In a large extent, the behavior of MSE structures on soft ground is controlled by the interaction between the reinforced mass and soft subsoil because the time dependent and differential settlements of the subsoil can cause the lateral displacement of the structure and modify the tension force distribution pattern in reinforcements. For describing the behavior of MSE structures on soft ground, the results from the AIT MSE test wall/embankment (described earlier) will be quoted.

7.7.1 Excess Pore Pressure and Settlement

The earth structures on soft ground usually have time dependent large settlements and lateral displacements. In the case of MSE structures, the foundation deformation will cause the variation of the stress/strain stages of reinforced soil mass. Therefore, understanding the consolidation process of the soft ground is important. The usual approach for analyzing the behavior of the soft ground under surface loading is to assume an undrained condition for short-term performance and drained condition for long-term response. However, all natural clays, as

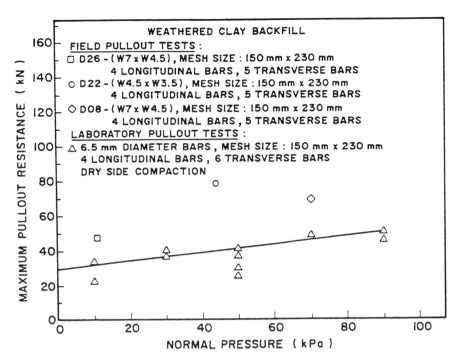

Fig. 7.39 Comparison of Pullout Resistance from Field and Laboratory Pullout Tests (Weathered Clay Backfill)

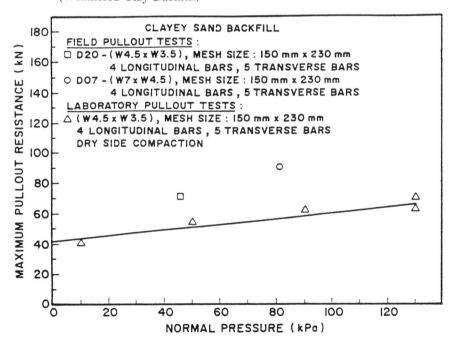

Fig. 7.40 Comparison of Pullout Resistance from Field and Laboratory Pullout Tests (Clayey Sand Backfill)

they occur in-situ, have developed some preconsolidation as the result of past loading and erosion, aging, cementation, water table fluctuation, etc. After studying the field observations of several embankments on soft clay foundation, Tavenas and Leroueil (1980) proposed that during the early stages of construction, the foundation soil behaves as overconsolidated soil which has a relatively high coefficient of consolidation. Thereby, significant consolidation can take place. When the embankment load reaches the preconsolidation pressure of the clay, the foundation soil becomes normally consolidated clay, and the foundation response becomes undrained with constant volume deformation. Figure 7.42 shows the typical settlement-time plots (AIT MSE wall/embankment). It can be seen that the settlements continued to 1 year after construction. The maximum surface settlement occurred at the center plate S5 below the middle section (lateritic soil) such that the overall settlement pattern at the surface indicate a dish-like configuration. The differential settlement will cause large tension as indicated in Fig. 7.43. It can be seen that during the construction period, the excess pore pressure increase was about 40% of the increase in total vertical load (elastic solution), and post construction excess pore pressure dissipation rate was slow and reduced with the increase in time.

7.7.2 Lateral Displacement

For the MSE wall, the lateral displacement is an important item because it controls the stability of the structure. The lateral displacement can be developed from the horizontal earth pressure of the soil within and behind the structure as well as from the different settlement of the foundation soil. The latter one is more significant for MSE wall on soft ground. Figure 7.44 shows the lateral displacement profiles of AIT MSE wall/embankment. It is indicated that the end of construction lateral displacements, specially the wall face lateral displacement increased significantly during the soft ground consolidation process. At 228 days after the construction, the maximum wall face lateral displacement reached 450 mm. Under the toe of wall face, there was stress concentration, which caused the differential foundation settlement and induced the large lateral wall face displacement. This is quite different from the case of the MSE wall on rigid foundation, where most of the deformation occurred during construction period. For AIT MSE wall, the wall face inclination was about 5 degrees. Therefore, it is recommended that for MSE wall construction on soft ground, an inclined wall face (about 80 degrees with ground surface) is more suitable than a vertical wall face.

7.7.3 Tension Force in Reinforcement

As mentioned previously, for MSE on soft ground, the reinforced mass and soft ground interaction is one of the important factors which controls the reinforcement tension force mobilization and variation. The differential settlement of the foundation caused bending effect on the MSE structure, i.e. top in compression and bottom in tension. The measured reinforcement tension forces of AIT MSE wall/embankment is depicted in Fig. 7.45 for the middle section of the test embankment, including the settlement profiles and base pressures. It can be seen that: (1) the reinforcement tension forces increased during consolidation process of the soft ground; (2) the tension forces in mats 1 and 2 in the lower portion were as high as the

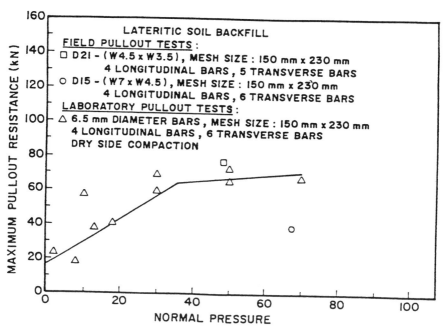

Fig. 7.41   Comparison of Pullout Resistance from Field and Laboratory Pullout Tests (Lateritic Soil Backfill)

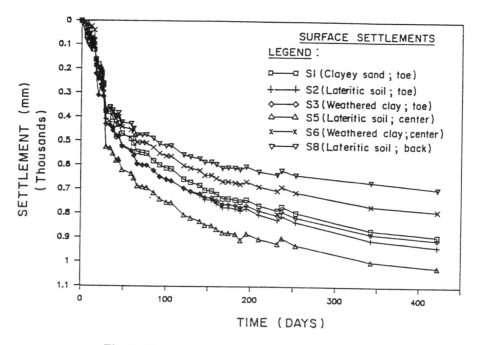

Fig. 7.42   Plots of Surface Settlements with Time

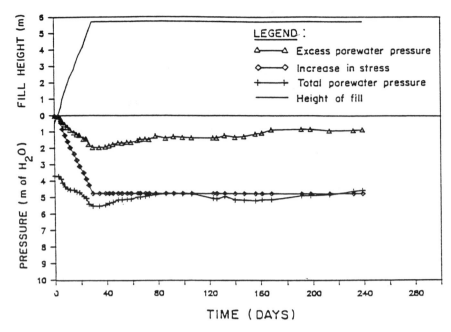

Fig. 7.43   Typical Plots of Excess Pore Pressure with Time

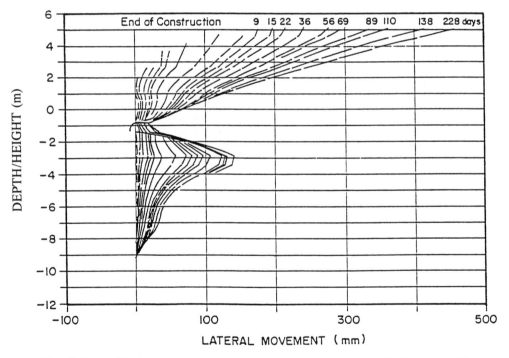

Fig. 7.44   Typical Plots of Wall Movement Against Depth (Inclinometer 2)

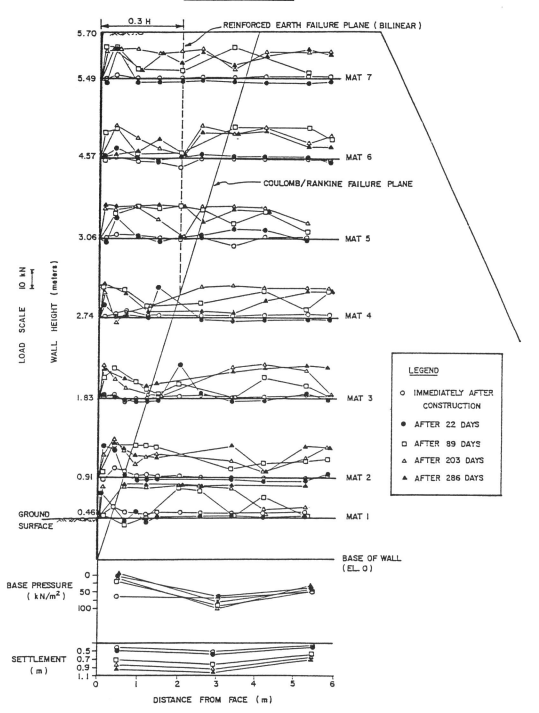

Fig. 7.45 Variations of Reinforcement Tension, Base Pressure and Settlements at Different Periods for the Middle Section

upper mats. For reinforced wall on rigid ground, the tension force in the base layer tends to be lower because of the friction between reinforced mass and the foundation soil (Adib et al. 1990). The maximum lateral pressures immediately after construction are plotted in Fig. 7.46 for the middle section, and were compared to existing earth pressure theories on reinforced soil structures as discussed by Jones (1985). The measured values immediately after construction were higher than the coherent and tie-back structure hypotheses but seem to be closely predicted by including the compaction effects proposed by Ingold (1983b).

7.7.4 Lateral Earth Pressure Coefficient, k

In the design of MSE structures, the lateral earth pressures in reinforced soil mass determine the required reinforcement force for internal stability analysis. For MSE wall on rigid foundation, the k values are varied with the height of wall and the types of reinforcement. Since the foundation settlement pattern can influence the reinforcement tension force mobilization process and distribution pattern, the k values of the MSE wall on soft ground will certainly be influenced by the reinforced mass/soft ground interaction with high k value at the bottom of MSE structure. Figure 7.47 indicates the variations of reinforcement tension during construction of AIT MSE wall/embankment. As shown, the k values are larger than active value, k_a, and closer to at rest value, k_o. Near the top of the wall, the k values were larger than k_o line due to the effect of compaction induced residual reinforcement tension force. Figure 7.48 shows the typical k value variation during the construction process for a reinforced mat. The k values varied from larger than 1.5 per line of low overburden pressure to about 0.4 at an overburden height of 6 m.

7.7.5 Maximum Tension Line

The response generated by the wall due to foundation compressibility created a unique situation wherein existing theories on earth pressures may not be directly applicable. For reinforced soil walls, the maximum tension line was reported to define a failure surface or wedge of a Coulomb/Rankine type failure plane, bilinear failure plane, or the log spiral failure plane. Any of these conditions may not be satisfied if there is large foundation settlement or deformation. It was found that the maximum tension line seems to conform closely to the log spiral failure plane in the lower half of the wall, and to the reinforced earth coherent gravity failure plane in the upper half of the wall.

The behavior of the MSE structure on soft ground is a function of properties of the reinforced earth, properties of the soft ground foundation, and their interactions. Normally, due to the large total and differential settlements of the soft ground, substantial wall face lateral displacement will be induced and maximum tension force will occur in the base reinforcement. These results should be taken into account for the design of MSE structures on soft ground.

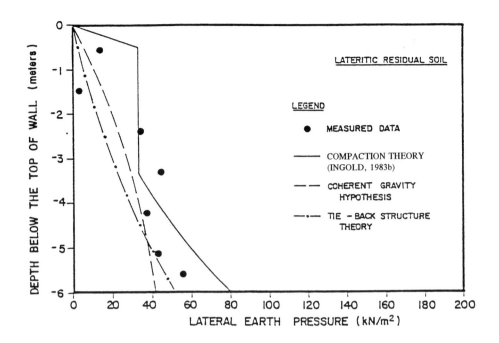

Fig. 7.46 Measured and Predicted Lateral Earth Pressure Along the Height of the Wall

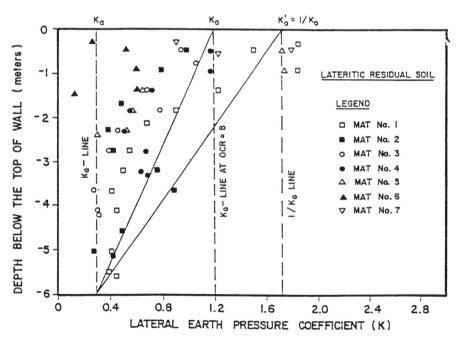

Fig. 7.47 Measured Lateral Earth Pressure Coefficient (K) During Construction of the Wall

7.8 MODELLING OF MSE STRUCTURES ON SOFT GROUND

The finite element method was first introduced in geotechnical engineering in the early 1960s. Since that time, it has been widely used in almost every aspect of geotechnical problems, due to its ability accommodate difficulties such as nonhomogeneous materials, nonlinear stress-strain behavior and complicated boundary conditions, etc. The behavior of the reinforced soil walls and embankments on soft ground, and its influence factors have been analyzed by using finite element methods (Hermann, 1978; Schaefer and Duncan, 1988; Hird and Kwok, 1989; Adib et al. 1990). Different soil behavior models, such as elastic perfect-plastic, nonlinear hyperbolic, and modified Cam clay have been used to represent the stress/strain behavior of the soils. Interface elements have been employed to simulate the interaction behavior between soils and reinforcements. Although the accuracy of the finite element analysis mainly depends on the accuracy of the models used and the correctness of the input parameters, the finite element method is a powerful tool to investigate the influence factors. Furthermore, if the model and the parameter determination method are validated, the finite element method can be used to predict the performance of the reinforced earth structure.

There are two general approaches to the analysis of reinforced soil system (Hermann, 1978), namely: discrete representation and composite representation. In the discrete representation, the reinforced system is treated as a heterogeneous body. The soil and every reinforcing member are discretely represented in the system. The elements can have different stress/strain relationships and boundary conditions. This approach can give detailed information about the interaction of the soil and the reinforcement, but needs more computer time.

In the composite representation, the reinforced system is modelled as a locally homogeneous cross-anisotropic material termed the "composite material". The body of the reinforced system is represented by an array of continuum elements whose boundaries need not bear a spatial relationship to the geometric arrangement of the reinforcement. The "composite material" properties assigned to the continuum members reflect the properties of the matrix material and the reinforcement, and their composite interaction. This approach can save computer time, but it does not directly yield detailed information about the stress and strain states at the interfaces of the soil and the reinforcement, nor about localized deformation near the edges of the reinforced mass.

At present in geotechnical engineering, the primary role of finite element analysis is to validate the simplified methods of analysis or to do the parametric study. This method is not used as a routine design tool. for reinforced soil, the soil/reinforcement interaction properties are key factors which control its performance. From this point of view, for the case of reinforced wall and embankment system, the discrete representation approach is superior to the composite representation approach.

7.8.1 Reinforcement and Wall Face Models

The reinforcement is usually modelled as an elastic-perfectly plastic material. Its yield stress is controlled either by its elastic limit or by its pullout resistance, whichever is smaller. A nonlinear elastic behavior can be numerically introduced. Interaction behavior between soil and grid or strip reinforcements is a truly three-dimensional situation. However, generally, in two-dimensional analysis, the reinforcement is treated as a continuous sheet and modelled as a bar element with axial stiffness but negligible flexure rigidity. The finite element representation of the reinforcement and the soil/reinforcement interface in two-dimensional analysis is already shown in Fig. 7.29. The model parameters are: cross sectional area per unit width of reinforcement, tangent modulus, and the yield stress.

The most commonly used facing units for reinforced wall construction are concrete panels and steel or polymer grids. In two-dimensional finite element analysis, the facing unit is commonly represented by a beam element with axial, shear, and bending stiffness. Normally, the facing unit is considered as elastic material, with the material properties of elastic modulus, shear modulus, cross-sectional area per unit width, and the moment of inertia. The connection between facing panels is modelled as a hinge in two-dimensional finite element modelling (Schmertmann et al. 1989).

7.8.2 Soil Models

Soil is a complex engineering material. It has nonlinear, non-homogeneous, elasto-plastic, and visco-plastic properties in most of the geotechnical engineering cases. Many attempts have been made to model the soil behavior. The commonly used models can be divided into two groups: nonlinear elastic models such as hyperbolic model with Mohr-Coulomb failure criterion, and the elasto-plastic models with Cam clay being typical.

7.8.2.1 Hyperbolic Model

The stress/strain curves of the soil are approximated by hyperbola (Kondner and Zelasko, 1963) as expressed in the following:

$$\sigma_1 - \sigma_3 = \frac{\epsilon_1}{a + b\epsilon_1} \tag{7.18}$$

where: σ_1, σ_3 = major and minor principal stress
ϵ_1 = axial strain
a, b = constants, whose value may be determined from conventional triaxial tests. 1/a is the initial tangent modulus, E_i, and 1/b is the ultimate principal stress difference.

Duncan and Chang (1970) improved this model by (a) obtaining the relationship for the initial tangent modulus, E_i. as given by Janbu (1963), from the following expression:

$$E_i = k.P_a \cdot \left[\frac{\sigma_3}{P_a}\right]^n \tag{7.19}$$

where: P_a = atmospheric pressure
K = modulus number
n = modulus exponent

and (b) utilizing the Mohr-Coulomb failure criterion for determining the value of b. The resulting tangent modulus, E_t, can be written as:

$$E_t = \left[1 - \frac{R_f(1-\sin\phi) \cdot (\sigma_1-\sigma_3)}{2.C.\cos\phi + 2.\sigma_3.\sin\phi}\right]^2 .k.P_a \left[\frac{\sigma_3}{P_a}\right]^n \tag{7.20}$$

where: C = cohesion of the soil
ϕ = frictional angle of the soil
R_f = failure ratio ($R_f = b(\sigma_1-\sigma_3)_{fail}$, $(\sigma_1-\sigma_3)_{fail}$ is the principal stress difference of the soil at failure)

The unloading and reloading modulus of the soil is usually higher than the virgin loading modulus, and under given confining pressure, it is approximately a constant for a given soil. The model expresses the unloading-reloading modulus as follows:

$$E_{ur} = k_w.P_a \cdot \left[\frac{\sigma_3}{P_a}\right]^n \tag{7.21}$$

where k_w is unloading-reloading modulus number. In most cases, the Poisson's ratio is taken as constant and determined by experience. For a constant Poisson's ratio, the model is a six-parameter model, k, n, k_w, R_f, C, and ϕ, as defined previously.

Subsequent studies have shown that the volume change behavior of most soil can be modelled with equal accuracy by assuming that the bulk modulus of the soil varies with the confining pressure and is independent of the percentage of mobilized strength (Duncan et al. 1980). Based on this assumption, the bulk modulus and Young's modulus have been used to formulate the stress/strain relationship of the soil. The value of bulk modulus, B, is defined by:

$$B = \frac{(\sigma_1+\sigma_2+\sigma_3)}{3.\varepsilon_v} \tag{7.22}$$

where: $\sigma_1, \sigma_2, \sigma_3$ = principal stresses
ε_v = volumetric strain corresponding to the stress condition.

Also, the bulk modulus may be found by an empirical equation if the same soil is tested at various confining pressure, as follows:

$$B = k_b \cdot P_a \left(\frac{\sigma_3}{P_a}\right)^m \qquad (7.23)$$

where: k_b = bulk modulus number
m = bulk modulus exponent

The tangent Poisson's ratio, ν_t, may be related to bulk modulus, B, and tangent modulus, E_t, as follows:

$$\nu_t = 0.5 - \frac{E_t}{6.B} \qquad (7.24)$$

When using E_t and B to express soil behavior, the hyperbolic model becomes an eight-parameter model, k, k_w, n, R_f, C, ϕ, k_b, and m, as defined previously. The model is illustrated in Fig. 7.49.

The nonlinear constitutive model is widely used in finite element analysis because it is simple, and the parameters can be easily determined from conventional triaxial tests. The drawback of this model is that is cannot consider the dilatancy properties of the soil during shear. Moreover, being based on the generalized Hooke's law, the relationships are not suited for analysis of stress and strain near the failure state where large plastic deformations occur.

7.8.2.2 Elasto-Plastic Models

For modelling the elasto-plastic behavior of the soil, three things need to be defined: (a) yield surface, (b) strain hardening law, and (c) flow rule.

For elasto-plastic materials, the strains due to any increment of stress are often divided into two components: elastic and plastic. A yield surface is defined in stress space as one which divides the regions of stress for which the strains are elastic from those for which the strains include a plastic component. For conditions of stress inside the yield surface, an infinitesimal increment of stress can cause only elastic strains. If the stress conditions correspond to a point on the yield surface and if the material is stable, an infinitesimal increment of stress is directed outside the yield surface and the material has strain hardening property. The strain hardening law is a relationship between the position of the yield surface in the stress space and the plastic strains experienced by the material in arriving at the specific stress point under consideration. The flow rule provides a relationship between the strain rate vector during plastic deformation and the imposed stress vector. For increment of plastic deformation to occur, the stress point should lie on the yield surface and the stress increment should be directed outside this surface.

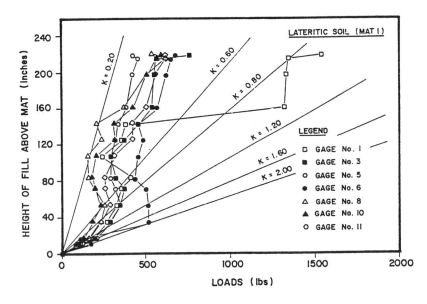

Fig. 7.48   Variation of Wire Tensions with Fill Height

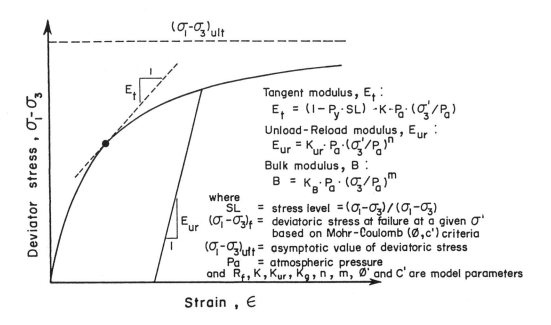

Fig. 7.49   Hyperbolic Soil Model of Stress-Strain Behavior

There are several elasto-plastic models for soils. The critical state theory is a typical one, such as the modified Cam clay model (Roscoe and Burland, 1968). The critical state model is a form of elasto-plastic, isotropic strain hardening law. It introduces a distinction between yield and ultimate collapse by using the concept of a critical state line in conjunction with strain-dependent yield surface. If a soil is at the critical state during continuous deformation, there is no change in both the void ratio and the effective stress components. In this model, a soil undergoing shear deformation can pass thorough a yield point without collapse and continue to deform until eventually the critical state line is reached, where perfect plastic conditions exist. In this state, the soil continues to deform without further change of void ratio or stress. Modified Cam clay model is widely used as a critical state model. Its yield surface is an ellipse in the p', q plane (p is effective mean stress, and q is deviator stress). The modified Cam clay model assumes an associated flow rule, so the yield function and plastic flow function are the same:

$$p' = \frac{p'_o . M^2}{M^2 + \eta^2} \tag{7.25}$$

where:
- p' = effective mean stress
- p'_o = isotropic preconsolidation pressure
- M = slope of the failure line in (q,p') plane
- η = stress ratio q/p'

The strain hardening law uses the consolidation parameters, λ and κ, obtained by isotropic loading and unloading of normally consolidated soil. The parameter λ is the slope at the void ratio versus ln(p') plot during loading, while κ is the initial value of the slope during rebound. The strain is expressed as volumetric strain and shear strain. In an elastic state, the elastic shear strain is neglected in the original theory, but included in the finite element formulation by using an additional parameter of shear modulus (G) or Poisson's ratio (ν) (Britto and Gunn, 1987). The elastic volumetric strain is given as follows:

$$dv^e = \frac{\kappa . dp'}{(1+e).p'} \tag{7.26}$$

where e is initial void ratio. The incremental plastic volumetric strain dv^p and shear strain $d\epsilon^p$ are given as follows:

$$dv^p = \left\{\frac{\lambda}{1+e}\right\} . \left[\frac{dp'}{p'} + \frac{2.\eta.d\eta}{M^2+\eta^2} . \left[1-\frac{\kappa}{\lambda}\right]\right] \tag{7.27}$$

$$d\epsilon^p = \left\{\frac{2.\eta}{M^2-\eta^2}\right\} . \left\{\frac{\lambda-\kappa}{1+e}\right\} . \left\{\frac{dp'}{p'} + \frac{2.\eta.d\eta}{M^2+\eta^2}\right\} \tag{7.28}$$

In finite element formulation, the modified Cam clay model only needs five fundamental constants, M, λ, κ, G, and specific volume on critical state line at unit mean pressure, Γ are illustrated in Fig. 7.50a,b. For the reinforced earth structure on soft ground system, usually the fill material can be modelled by nonlinear elastic model. For soft foundation soil, the elasto-plastic model is widely used.

7.8.3 Finite Element Analysis of Consolidation

Consolidation analysis corresponds to a coupling between the laws governing the behavior of the skeleton of the soil and the flow of the pore fluid. In a variational form, the analysis of this problem leads to the search for a field of displacements and a distribution of hydraulic heads satisfying the simultaneous equations. Biot's consolidation theory (Biot, 1941) is a rigorous solution for this problem, when the soil skeleton is linear elastic and the pore fluid is incompressible. However, soil is not an elastic material, especially, for soft clays, wherein plastic deformation is significant. In finite element analysis, the consolidation of elasto-plastic soil is treated by incremental form. The basic matrix equations of consolidation for finite element analysis can be obtained using the standard isoparameter finite element formulation procedure (e.g. Britto and Gunn, 1987).

The reinforced embankment and wall on soft ground usually causes large settlement of soft ground. Conventional stress/strain relationship is developed based on the infinitesimal deformation assumption, and it is not suitable for analyzing the large deformation problem. Theoretically, large deformation problems can be solved by the rate type constitutive law (Carter et al. 1977, 1979). For defining the stress rate, the rotation effect is taken into account. However, in the geotechnical engineering area, for most cases, the rotation effect is not serious. In finite element analysis, the large settlement problem is treated mainly by updating the nodal coordinate (Rowe, 1984; Hird and Kwok, 1986, 1989). Although the approach is only approximate since it neglects rotational effects, the approach was checked against the rigorous large deformation analysis (Carter et al. 1979) and was found to provide good results for the embankment on soft ground problem (Rowe, 1984)

7.8.4 Soil/Reinforcement Interaction Model

The behavior of junctions or interfaces between structural and soil influences the performance of reinforced earth structures. The motions at the interface involve relative translational and rotational under loading. In the context of numerical methods, such as finite element method, special interface or joint elements are used in order to account for the relative motions and associated deformation modes, and eventually take care of the effects of interface properties on the behavior of the structures.

7.8.4.1 Physical Models

One of the main purposes of theoretical modelling is to apply the model to numerical analysis, such as finite element analysis, of geotechnical problems. Two basic physical models or basic elements have been evolved for modelling the interface behavior. The first one involves the insertion of distinct "joint" or "interface" elements with zero thickness. This approach was pioneered by Goodman et al. (1968) and has been used extensively. The second involves the use of "thin" continuum elements, with or without special constitutive relations (Desai et al. 1984).

The joint element formulation is derived on the basis of relative nodal displacements of the solid elements surrounding the interface element as shown in Fig. 7.51. For two-dimensional analysis, the constitutive or stress/relative displacement relation is expressed as:

$$\begin{Bmatrix} \sigma_n \\ \tau \end{Bmatrix} = \begin{bmatrix} k_n & 0 \\ 0 & k_s \end{bmatrix} \begin{Bmatrix} v_r \\ u_r \end{Bmatrix} \quad (7.29)$$

where:
- σ_n = normal stress at interface
- τ = shear stress at interface
- k_n = normal stiffness on interface
- k_s = shear stiffness of interface
- v_r = relative normal displacement of interface
- u_r = relative shear displacement of interface

For application to soil/structure interaction problems, the thickness of the "joint" element is assumed to be zero. Based on the assumption that the structure and soil media do not overlap at interfaces, a high value of the order of 10^8 to 10^{12} units is assigned for the normal stiffness, k_n.

In the thin layer elements, the actual shear zone between two solid materials is a thin layer. The idea of using a thin layer element to simulate the behavior of interfaces appears logical. Zienkiewicz et al. (1970) is the first investigator to use a solid element as an interface element. The constitutive law is expressed as follows:

$$\{d\sigma\} = [c]_i \cdot \{d\varepsilon\} \quad (7.30)$$

where:
- $\{d\sigma\}$ = the vector of increments of stresses
- $\{d\varepsilon\}$ = the vector of increments of strains
- $[c]_i$ = the constitutive matrix.

In this case, the normal properties of the thin layer element depend on the characteristics of the thin interface zone as well as the state of stress and properties of the surrounding elements. However, it is very sensitive to the thickness of the interface zone. It is found that satisfactory simulation of interface behavior can be obtained for the ratio of t/B (t is thickness of the thin layer element, and B is width of the element) in the range of 1.01 to 0.1 (Desai et al. 1984).

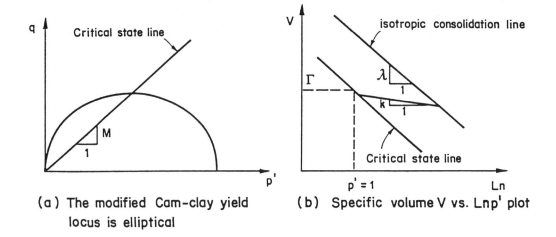

Fig. 7.50   Modified Cam-Clay Model

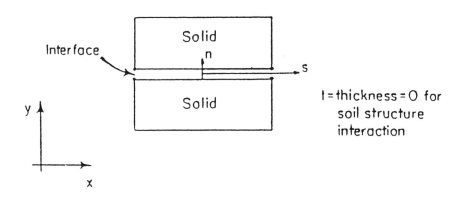

Fig. 7.51   Zero Thickness Interface Element

These two elements are based on different methodology. However, in actual practice, there seems to be not much difference in the sense of the thickness of the element. It is assumed that the thickness of the joint element is zero, but the thickness of the joint layer is considered in determining the shear stiffness. The thickness of the thin layer element is not the real thickness of the interface layer. It is related to the width of the element to fit the actual interface behavior.

7.8.4.2 Constitutive Models for Interface

For both joint element and thin layer element, the most important thing is to specify a constitutive law for shear stress/relative displacement relationship. Some models also give special normal stress/relative normal displacement relationship. There are several constitutive laws that have been developed which can be classified as: linear elastic-perfectly plastic, hyperbolic, and elasto-plastic with hardening law. In the case of linear elastic-perfect plastic model, the shear strength τ_f of the interface is governed by Mohr-Coulomb failure criteria:

$$\tau_f = c + \sigma_n \cdot \tan\delta \tag{7.31}$$

where:
c = cohesion at interface
δ = friction angle of interface
σ_n = normal stress acting on interface

Before the shear stress reaches the shear strength, the shear stress, τ, is relative displacement, u_r, by shear stiffness, k_s, as follows:

$$\tau = k_s \cdot u_r \tag{7.32}$$

When the shear stress reaches the strength, τ_f, the shear displacement will develop without increase in shear stress. This model is simple and the parameters, c, ϕ, and k_s, can be easily determined by direct shear test. However, it cannot simulate either hardening softening behavior of interface.

For the hyperbolic model, the hyperbolic stress/strain relationship of soil developed by Duncan and Chang (1970) has been extended to simulate the shear stress/relative displacement relationship of Clough and Duncan (1971) as follows:

$$\tau = \frac{\delta u_r}{b_1 + b_2 \cdot \delta u_r} \tag{7.33}$$

where:
τ = shear stress at interface
δu_r = relative displacement at interface for a given normal stress σ_n
b_1, b_2 = constants equal to $(1/k_1)$ and $(1/\tau_{ult})$, respectively, where k_1 is the initial tangent shear stiffness per unit shear area and τ_{ult} is the ultimate value of the shear stress

In a manner similar to that for tangent modulus of soil, the tangent shear stiffness, k_t, of the interface can be expressed as:

$$k_t = \left[1 - \frac{R_{fl} \cdot \tau}{c_a + \sigma_n \cdot \tan\delta}\right]^2 k_1 \cdot \gamma_w \left[\frac{\sigma_n}{P_a}\right]^{n_1} \tag{7.34}$$

where: R_{fl} = failure ratio which relates the ultimate shear stress on the interface to failure shear stress
c_a = interface adhesion
δ = friction angle at interface
σ_n = normal stress on interface
k_1 = shear stiffness number
n_1 = shear stiffness exponent
γ_w = unit weight of water

This model is also simple. All parameters have physical meanings, and it can treat the nonlinear and strain hardening behavior of the interface. But like the linear elastic-perfect plastic one, it cannot take into account strain softening behavior.

In the elasto-plastic models, as for the soil elements models, there are several elasto-plastic models for soil/structure interface behavior. Some of them just use the same mathematical structure as for soil elements, but use the interface relative displacement instead of strain, such as Boulon's model (Boulon, 1989). A flexible constitutive law proposed by Gens et al. (1990) for interface is typical. The yield surface is a hyperbolic function in shear stress, τ, and normal stress, ν_n, space. However, plastic potential function needs to be determined from test results. The key point is to assume that the factors controlling the hardening/softening, dilatancy angle of interface, and the friction angle of interface are varied with plastic displacement and follow the same function which is flexible.

Ideally, the elasto-plastic model can represent the interface stress/displacement behavior better than other models. However, it is difficult to obtain the model parameters and need special tests to determine them. Therefore, at present, for modelling the soil/reinforcement interface, the elastic, perfectly plastic or hyperbolic models are still widely used (Adib et al. 1990; Hird and Pyrah, 1990). These models are simple and the parameters can be easily determined by conventional tests. Also, the stress level dependency, nonlinear interface shear stress/shear displacement relationship, and strain hardening behavior still can be modelled correctly.

7.8.4.3 Modelling Different Soil/Reinforcement Interaction Modes

The soil/reinforcement interaction mode can be either direct shear or pullout. For grid reinforcement, these two different interaction modes yield different interface strengths and deformation parameters. The interface elements above and below reinforcement work as pair elements and the direct shear (the same sign of shear stresses) and pullout (different sign of shear stresses) soil/reinforcement interaction modes, as shown in Fig. 7.52, should be automatically adopted according to their relative shear displacement pattern.

Pullout of reinforcement, especially, the grid reinforcement, from the soil is a truly three-dimensional problem and it can only be approximately modelled in a two-dimensional analysis. It is assumed that the pullout resistance is uniformly distributed over the entire interface areas. Pullout interface shear stiffness consists of stiffness from skin friction resistance, k_{sf}, stiffness from passive bearing resistance, k_{sp}. The total equivalent tangential shear stiffness k_s is the sum of k_{sf} and k_{sp} as follows:

$$k_s = k_{sf} + k_{sp} \tag{7.35}$$

For both direct shear and pullout interaction modes, when the normal stress at the interface is in tension, very small (e.g. 100 kN/m^3) normal and shear stiffness are assigned to allow the opening and slippage at interface.

7.8.5 Simulating the Actual Construction Process

The actual embankment construction is carried out by placing and compacting the fill material layer by layer. Therefore, in finite element analysis, the incremental load should be applied by placement of embankment elements one layer after another. In analyzing the problems, such as embankment on soft ground, the large deformation phenomenon can be considered by updating the node coordinates during the incremental analysis. In this case where considerable deformation of the current construction level is also corrected based on the following assumptions: (a) the original vertical lines are kept at vertical direction and the horizontal lines remained straight; (b) the incremental displacements of the nodes above current construction top surface are linearly interpolated from the displacements of the two end nodes of current construction top surface according to their x-coordinates (horizontal direction).

As shown in Fig. 7.53, node $C(x_c, y_c)$ is above the current construction top surface. The incremental displacements of node $C(\Delta x_c, \Delta y_c)$ is calculated using the incremental displacements at node $A(\Delta x_a, \Delta y_a)$, node $B(\Delta x_b, \Delta y_b)$, and their coordinates of $A(x_a, y_a)$ and $B(x_b, y_b)$ as follows:

$$\Delta x_c = \Delta x_a + \frac{x_c - x_a}{x_b - x_a}(\Delta x_b - \Delta x_a) \tag{7.36}$$

$$\Delta y_c = \Delta y_a + \frac{x_c - x_a}{x_b - x_a}(\Delta y_b - \Delta y_a) \tag{7.37}$$

The operation ensures that the applied fill thickness is the same as the actual value and, thus, the actual construction process is closely simulated.

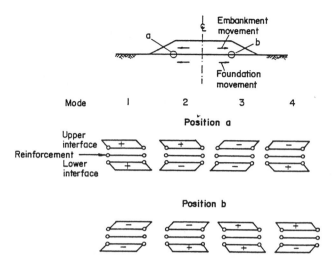

Fig. 7.52   Soil/Reinforcement Interaction Modes for Reinforced Embankment on Soft Ground (Hird and Kwok, 1989)

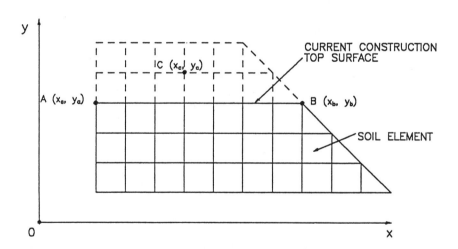

Fig. 7.53   Configuration of the Embankment During Construction Showing the Global Node Coordinates

7.8.6 Variation of Permeability

The settlement rate and pore pressure dissipation rate are mainly controlled by the permeability of soil. Both laboratory and field tests show that the permeability is varied during the loading and consolidation process (Tavenas et al. 1980; 1983). The formula proposed by Taylor (1948) and verified by Tavenas et al. (1983) can be used to represent variation of the permeability of soft clay during the consolidation.

$$k = k_o \cdot 10^{\left[-\frac{(e_o-e)}{c_k}\right]} \tag{7.38}$$

where:
- e_o = the initial void ratio
- e = the void ratio at the condition under consideration
- k = the permeability
- k_o = the initial permeability
- c_k = constant, which is equal to $0.5e_o$ (Tavenas et al. 1983).

The computer program, CRISP-AIT, incorporated above in the modelling techniques, which was developed based on CRISP computer program (Britto and Gunn, 1987).

7.8.7 Modelling the Compaction Operation

The strength and stress/strain behavior of a soil depends largely on the levels of stresses within the soil mass, and the compaction operation can significantly increase these stresses. For reinforced earth structure, especially reinforced wall, the compaction first induce the lateral displacement of the soil. For inextensible reinforcement, the tensile force in the reinforcement can be mobilized by a small lateral displacement, and in return, the tensile force in reinforcement will restrain further lateral movement of the soil and increase the lateral stress. Because most compaction-induced lateral displacement is plastic displacement, the compaction induced lateral stress cannot be totally released after the removal of compaction equipment.

Several investigators have studied the compaction induced stresses especially for earth retaining structures. Duncan and Seed (1986) proposed a useful analytical model for evaluation of peak and residual compaction-induced lateral earth pressure in the compacted soil mass. The model is a bilinear and hysteretic based on the k_o compression test results, and then extended to non-k_o conditions by using an equivalent vertical effective stress. The increment of equivalent vertical stress can be calculated by elastic solution of loading. The application of this model to a single loading/unloading cycle is illustrated in Fig. 7.54, and the horizontal axis is the equivalent vertical stress. The compaction-induced stresses can be calculated according to different loading conditions. The virgin load stress path follows the k_o line, and reloading and unloading follow the k_2 and k_3 lines, respectively. The term k_o is the lateral earth pressure coefficient at-rest condition, while k_2 and k_3 are reloading and unloading lateral earth pressure coefficients, respectively. Practically, k_2 is equal to k_3. The unloading stress path limited by k_1 is lateral earth pressure coefficient for passive failure condition. The overall lateral effective stress (σ_h') at any point is then the sum of the geostatic (σ_{ho}) and compaction-induced (σ_{hc}') stresses.

$$\sigma_h' = \sigma_{ho}' + \sigma_{hc}' \qquad (7.39)$$

In the incremental finite analysis, the general hysteretic model plays two roles: (a) providing a basis for the controlled introduction of compaction-induced stresses; (b) controlling the interaction between geostatic and compaction-induced stress fractions in order to account for hysteretic effects associated with changes in the geostatic stress fractions. The model retains the following characteristics:

a) Unloading from a peak loading condition results in partial relaxation of peak lateral stress as some fraction $(1-k_2/k_o)$ of the peak lateral stress is retained as residual stress.

b) Reloading results in a smaller increase in lateral stress than does virgin loading.

c) Geostatic stresses increase due to the increase of overburden, reloading eventually "overcomes" prior compaction-induced stresses, and further loading corresponds to primary loading.

d) Unloading is subject to a k (passive earth pressure coefficient) type of limiting condition.

e) No loading/unloading cycle to the same initial/final vertical stress results in a decrease in lateral stress unless the decrease is due to the lateral strain in the soil.

Multiple passes of surface compaction equipment continually reintroduced the lateral stresses relaxed by deflection resulting in progressive arrangement of soil particles. In order to approximate this process with a single solution increment, a specific depth, termed softening depth, h, is introduced. All soil elements within softening depth are assigned negligible moduli during the compaction increment. This modelling procedure results in the calculation of displacements at all locations as a result of compaction-induced lateral forces, but no changes in compaction-induced stresses and node forces within the softening depth (Seed and Duncan, 1986). The softening depth is one of the key parameters controlling the compaction effects, especially the lateral wall displacement. Detailed finite element analysis shows that a softening depth of 0.38 m produced reasonable lateral displacements (Schmertmann et al. 1989).

7.9 FINITE ELEMENT RESULTS OF MSE WALL/EMBANKMENT ON SOFT GROUND

The finite element modelling techniques disrobed in the previous section have been used to analyze the MSE wall/embankment on soft ground under strain condition. Two examples are given; one is the AIT MSE wall (Bergado et al. 1991), and another is the reinforced Malaysian test embankment at Muar Flats (MHA, 1989; Chai, 1992). The effect of foundation soil permeability and soil/reinforcement interaction modes (direct shear or pullout) have been investigated.

7.9.1 Analysis of AIT Wall

The dimensions of the AIT MSE wall have been shown in Fig. 7.33 and 7.34. The finite element mesh and the boundary conditions are shown in Fig. 7.55. The bar and beam elements are indicated by a darker solid line. For clarity, the interface elements are not shown in the mesh.

7.9.1.1 Model Parameters

The linear elastic-perfectly plastic model parameters for the topmost 1.0 m thick heavily weathered clay layer and modified Cam clay parameters for soft to medium stiff clay layers are shown in Table 7.3. The parameters were determined based on actual test data (Balasubramaniam et al. 1978; Asakami, 1989). Since there is uncertainty of the permeability of the foundation soil, 3 sets of permeability parameters, namely: high, middle, and low permeabilities (Ahmed, 1977; Bergado, 1990), are indicated also in Table 7.3. The top 2.0 m weathered clay is overconsolidated with an average overconsolidation ratio (OCR) of 5 and the underlying soil layers are slightly overconsolidated with an average OCR of 1.2.

The hyperbolic, non-linear elastic soil model parameters for compacted lateritic fill material (middle section of the embankment) are tabulated in Table 7.4 based on triaxial unconsolidated undrained (UU) test results (Bergado et al. 1988) and followed the technique established by Duncan et al. (1980).

The interface hyperbolic direct shear model parameters were determined from direct shear test results of the fill material (Macatol, 1990). The adopted parameters were: friction angle, ϕ, of 32.5 degrees, cohesion, C, of 60 kPa, shear stiffness number, k_1, of 10500, shear stiffness exponent, n_1, of 0.72, and failure ratio, R_{fl}, of 0.85. The skin friction parameters between reinforcement frictional surface and lateritic soil were determined from test results of Shivashankar (1991) with adhesion of 50 kPa, skin friction angle of 9 degrees. The spacing between the grid reinforcement bearing member was 225 mm and the diameter of the bearing member 5.4 mm. For both direct shear and pullout models, the normal stiffness of the interface was defined as 10_7 for compression case and 10^3 for tension case.

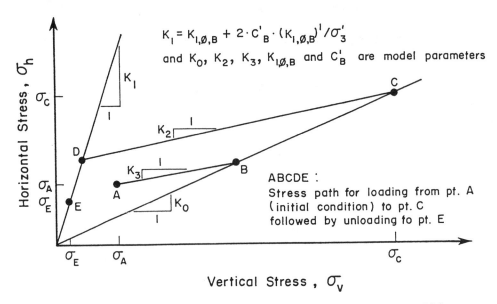

Fig. 7.54   Hysteretic Compaction Model (Duncan and Seed, 1986)

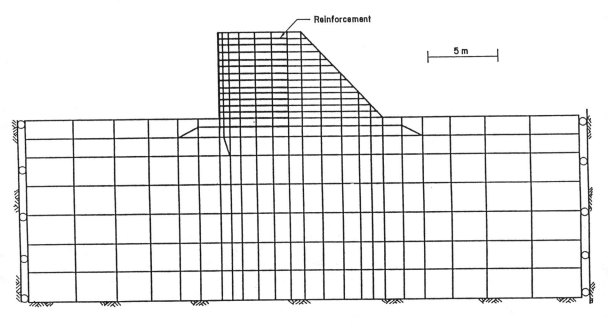

Fig. 7.55   Finite Element Mesh Used for AIT Test Reinforced Wall/Embankment

For welded wire reinforcement including the wall face, the Young's modulus was 2×10^8 kPa and the cross-sectional area of longitudinal bar per meter width was 180 mm^2. For the reinforcement, the yield stress was $6 \times 10_5$. For the wall face, the shear modulus was $8.3 \times 10_7$ kPa, and the moment of inertia of cross-sectional area was 45 mm^4 which was the sum of the moment of inertia of individual bars within 1.0 m width. The shear and normal stiffness for nodal link were assigned as 1.5×10^4 kN/m and 5×10^6 kN/m, respectively.

Table 7.3 Soil Parameter of Soft Bangkok Clay

Parameter	Symbol	Soil Layer				
		1	2	3	4	5
	Depth (m)	0-1	1-2	2-6	6-8	8-12
Kappa	κ		0.04	0.11	0.07	0.04
Lambda	λ		0.18	0.51	0.31	0.18
Slope	M		1.10	0.90	0.95	1.10
Gamma (P'=1 kPa)	Γ		3.00	5.12	4.00	2.90
Poisson's ratio	ν	0.25	0.25	0.30	0.30	0.25
Modulus, (kPa)	E	4000				
Friction angle (degree)	ϕ	29.0				
Cohesion, (kPa)	C	29.0				
Unit weight, (kN/m^3)	γ	17.5	17.5	15.0	16.5	17.5
Horizontal permeability (m/sec), (10^{-8})	High — K_h	69.4	69.4	10.4	10.4	69.4
	Middle — K_h	34.7	34.7	5.2	5.2	34.7
	Low — K_h	13.9	13.9	2.1	2.1	13.9
Vertical permeability (m/sec), (10^{-8})	High — K_v	34.7	34.7	5.2	5.2	34.7
	Middle — K_v	17.4	17.4	2.6	5.2	17.4
	Low — K_v	6.9	6.9	1.0	2.6	6.9

NOTE: High: K_v = 50 times of estimated average test value
Middle: k_v = 25 times of estimated average test value
Low: k_v = 10 times of estimated average test value
Horizontal permeability is always 2 times of the vertical value.

Table 7.4 Hyperbolic Soil Parameters Used for Lateritic Backfill Material

Parameter	Cohesion C, (kPa)	Friction Angle ϕ (deg)	Modulus Number (k)	Modulus Exponent (n)	Failure Ratio (R_f)	Bulk Modulus Number (k_b)	Bulk Modulus Exponent (m)	Unit Weight (kN/m³)
Value	60	32.5	1078	0.24	0.96	1050	0.24	20

The wall/embankment above the ground surface was simulated by 13 incremental layers. For each layer, the gravity force was applied in two increments. Except the selected three sets of foundation soil permeability values, the varied permeability analysis with initial values of middle permeability has also been conducted. A computer program named CRISP-AIT which was developed by modifying the CRISP computer program (Britto and Gunn, 1987), was used for the analyses. All the analyses were consolidation analyses. The presentation of the results of finite element analysis together with field data is made in terms of excess pore pressures, vertical settlements, wall face and subsoil lateral displacements, tension forces in the reinforcements, and the wall/embankment base pressures.

7.9.1.2 Excess Pore Pressure

Figure 7.56 shows the typical calculated excess pore pressure variations with time together with the field data at piezometer point 7 m below the ground surface. From the figure, it can be seen that the excess pore pressures are strongly influenced by the foundation soil permeability and the middle permeability analysis fit the field data better from overall point of view. The varied permeability analysis predicts the excess pore pressure dissipation rate well. This implies that by using constant permeability, the actual excess pore pressure change may not be predicted well.

7.9.1.3 Settlement

Calculated and measured surface settlements under the center point of reinforced mass are compared in Fig. 7.57. The locations of settlement plate are also shown by the key sketch in the figure. It can be seen that the calculated values using middle permeability have remarkable agreement with measured data. The results from varied permeability analysis are somewhat between the results of using middle and low permeabilities but closer to that of using middle permeability. For the sake of clarity of the figures, the settlement of varied permeability analysis was omitted.

7.9.1.4 Lateral Displacement

Figure 7.58 is the comparison of lateral displacement fora both end of construction and 7 months after construction only reach down to 3 m depth because the inclinometer probe could not be inserted into the deformed casing below 3 m depth. At the end of construction, the

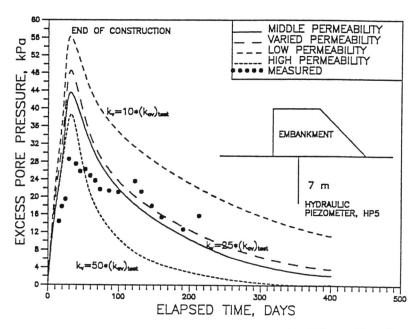

Fig. 7.56 Comparison of Measured and Predicted Excess Pore Pressure

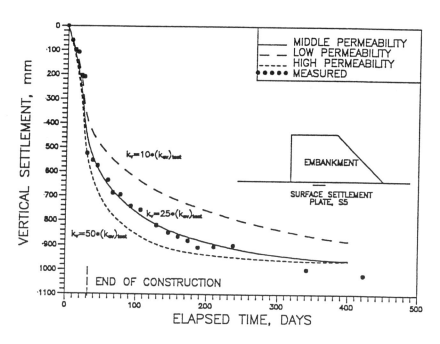

Fig. 7.57 Comparison of Measured and Predicted Surface Settlements

calculated wall face lateral displacements are twice as large as that of measured data. At 7 months after construction, the calculated subsoil and wall face lateral displacements reasonably agreed with the measured values. However, at the top of the wall face, the calculated values are less than the measured ones and the calculated maximum subsoil lateral displacements are still larger than the field data. There are two reasons for the differences obtained between the measured lateral displacements and those calculated by the finite element analyses, namely: (1) the deficiency of the analytical method (Poulos, 1972); and (2) the influence of inclinometer casing stiffness which may result in relative displacements between the soil and the casing.

7.9.1.5 Tension Force in Reinforcement

The maximum tension forces in reinforcements at immediately after construction and 1 year after construction are shown in Fig. 7.59, together with the measured data at immediately after construction. Also shown are the active and at-rest earth pressure lines without considering the cohesion in drawing the active earth pressure line. The data are presented in terms of per meter width and per reinforcement layer (0.45 m). Both measured and calculated data show that at the end of construction, the maximum tension forces in the reinforcements at the top half of the wall are much larger than k_o line. At the middle wall height, the data are close to k_o line. At the bottom of the wall, the data are much higher than k_o line. The maximum tension forces in the reinforcements increased during the foundation soil consolidation process. At the top half of the wall, the maximum reinforcement tension forces at 1 year after construction are twice as large as those immediately after construction due to the large lateral displacement of the wall face.

The tension force distributions along the reinforcements are shown in Fig. 7.60 for the time immediately after construction. It can be seen that the soil/reinforcement interaction mode has strong influence on reinforcement tension force distribution. The pullout interaction mode has weaker interface stiffness and yields large tension force and longer length of the reinforcement in tension in the lower half of the reinforced wall. In the upper half of the wall, the difference is not significant because of the bending effect on reinforced mass caused by the differential settlements of foundation. The difference between direct shear/pullout interaction mode and direct shear interaction mode is not evident because for this particular reinforced earth structure, the soil/reinforcement pullout interaction mode zones are small and located near the wall face and the bottom of the reinforced mass as shown in Fig. 7.61.

7.9.1.6 Wall/Embankment Base Pressure

The base pressure is an important item for design of reinforced wall and embankment on soft ground because it controls the safety factor of bearing capacity of the foundation. Figure 7.62 shows the predicted and measured total earth pressures at the base of the reinforced mass. There are two points that can be made from the predicted total base pressure distribution. Firstly, the predicted earth pressure distribution is more likely a trapezoidal pattern, even though there is a stress concentration under the wall face. Secondly, during the foundation consolidation process, there is a reduction in the total base pressure, and an increase in stress

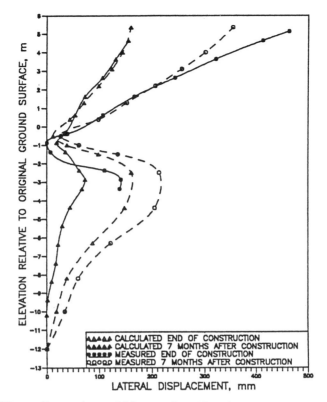

Fig. 7.58 Comparison of Measured and Predicted Lateral Displacement

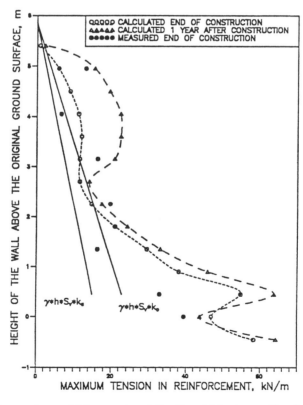

Fig. 7.59 Comparison of Measured and Predicted Maximum Tension in Reinforcement

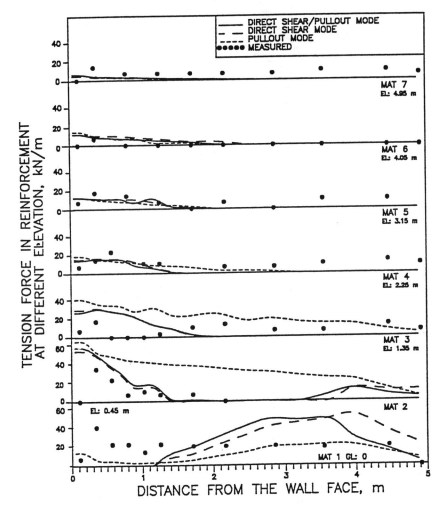

Fig. 7.60 Soil/Reinforcement Interaction Mode on Tension Force in Reinforcement Immediately After Construction

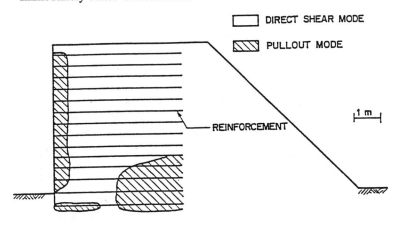

Fig. 7.61 Soil/Reinforcement Pullout and Direct Shear Interaction Mode Zones

concentration under the wall face. This is because during the foundation consolidation process, the reinforced mass sinks into the ground, and the vertical load is distributed into a larger horizontal area. At the same time, the overturning movement of the reinforced mass also increased due to the differential settlement. The overall effect is the increased stress concentration under the wall face and the reduction of the base pressure at other locations.

7.9.1.7 Effect of Properties of Reinforced Mass on Soft Ground

The response of the soft foundation soil under the reinforced wall/embankment load can be classified into two extremes, namely: (1) the same as that of under rigid footing, (2) the same as that under flexible surface loading. Any factor tending to increase the rigidity of the reinforced mass will result in larger settlement under the toe of the reinforced wall and smaller lateral spreading of the foundation soil. Figure 7.63 shows the influence of the different soil/reinforcement interaction modes on the foundation deformation pattern. From the figures, it can be seen that the direct shear interaction mode has a stronger tangent shear stiffness, and the whole reinforced mass deformed more like a rigid body, resulting in larger settlement occurring under the wall face. On the other hand, the pullout interaction mode has a weaker tangent stiffness making the soil under the wall/embankment to squeeze out easier, resulting in less settlement under the wall face, more heave at the free surface, and more settlement at the centerline of the reinforced mass. In return, this kind of foundation settlement pattern will induce more bending effect on reinforced mass, i.e. bottom in tension and top in compression. Reducing the backfill soil strength and reinforcement stiffness results in similar effects on foundation response as reducing the soil/reinforcement interface shear stiffness.

7.9.1.8 Effect of Foundation Properties on Behavior of Reinforced Mass

Generally, more compressible foundation soil means more foundation lateral displacement, more wall face lateral displacement, and more tension force in the reinforcement at the lower part of the reinforced wall would occur. The foundation soil consolidation rate also influences the interaction behavior between the reinforced mass and soft ground. Figure 7.64 shows the comparison of maximum tension force distribution along the wall height at approximately 90 degrees of average foundation soil consolidation condition for the cases of different foundation soil permeabilities. It shows that the higher the permeability, the larger the maximum tension force developed in the lower half of the wall. This is because the reduction of the foundation soil permeability increased the foundation soil lateral spreading and the bending effect on the reinforced mass which is top in compression and bottom in tension.

7.9.2 Analysis of Malaysian Reinforced Embankment (Scheme 6/8)

7.9.2.1 Description of the Embankment

The embankment was located on Muar Flats, 50 km due east of Malacca on the southeast coast of west Malaysia and was constructed with a base width of 88 m and length of 50 m initially to a fill thickness of 3.9 m. Then a 15 m berm was left on both sides and the

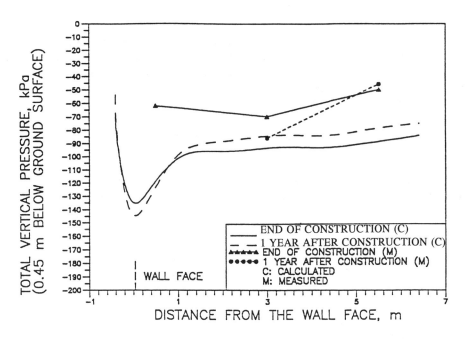

Fig. 7.62   Comparison of Measured and Predicted Base Pressures

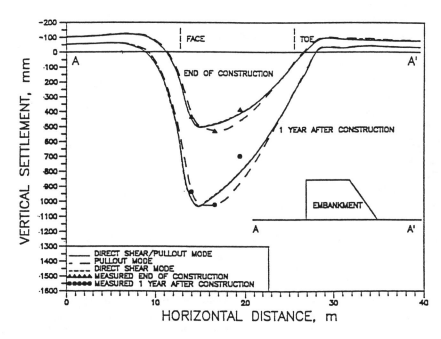

Fig. 7.63   Influence of Different Soil/Reinforcement Interaction Made on Surface Settlement

embankment was constructed to a final fill thickness of 8.5 m. The construction history is shown in Fig. 7.65. Two layers of Tensar SR110 geogrids were laid at the leveled ground surface in a 0.5 m thick sand blanket with 0.15 m vertical spacing between them. The vertical drains (Desol) were installed in a square pattern with 2.0 m spacing to a 20 m depth (MHA, 1989).

The soil profile at the test site consisted of weathered crust at the top 2.0 m which is underlain by about 5 m of very soft silty clay. Below this layer lies a 10 m thick layer of soft clay which in turn is underlain by 0.6 m of peat with high water contents. Then, a thick deposit of medium dense to dense clayey silty sand is found below the peat layer.

The finite element meshes together with the boundary conditions for analyzing the Malaysian reinforced trial embankment is shown in Fig. 7.66. The bar elements representing the reinforcements and interface elements are indicated by a coarse solid line. The two layers of SR110 geogrids are simulated by a single layer of bar elements, because the distance between the two layers is very small, only 0.15 m.

7.9.2.2 Input Parameters

The modified Cam clay model parameters for foundation soil are listed in Table 7.5. The parameters, M, λ, and κ, are obtained directly from test results. The Poisson's ratio for soft and stiff clay layers in 0.25 m, and for weathered clay layer and dense clayey silty sand layer is 0.2 (Balasubramaniam et al. 1989; Magnan, 1989). Based on existing information (e.g. Poulos et al. 1989), two sets of the permeability values are selected for finite element analysis as shown in Table 7.5. The first set of parameters does not consider the effect of the vertical drain. The second set of parameters considers the vertical drain effect as vertical seams which increased the vertical permeability and the vertical permeability in the zone with vertical drain is twice as large as that without vertical drain.

The backfill material is cohesive-frictional soil consisting of decomposed granite with consistency of sandy clay. The hyperbolic soil model parameters for backfill material and sand blanket are given in Table 7.6, which are from test results or selected from the parameters collected by Duncan et al. (1980).

Tensar grid SR110 has the spacing of 150 mm between transverse members and 22.7 mm between longitudinal members. The average cross-section of the transverse member is 5.7 mm in thickness and 16.0 mm in width, and the average cross section of the longitudinal member is 2.1 mm in thickness and 10 mm in width. The constant stiffness of 650 kN/m were used in the analyses for Tensar SR110. For parametric study, steel grid reinforcement with 12.7 mm bar diameter and 152 mm by 225 mm spacing (longitudinal by transverse) is assumed.

The adopted interface hyperbolic direct shear model parameters were: interface frictional angle, ϕ, of 35 degrees; cohesion, C, of zero; shear stiffness number, k_1, of 4800; shear stiffness exponent, n_1, of 0.51; and failure ratio, R_{f1}, of 0.86. The skin friction angle between

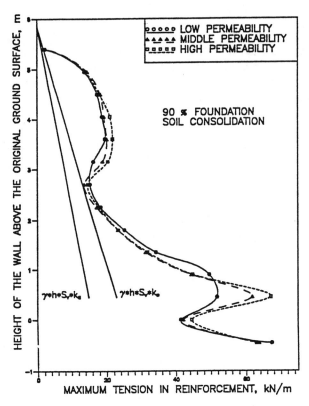

Fig. 7.64 Influence of Foundation Soil Permeability on Maximum Tension in Reinforcement

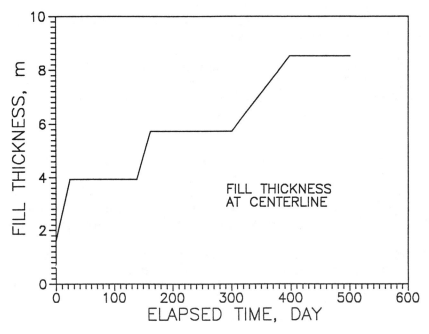

Fig. 7.65 Construction History of a 6 m Malaysian Reinforced Embankment

Tensar grid plane surface and the sand was 10 degrees and adhesion was zero. The maximum relative displacement for mobilizing the peak skin friction was 2 mm.

All the finite element analyses conducted are consolidation analyses. Except for the two sets of selected foundation soil permeability values, varied permeability analyses are also conducted with the initial value of high permeability (set 2 in Table 7.5). The results of the analysis include the response of the soft foundation, the reinforcement tension force, the interface shear stress, and the influence of reinforcement on the foundation soil stress state.

Table 7.5 Soil Parameter of Soft Bangkok Clay

Parameter	Symbol		Soil Layer				
			1	2	3	4	5
	Depth (m)		0-2	2-7	7-12	12-18	18-22
Kappa	κ		0.06	0.10	0.06	0.04	0.03
Lambda	λ		0.35	0.61	0.28	0.22	0.10
Slope	M		1.20	1.07	1.07	1.07	1.20
Gamma (P'=1 kPa)	Γ		4.16	5.50	3.74	3.45	2.16
Poisson's ratio	ν		0.20	0.25	0.25	0.25	0.20
Unit weight, (kN/m³)	γ		15.5	14.5	15.0	15.5	17.0
Horizontal permeability (m/sec), (10^{-8})	SET 1	K_h	2.78	1.40	1.04	0.70	14000
	SET 2	K_h	2.78	1.40	1.04	0.70	14000
Vertical permeability (m/sec), (10^{-8}) Drain/No Drain	SET 1	K_v	1.39	0.70	0.52	0.35	7000
	SET 2	K_v	2.78/ 1.39	1.40/ 0.70	1.04/ 0.52	0.70/ 0.35	14000/ 7000

Drain/No Drain: In the zone installed with vertical drain, the vertical permeability is equal to the horizontal value and two times of that the zone without vertical drain.

7.9.2.3 Response of the Foundation Soil

The typical variation of the excess pore pressure with elapsed time for a piezometer point 4.5 m below ground surface and on the embankment centerline is shown in Fig. 7.67. It can be seen that finite element results, especially the varied permeability analysis results, predict both excess pore pressure build up and dissipation for stage constructed embankment well.

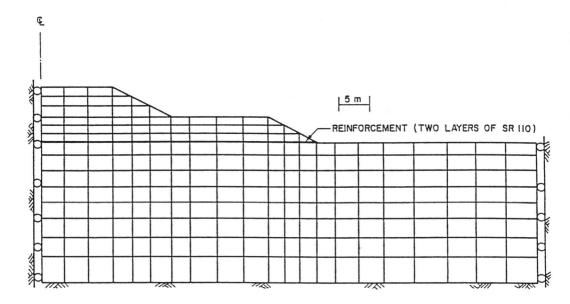

Fig. 7.66   Finite Element Mesh Used for 6 m High Malaysian Reinforced Embankment

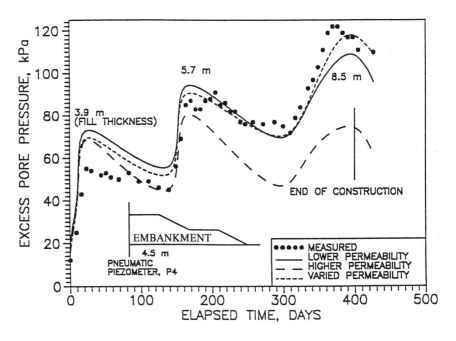

Fig. 7.67   Comparison of Measured and Predicted Excess Pore Pressures

The comparison of surface settlement profiles is shown in Fig. 7.68. An interesting factor is that at the early stage of the embankment construction (3.9 m fill thickness), at the zone near the embankment toe, the finite element analyses yielded large settlement than the center point of the embankment, and the measured data also show this tendency because this zone is high shear stress zone.

Figure 7.69 compared the finite element results and measured lateral displacement profiles at inclinometer position. It shows that the agreement between the calculated and the measured data is good. However, at the early stage of the construction, the calculated values considerably overestimated the lateral displacements. At the end of construction, the calculated values slightly underestimated the lateral displacements. This lateral displacement variation tendency indicates that the consolidation process at a very early stage of the construction is not simulated well by the finite element analysis.

Table 7.6 Hyperbolic Soil Parameters for Backfill Materials and Sand Blanket of Malaysian Embankment

Parameter	Symbol	Backfill	Sand Blanket
Cohesion (kPa)	C	19	0
Friction Angle (°)	ϕ	26	38
Modulus Number	k	320	460
Modulus Exponent	n	0.29	0.50
Failure Ratio	R_f	0.85	0.85
Bulk Modulus Number	k_b	270	392
Bulk Modulus Exponent	m	0.29	0.50
Unit Weight (kN/m³)	γ	20.5	20.5

7.9.2.4 Reinforcement Tension Force

The calculated reinforcement tension force distributions for different fill thickness are shown in Fig. 7.70. It shows that at the early stage, the higher tension force developed near the embankment toe at the location of higher shear stress level zone. Later on, since the berm are placed on both sides of the embankment and also due to consolidation effect, the tension force increased at the embankment center position and decreased under the berm. At the end of the construction, the maximum tension force in each SR110 geogrid is 13 kN/m, equivalent to 2% of axial strain in the reinforcement. For 3 years after construction condition, the maximum tension force in each SR110 geogrid is 20 kN/m or 40 kN/m in two layers of SR110 geogrids. This indicates that during the consolidation process, the maximum tension force in the

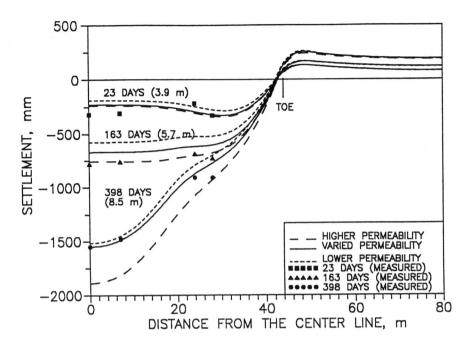

Fig. 7.68   Comparison of Measured and Predicted Surface Settlements

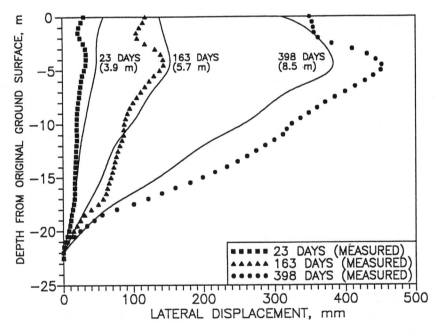

Fig. 7.69   Comparison of Measured and Predicted Lateral Displacements

reinforcement increased.

7.9.2.5 Interface Shear Stresses

Figure 7.71 shows the shear stress distributions at soil/reinforcement upper and lower interfaces at different fill thicknesses. The convention is also shown in the figure by key sketch. It can be seen that for Tensar polymer grid, the signs of shear stress at upper and lower interfaces are the same for most interface areas, i.e. the direct shear interaction mode is applicable for this case. In the zone near the toe of embankment and the intersection point between the berm and the main embankment, because of the free face of the embankment fill, the lateral displacement of the fill is large, and the interface shear stress has negative sign. In other zones, the lateral squeezing of the foundation soil causes the interface shear stresses to have a positive sign. Parametric study has shown that only for high stiffness base reinforcements, such as steel grid, the reinforcement tends to hold both lateral spreading force from fill material and lateral squeeze force from soft ground and pullout soil reinforcement interaction mode controls the interface properties.

7.9.2.6 Influence of Reinforcement on Soft Ground Stress State

Figures 7.72a,b show the effective stress path followed by the soil elements located at 1.0 m and 5.5 m, respectively, below the reinforcement layer and on the selected vertical line. Both of the figures show that the stiff steel grid reinforcement reduced the stress ratio (q/p', where q is the deviator stress and p' is the effective mean stress) about 0.1 to 1.15 for corresponding fill thickness. The steel grid reinforcement yields at the end of construction, which caused certain reduction of foundation confining pressure and increase of shear stress in foundation soil.

7.10 FIELD EVALUATION OF EARTH PRESSURES ON RSE WALLS

Current design procedures for welded wire mesh reinforced soil walls are derived mainly from the results of fully instrumented welded wire walls. These walls consist of welded wire mats placed between successive layers of backfill and bent up at the front to form the face of the wall. One of the recent innovations to the wire mesh reinforced soil wall is the Reinforced Soil Embankment (RSE) system. The wall system uses either flat-faced or shadow panel facing elements, which are made up of precast reinforced concrete. The reinforcing mats of the flat-faced system are connected to the facing by means of steel anchors embedded on the concrete panels. The restraint conditions imposed by the facing panels and the panel/reinforcement connection can influence the value of the lateral earth pressure coefficient, K, for the use in the design.

Reinforced Soil Embankment (RSE) walls use either standard flat face or shadow panel facing elements. The standard flat face panels are made up of precast reinforced concrete and have standard dimensions of 0.76 m (2.5 ft) high by 3.81 (12.5 ft) long by 0.13 m (5 in) thick. Panel anchors that provide connection between the facing and reinforcing elements are made of

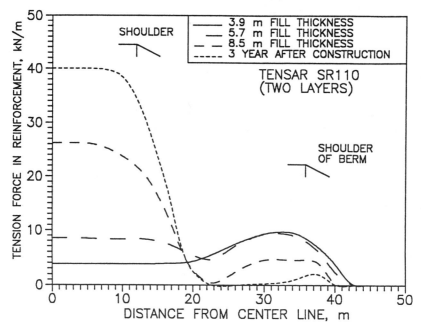

Fig. 7.70   Distribution and Variation of Calculated Tension Forces in Tensar SR110 Reinforcement

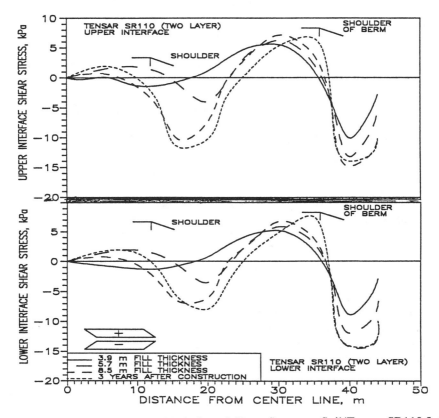

Fig. 7.71   Distribution and Variation of Shear Stress on Soil/Tensar SR110 Interface

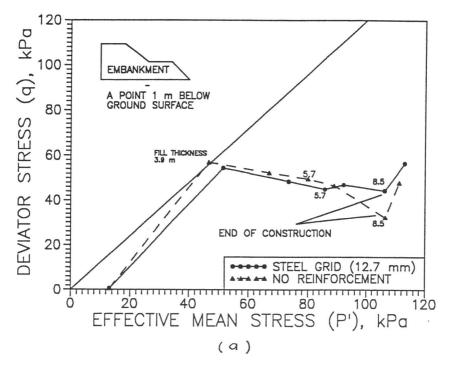

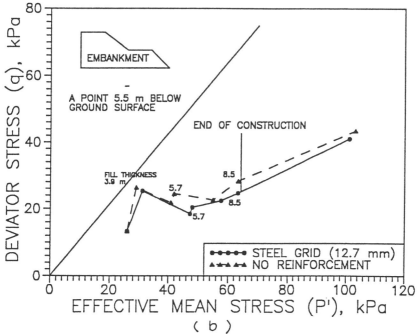

Fig. 7.72 Stress Paths Followed for Both With and Without Steel Grid Reinforcements

a 0.01 m dia (W12) steel wire bent in an elongated loop and embedded in the concrete panel.

The reinforcing mats are made of steel longitudinal wires spaced at 0.15 m (0.5 ft) on centers and welded to transverse wires. The mats are spaced at 0.76 m (2.5 ft) along the vertical direction corresponding to the standard panel height. Standard conditions require two 1.22 m (4 ft) wide reinforcing mats per panel but continuous reinforcements are also available. The length and size of the longitudinal wires and the size and spacing of the transverse wires depend on the stability requirements of the wall. The longitudinal reinforcement wires are attached to anchors that are embedded in the concrete facing panel. There is some flexibility in the connection.

7.10.1 Stability of Wire Mesh Reinforced Walls

As in other types of mechanically stabilized earth (MSE) walls, wire mesh systems such as welded wire and RSE walls must be designed to meet both external and internal stability criteria. External stability is evaluated by considering the reinforced mass as a gravity retaining wall and conventional stability criteria for overturning, sliding, bearing capacity, and deep stability are then applied.

Internal stability evaluation involves provisions for a sufficient safety factor against rupture as well as pullout failure of the mats. For the case of inextensible wire mesh systems, the failure plane is considered to correspond closely to the bilinear failure plane assumption (Mitchell and Christopher, 1990). This differs from extensible reinforcement systems in which the maximum tension line closely conforms to the Coulomb failure plane. The longitudinal members of the reinforcing mats restrain a failure mass from sliding along the potential failure surface. Thus, the embedment length of the mats behind the failure plane must extend far enough behind the potential failure surface to develop sufficient pullout resistance to restrain the potential failure mass (Mitchell and Christopher, 1990).

The maximum tension acting on the failure plane can be calculated by assuming that each longitudinal wire in a mat provides the lateral restraint for an area that extends half the distance to the adjacent wires above, below, and on both sides. Generally, the maximum tension can be expressed as:

$$T_{max} = \sigma_n + S_h + S_v \qquad (7.40)$$

where σ_n is the horizontal stress, S_h is the average horizontal spacing of longitudinal wires, and S_v is the vertical spacing between layers of reinforcement. The horizontal stress at a depth z from the top of the wall is (Christopher et al. 1989):

$$\sigma_h = K(\gamma z + q + \Delta\sigma_v) + \Delta\sigma_h \qquad (7.41)$$

where K is the lateral earth pressure coefficient, γ is the unit weight of backfill material, q is the surcharge load, $\Delta\sigma_v$ is the increase in vertical stress due to concentrated vertical loads, and $\Delta\sigma_h$ is the increment of horizontal stress due to horizontal concentrated surcharges (if any).

The value of lateral earth pressure coefficient, K, differs for each type of soil reinforcement system and depends on the amount of yielding that can take place. Recommended values of K were based mainly on fully instrumented reinforced soil walls. For welded wire walls, Anderson et al. (1987) developed a relationship between the lateral earth pressure coefficient (K) and the wall height based on measurements of a 15.24 m (50 ft) high welded wire wall which reportedly agrees with the earlier findings of Bishop and Anderson (1979) and Anderson and Wong (1989).

7.10.2 RSE Wall Construction, Houston, Texas

The highway expansion program that was initiated by the Texas State Department of Highways and Transportation for the city of Houston involves the construction of nine double-faced Hilfiker RSE walls with segmented, flat-faced concrete panels along State Highway 225 in Harris County, Texas. Since the reinforcement for the double-faced walls was independent for each face (not connected), the project actually involved eighteen separate back to back RSE walls. One of the highest sections of these walls was chosen for study, which is a segment of Wall 17 located at the junction between Highway 225 and Battleground Road and constructed back to back with Wall 18.

The section of the wall is 10.06 m high and consists of fourteen reinforcement layers as shown in Fig. 7.73. The 7.32 m long reinforcement mats overlapped the reinforcements for Wall 18 as much as 0.61 m in some locations. However, since the reinforcement was not continuous along the panel, as shown in Fig. 7.74, the mats were generally staggered and did not physically overlap. The backfill material was non-plastic, poorly graded, fine brown sand having about 11% to 14% finer than the No. 200 sieve and a D_{50} of 0.15 mm. The Standard Proctor maximum dry density was 18 KN/m³ corresponding to an optimum water content of about 11%. The friction angle (ϕ) obtained from direct shear test was about 30°. The backfill material was spread by a bulldozer in 0.23 to 0.30 m compacted lifts. Hand vibrators were used to compact the soil within 0.91 to 1.22 from the wall face while smooth wheel rollers were used for that distance and farther away from the face. Backfill compaction ranged from 90 to 95% of the Standard Proctor densities with the measured placement moisture content averaging the optimum condition. Several boring holes were made at the site of Wall 17 to depths of about 9.0 m. These boring holes revealed the presence of dark brown to gray stiff silty clays. An elevation view of the instrumented section of Wall 17 is shown in Fig. 7.73.

7.10.2.1 Instrumentation Program

A total of eight layers were instrumented with strain gages. For each instrumented layer, sixteen strain gage points were chosen for the reinforcing mats (eight gage points for each mat) and twelve gage points for the panel anchors (Fig. 7.74). Each instrumentation point consists

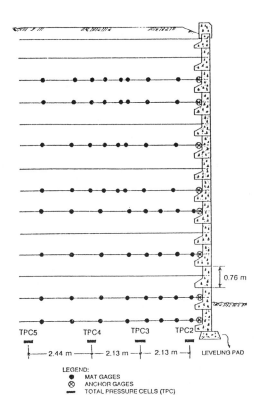

Fig. 7.73 Elevation View of the Instrumented Section of Wall 17

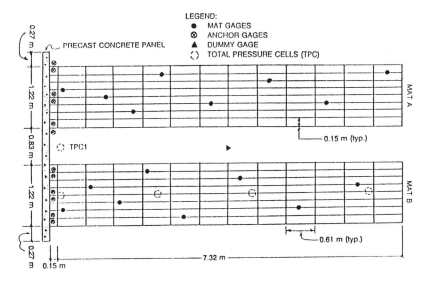

Fig. 7.74 Plan View of the Instrumented Section Showing Strain Gauge and Pressure Cell Locations

of two strain gages which were glued diametrically opposite each other on top and bottom of the longitudinal members. A wiring scheme was developed such that readings could be obtained for both axial tension and bending stresses acting on the wires. Dummy gages for temperature compensation were also embedded on each layer as shown in Fig. 7.74. These gages were attached on the same grade of material and size of longitudinal bars as the reinforcing mats for that layer.

Initial readings were taken at the time the mats were laid in place to indicate the strain reading corresponding to zero tension in the wires. Subsequent readings were then taken as the wall was constructed. The strain induced by the backfill placed over the mat at the time of the readings was obtained from the difference between the initial and the subsequent reading. From the strain, the tension (T) in wire was calculated using the following equation:

$$T = E \cdot \epsilon \cdot A \tag{7.42}$$

where E is the modulus of elasticity of steel, ϵ is the axial strain, and A is the cross-sectional area of the longitudinal wire.

To measure the vertical stress at the base of the wall for each layer of the backfill, five 0.23 m diameter SINCO pneumatic total pressure cells were installed at the same level as the leveling pad. Four of these cells were aligned at different distances behind the facing and below the reinforcing mats (Fig. 7.73) while the other one was installed at 0.15 m behind the wall face but below the unreinforced zone as shown in Fig. 7.74.

7.10.2.2 Results

The tension in the longitudinal wires was measured as the wall was constructed. Figure 7.75 shows a typical plot of the measured tension as a function of the distance from the face of the wall for different heights of the backfill above the mat. Similar trends were observed for the other mats.

The vertical pressure at the base of the wall was measured by total pressure cells and the results are plotted as a function of the distance from the face of the wall in Fig. 7.76. The overburden pressure (γH), computed using the average measured unit weight, γ, of 19 kN/m$_3$, is also plotted on the figure for a wall height of 9.3 m. The pressure cells closest to the face (TPC1 and TPC2) yielded the lowest measured values which are much less than would be produced by the weight of overburden. Both pressure cells (TPC1 and TPC2) measured approximately the same magnitudes of vertical base pressure.

The low vertical pressure measured at the face of the wall suggested that the reinforced mass did not behave like a gravity retaining wall subjected to lateral overturning forces. This can be partially explained because the wall was constructed back to back with another RSE wall, eliminating much of the active thrust. Similar base pressure distributions have been reported in the literature for instrumented reinforced soil walls on foundations that were somewhat more compressible than that of Wall 17 (Murray and Farrar, 1990; Bergado et al. 1991). The

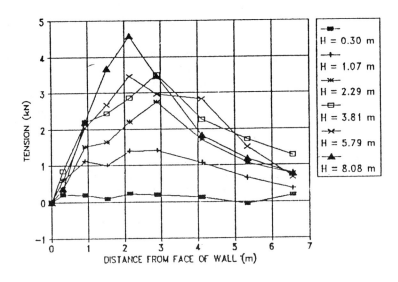

NOTE: (1) Data for Mat 2B located at 9.30 m below the final top elevation of Wall 17
(2) H is the height of fill above the mat

Fig. 7.75 Typical Plot of Tension as a Function of Distance from the Face of the Wall

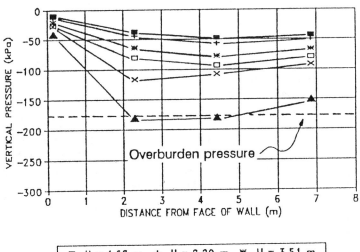

NOTE: (1) Overburden pressure is for height of fill of 9.3 m above the cells
(2) Vertical pressure closest to the wall face is for TPC2

Fig. 7.76 Measured Vertical Pressure Distribution at the Base of the Wall

behavior of these other walls was attributed to the influence of backfill settlement in association with interface friction on the facing.

The maximum tension was obtained for each mat at different heights of the wall and the corresponding maximum K values were calculated from:

$$K = \frac{T_{max}}{\gamma z S_h S_v} \qquad (7.43)$$

Equation 7.43 only includes the weight of the compacted backfill above the mat. This method of back-calculating K has been used by Anderson et al. (1987) to develop an envelope of the lateral earth pressure coefficient, K, for a welded wire retaining wall.

Figure 7.77 shows the back-calculated values of K as a function of the wall height. Also shown in the Figure is the envelope of K values measured by Anderson et al. (1987) for a 15.24 m high welded wire wall in Seattle, Washington. Except for the top 1.22 m of the wall, the K values for the RSE wall are somewhat lower than for the Seattle wall and the measured values for other welded wire walls (Anderson and Wong, 1989). The backfill material that was used in the Seattle welded wire wall was also poorly graded sand but has a higher D_{50} and slightly higher friction angle of 0.70 mm and 31°. The high K values near the top of the RSE wall and welded wire walls were due to compaction stresses. It should be acceptable to design the reinforcement near the top of the wall using a K value that is less than the K envelope.

It might seem that RSE walls may have more rigidity due to the reinforced concrete facing and, therefore, may impose higher lateral earth pressures than welded wire walls. However, the yield conditions of the reinforcement (flexibility of the wall system) will be a function of the pullout resistance of the reinforcing elements and the stiffness of the connection between the facing panels and the welded wire mats. The length of mats used in this study was 7.32 m, in contrast to the 13.41 m long mats used in the Seattle welded wire wall. The maximum tension line for the Seattle wall occurred within 1.22 to 3.05 m from the face and the maximum tension line obtained in this study was similarly in the vicinity of 1.52 to 3.05 m from the face (Sampaco et al. 1992). Thus, the ratio of the available embedment lengths to height of the wall for the Seattle welded wire wall were much greater than those of the RSE wall described in this paper. Furthermore, the transverse wires on the Seattle wall were spaced 0.23 m on centers and for this study the spacing was 0.61 m. Considering the transverse wire spacing and wire size, as well as the soil type and the embedment length, the pullout resistance of the Seattle wall was much higher. It seems, therefore, that the yield conditions of the reinforcing mats (flexibility of the system) will influence the resulting lateral earth pressure coefficient.

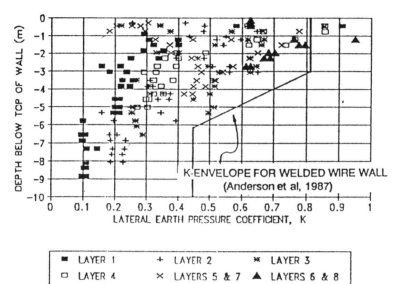

Fig. 7.77   Back-Calculated K Values Plotted as a Function of the Height of Wall

7.11 PERFORMANCE OF GEOTEXTILE REINFORCED EMBANKMENT ON SOFT BANGKOK CLAY

In order to study the improvement of embankment stability using geotextile reinforcement, two test embankments on soft Bangkok clay were constructed to failure. One test embankment was reinforced with geotextile as base reinforcement. The reinforcement consists of one layer of high-strength, woven-nonwoven geotextile placed directly on the natural ground surface at the bottom of the embankment fill. For comparison, another full-scale, unreinforced embankment using the same fill material as that of the reinforced one was also constructed to failure at the adjacent site. The details of these test embankments have been described by Bergado et al. (1994).

7.11.1 Test Location and Soil Profile

The test embankments were located on the AIT campus. Layout of the test embankments is given in Fig. 7.78, including the canal excavation. The locations of field vane tests and boreholes are also shown. The general soil profile and soil properties of the test site are presented in Fig. 7.79. The uppermost 12 m can be divided into 4 layers. The weathered crust consisting of heavily overconsolidated reddish-brown clay forms the uppermost 2 m. This layer is underlain by a soft, grayish clay down to about 8 m depth. The medium stiff clay with silt seams and fine sand lenses is found at the depth of 8 to 10.5 m depth. Below this layer is the stiff clay layer. The undrained shear strength profile measured by field vane shear tests is also presented in Fig. 7.79.

7.11.2 Instrumentation Program

Foundation instrumentation of both embankments is given in Fig. 7.80. The instruments consisted of settlement gages, piezometers, inclinometer, and earth pressure cells. The instrumentation program for the geotextile consisted of wire extensometers, Glotzl extensometers, strain gages, and dog bone load cells. The side view and layout of geotextile instrumentation are shown in Fig. 7.81.

7.11.3 Construction of Embankment to Failure

The layout of the test embankments is presented in Fig. 7.78, including the canal excavation. Before construction of the embankments, the ground surface was excavated to about 20 cm to 25 cm depth and levelled. The canal of 2 m deep and 7.5 m wide at the bottom was excavated along the pre-determined failure side of all embankments. The ground levelling and canal excavation were carried out in 10 days and finished on January 8, 1993.

The first layer of control embankment was placed on January 8, 1993. The embankment was constructed in layers with compaction lift thickness of about 33 cm. The main equipment used for embankment construction consisted of one D4 bulldozer, one backhoe, and one vibrating roller. Compaction at the area nearby the instruments were performed by a Walker

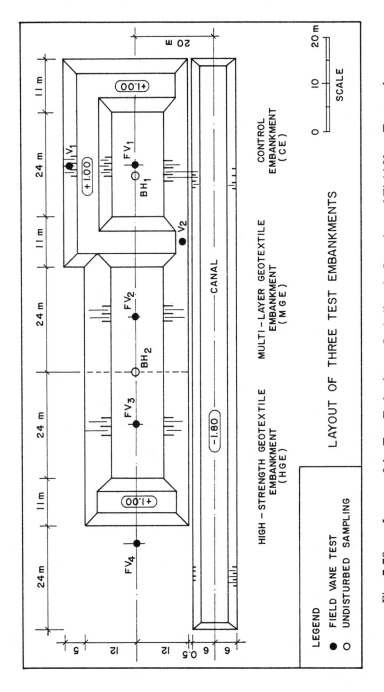

Fig. 7.78 Layout of the Test Embankment Including the Locations of Field Vane Tests and Boreholes

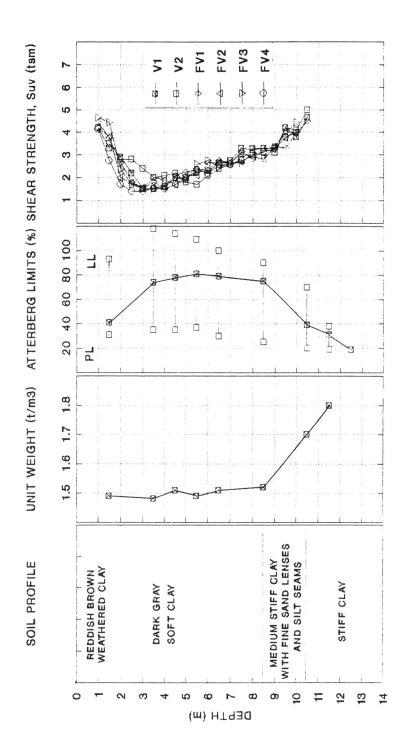

Fig. 7.79 Soil Profile and Soil Properties at the Site

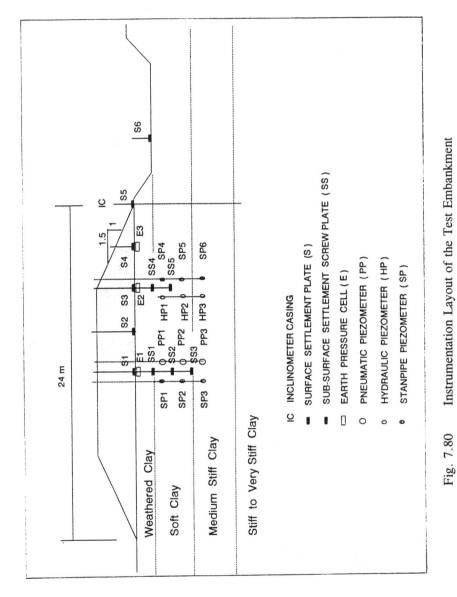

Fig. 7.80   Instrumentation Layout of the Test Embankment

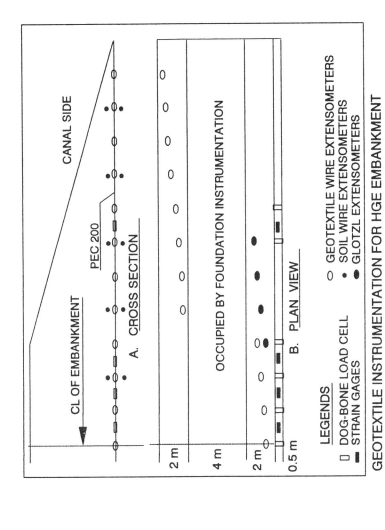

Fig. 7.81   Geotextile Instrumentation of HGE Embankment

hand compactor. Density and moisture content of each layer were controlled by field nuclear tests and sand cone tests. The dry density of 1.7 t/m^3 and moisture content of 9% were maintained. The embankment reached the height of 3 m without any signals of significant deformations. The height of 4 m was reached in the afternoon of February 4, 1993. In the early morning of February 5, 1993, a crack of about 5 mm wide and 3 m long was observed near the settlement plate S6 in the canal. The width of the crack opened quickly to about 1 cm wide, and the embankment completely failed. The cross-section of the embankment just after failure is given in Fig. 7.82.

Two months after the failure of the control embankment, two reinforced embankments were constructed adjacent to the unreinforced one. One embankment was reinforced by 4 layers of low-strength, non-woven geotextile (MGE). The other was reinforced by one layer of high-strength, woven-nonwoven geotextile (HGE). Both embankments were built at the same time and the same rate of filling. Construction procedure and quality control were kept to be the same as that of the control embankment. At 3 P.M. of May 28, 1993, when the two embankments reached 3.75 m high, one small crack of about 5 mm width was observed near the surface settlement plate S_6 in the canal. The crack developed into two larger cracks on the next day. Construction of embankments were continued to reach 4.2 m high on May 30, 1993. Failure of the MGE embankment and induced failure of HGE embankment occurred during the night of May 30, 1993. The cross-section of the HGE embankment measured on May 31, 1993 is given in Fig. 7.82 together with the failure section of the control embankment. After this event, all instruments of HGE embankment were still working except the inclinometer, strain gages, and dog-bone load cells. The measured data indicated that the geotextile in HGE embankment was not yet broken. The HGE embankment construction was then continued starting on June 19, 1993. The HGE embankment reached 6 m height on June 30, 1993 and failed during the night of that day. The cross-section of the embankment after failure are presented in Fig. 7.83.

It is noted that there were 35 rainy days during construction of MGE and HGE embankments. Consequently, the average field moisture content of MGE and HGE measured at the end of construction was about 13%, and was about 4% higher than that of the control embankment, resulting in higher surcharge loading.

7.11.4 Measured Data

The main results obtained during construction of the CE and HGE embankments are given in Figs. 7.84 to 7.90. The measured results are summarized as follows.

Lateral displacement Lateral displacements of two embankments are plotted together in Fig. 7.84. Up to the height of 3 m, the displacements in both embankments are nearly the same, and the maximum displacement was observed at the depth of 3 m. The maximum horizontal displacement of CE embankment at the day before failure on February 4, 1993 is 17 cm, and occurred at the ground surface. The displacement of ground surface of HGE embankment at 4.2 m height is 6 cm smaller that of CE embankment on the day before failure (H = 4m). However, the maximum displacement of HGE embankment still occurred at the depth of 2.5 m with the

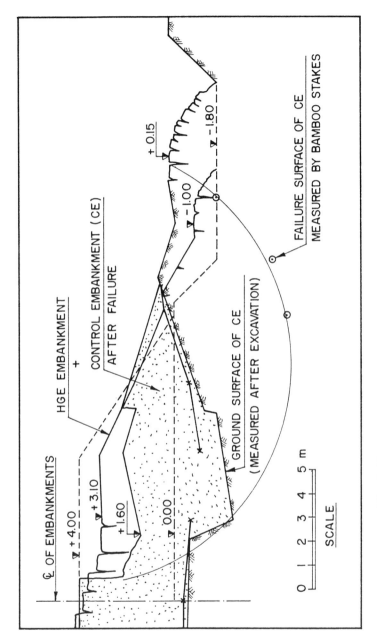

Fig. 7.82 Cross Section of the CE Embankment After Failure and Induced Failure of HGE Embankment

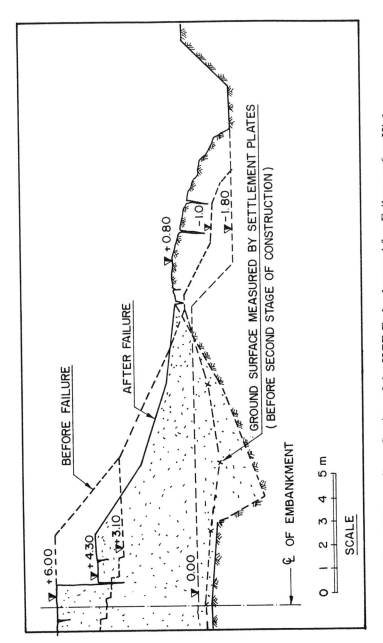

Fig. 7.83　Cross Section of the HGE Embankment After Failure at 6 m High

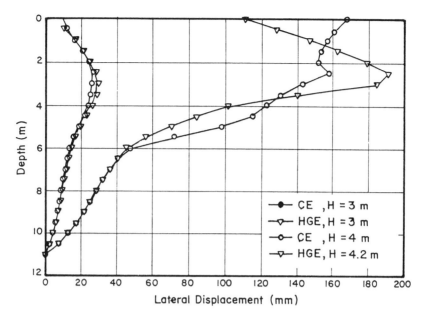

Fig. 7.84　　Lateral Displacement of CE and HGE Embankments

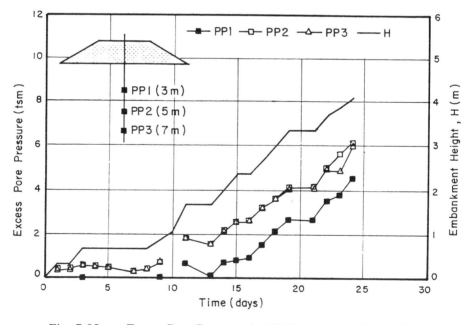

Fig. 7.85　　Excess Pore Pressures in CE Embankment Foundation

magnitude of 19 cm, and was about 3 cm larger than that of CE embankment. This event can be explained by (a) the vertical load acting on the ground surface of HGE embankment was larger than that of CE embankment because of the extra 0.2 m height and the increase of total unit weight due to rainfall during construction of HGE, and (b) the reinforcement has no significant effects on the lateral displacement to the depth of the soft clay layer.

Excess pore pressures The excess pore pressures measured at the depth of 3 m, 5 m, and 7 m of CE and HGE embankment foundations are plotted versus the embankment height in Figs. 7.85 and 7.86, respectively. The excess pore pressure in HGE embankment foundation were larger than that of CE embankment at the same embankment height, e.g, at the height of 4 m the excess pore pressures of 6 tsm and 6.5 tsm were observed in CE and HGE embankment foundations, respectively. This can be explained by the effects of rainfall during HGE embankment construction.

Geotextile displacements The displacements of geotextile measured by wire extensometers are presented in Figs. 7.87 and 7.88. At 4 m high, the maximum displacement of geotextile at the location near the toe of the embankment is about 10 cm, which corresponds to lateral displacement of ground surface measured by inclinometer at 4 m high (Fig. 7.84). Before the failure at 6 m high, the maximum displacement of geotextile was about 60 cm.

Strain measurement Strain in geotextile measured by wire extensometers, Glotzl extensometers, and strain gages is plotted together in Figs. 7.89 and 7.90. There were no significant strains in geotextile at the embankment height of lower than 3.5 m. The strain of about 2% was obtained at the height of 4.2 m. After the induced failure, the strain in geotextile was about 8%. After increase of the embankment height, the maximum strain of 12% was measured before embankment failure at 6 m high. The distribution of strains in geotextile does not correspond to the expected one with the classical design method. In fact, the maximum strain did not occur at the center of embankment and some compression was measured under the slope. Similar behavior was also reported by Delmas et al. (1992). The validity of strain measurements can be found by the comparable results obtained from wire extensometers, Glotzl extensometers, and strain gages, as shown in Figs. 7.89 and 7.90.

7.11.5 Slope Stability Analyses

Limit equilibrium analyses with circular slip surfaces have been commonly used in conventional design of geotextile reinforced embankments. In this type of analysis, two main assumptions have to be made: (a) the mobilized tensile force in geotextile at limit state, and (b) the orientation of the reinforcement which is often presented by the inclination factor, I_f. The mobilized tensile force may be taken from 2% to 10% (Bonaparte and Christopher, 1987). The values of I_f range from 0 to 1 corresponding to horizontal and tangential direction of the reinforcement, respectively. The tensile force has been considered to act horizontally by some investigators (Fowler, 1982; Jewell, 1982; Ingold, 1983; Milligan and La Rochelle, 1984), tangentially to the failure surface by the others (Binquet and Lee, 1975; Quast, 1983; Delmas et al. 1992) and between the above two directions (Huisman, 1987).

The actual failure surfaces of the test embankments presented in this study can be fitted well by the circular surfaces, as shown in Figs. 7.82 and 7.83. Hence, Bishop's simplified method using SB-SLOPE and STABL6 software is used for stability analyses of these unreinforced and reinforced embankments, respectively.

7.11.5.1 Shear Strength of Backfill Material

The locally silty sand was used as backfill material in these test embankments. The shear strength parameters of backfill material used in this analysis were determined by large direct shear tests. The values of $\phi = 30°$ and c = 15 kPa which correspond to the moisture content of 9% were applied for the case of the control embankment. In the case of the reinforced embankment (HGE), the measured moisture content at the field at the end of construction was 13% due to the rainfall, as mentioned in previous section. Corresponding to this moisture content, the values of $\phi = 30°$ and c = 10 kPa were obtained.

7.11.5.2 Results of Analyses

The results of back analyses indicated that in the case of the control embankment, using the field vane shear strength with the correction factor $\mu = 0.8$ for the soft clay layer, the factor of safety (FS) equal to 1 was obtained. In the case of reinforced embankment (HGE), the maximum strain measured in geotextile at 4.2 m high (before the first failure) was 3.5% which corresponds to the mobilized tensile force of 60 kN/m (Fig. 7.91). Using this value of tensile force, the FS of 1 and 1.08 were obtained for the case of the inclination factor, I_f, of 0 and 1, respectively. The results implied that the reinforced embankment was in the stable equilibrium state during mobilization process of tensile strain in the geotextile. For comparison, taking the same properties of backfill sand as used in HGE embankment, the factor of safety of unreinforced embankment must be 0.94 and 1.0 corresponding to the height of 4.2 m and 3.7 m, respectively. The results indicated that at low strain level of less than 3.5%, the reinforcement increased about 10% of factor of safety or 0.5 m height of the unreinforced embankment.

At the height of 6 m of the HGE embankment, using the tensile force of 200 kN/m which corresponds to the measured strain of 13% together with the residual strength of foundation soils, the calculated factor of safety were 1.05 for $I_f = 1$. The factor of safety of 0.65 were obtained for the equivalent embankment without reinforcements. Thus, the geotextile increased 60% the factor of safety or at least 2 m in the ultimate height of the unreinforced embankment.

7.12 FULL-SCALE TEST EMBANKMENT ON SOFT BANGKOK CLAY WITH TENAX GEOGRID REINFORCEMENT

A full-scale and fully instrumented test embankment with polymer Tenax TT 201 SAMP geogrid reinforcement was built to 6 m high on soft Bangkok clay at the AIT campus. The plan and section views of the test embankment together with the locations of the instrumentations are shown in Fig. 7.92. The pneumatic piezometers, inclinometers, and stains in the geogrids were

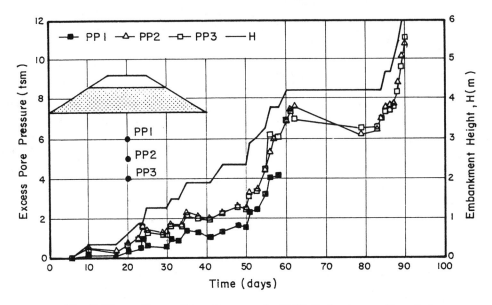

Fig. 7.86   Excess Pore Pressures in HGE Embankment Foundation

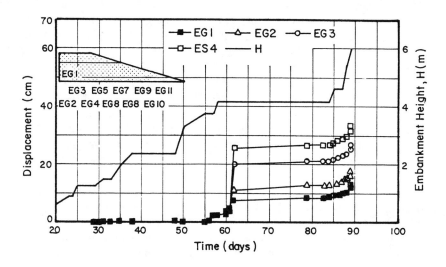

Fig. 7.87   Displacement of Geotextile Measured by Wire Extensometers EG1 to EG4

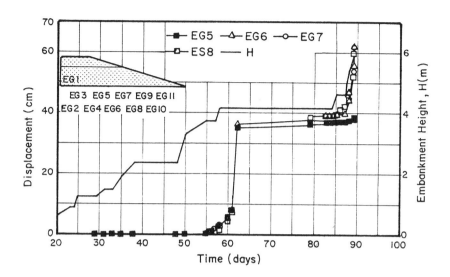

Fig. 7.88 Displacement of Geotextile Measured by Wire Extensometer EG5 to EG8

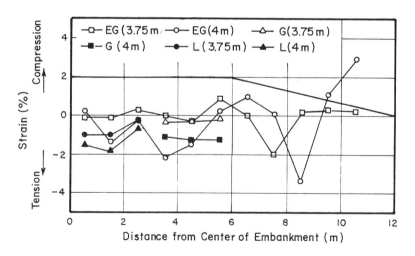

Fig. 7.89 Strains in Geotextile at Embankment Heights of 3.75 m and 4 m

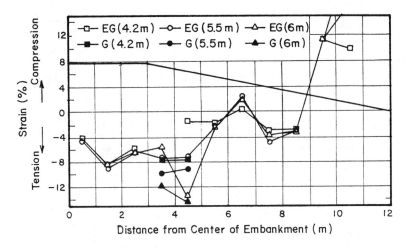

Fig. 7.90   Strains in Geotextile at Embankment Heights of 4.2 m and 6 m

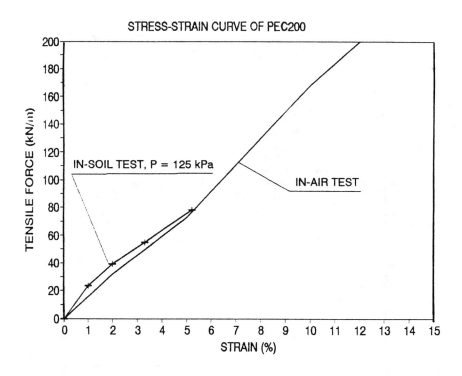

Fig. 7.91   Stress-Strain Curve of PEC200

monitored using standard portable readout boxes. Settlement plates and settlement markers were located and monitored using standard surveying techniques. Fluctuations in water table and dummy piezometers were measured using a galvanometer electric wire system.

7.12.1 Excess Pore Pressure

Figure 7.93 shows the typical relationship between increasing embankment height and the resulting excess pore water pressures measured by pneumatic piezometer at 6.0 m depth (P2) beneath the embankment. All the piezometers demonstrated an increase in the excess pore pressures during the full period of construction which was developed due to the increase in the surcharge load. Both the hydraulic and pneumatic piezometers showed the same response. Two of the pneumatic piezometers located at P1 and P3 ceased to function after construction, probably due to the disconnection of the tubing. Consequently, hydraulic piezometers were relied upon for the subsequent readings after construction. It is also noted that both total and excess pore water pressures showed a tendency to decrease after construction indicating the subsoil consolidation. Vertical stress increases due to embankment loading in all directions are also incorporated in Fig. 7.93. The methods of Gray (1936) and Poulos and Davis (1974) were adopted to determine the stress increase. The former assumed a semi-infinite layer while the latter assumed a finite layer. The concept was based on a uniformly loaded rectangular area. Both methods also assumed that the subsoil was elastic, homogeneous, and isotropic.

7.12.2 Settlements

Surface and subsurface settlement curves followed classical behavior, with an initially high rate of settlement during construction, slowing down thereafter. About 200 days after the end of construction, the rate of settlement considerably decreased. Correspondingly, about 70% degree of consolidation was attained. The surface settlement indicated by S_1 and S_6 were observed to be almost identical at about 250 mm at the end of construction and continue to increase thereafter. Figure 7.94 showed that at the end of 270 days, the magnitude of surface settlements in all locations is almost the same, signifying near uniform distribution of vertical stress. The last monitored reading showed maximum value at S_3 and S_4, both the center followed by S_5, S_2, and S_1 at near center and at front with the lowest value at S_6 near the back slope.

7.12.3 Lateral Displacement

The distribution of the horizontal movements at the face and at the back, both in the embankment and the subsoil, were monitored by means of a biaxial inclinometer. The inclinometer showed that the lateral movement in the subsoil occurred mainly between 3 to 4 m below the ground surface. On the other hand, the maximum lateral movement at the face wall occurred at the top of test embankment. The recorded lateral movement of the soft clay subsoil at the end of construction is about 50 mm, which is approximately 35% of the maximum recorded value of 160 mm after 223 days. Inclinometer 2, which is located at the back slope, showed outward movement with smaller magnitude than the front. The direction of the lateral

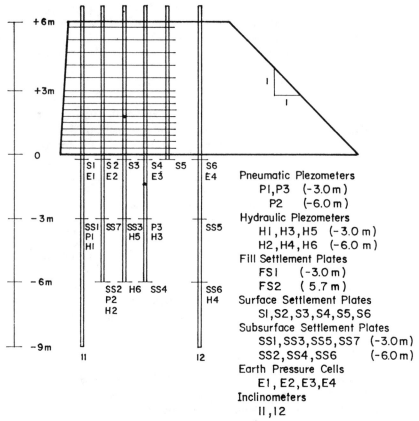
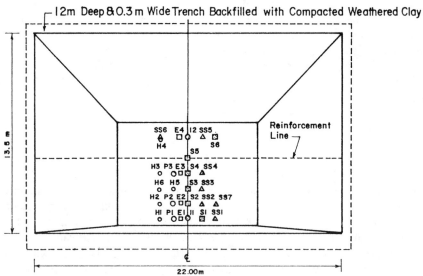

Fig. 7.92   Cross Section and Instrumentation Layout of Test Embankment

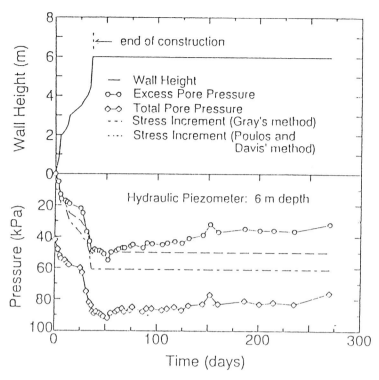

Fig. 7.93   Relationship Between Increasing Embankment Height and the Resulting Pore Water Pressure

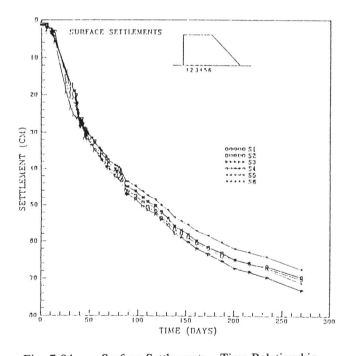

Fig. 7.94   Surface Settlements - Time Relationship

movement of the embankment and subsoil at the back slope is opposite to that of the front wall.

7.12.4 Vertical Soil Pressure

To determine the vertical pressure distribution at the base of the embankment, four pneumatic pressure cells (SINCO) were installed. The measured distribution of the vertical stress at the base, and therefore on the geogrid laid at the base, is slightly nonlinear. This variations of base pressure is due to the subsoil compression beneath the embankment. The maximum value recorded at the end of construction occurred at the front and decreased farther from the wall face with the lowest pressure measured at a distance of about 6.5 m away from the wall face. The uniform distribution of vertical stresses is reflected by the uniformity of measured settlements (see Fig. 7.95).

7.12.5 Geogrid Strain

Strains in the geogrid reinforcement were measured at seven different elevations, each instrumented fully with 10 resistance strain gages (Type YFLA-5). This distribution of strains along the geogrid length after construction at different elevations is presented in Fig. 7.95 together with measured settlements. The values plotted were chosen from different periods of time with noticeable change in magnitude, including the maximum recorded value. The tensile forces were obtained using the calibration curve derived from laboratory pullout tests at different applied normal pressures. What is worth noting is that all the observed tensile strains are in the range of 0.4% to 1.1%, corresponding to 0.84 to 2.3 kN/m load in the geogrids. Comparing this load to the maximum tensile strength of geogrid, which is 55 kN/m, the grids were only loaded to between 1.5% and 4.2% of the ultimate strength.

7.12.6 Maximum Tension Line

Previous research on reinforced embankments supported on good foundation showed that maximum tension line can either be defined by Coulomb/Rankine type linear failure plane (Juran and Christopher, 1989) or a bilinear/logarithmic spiral failure plane (Schlosser and De Buhan, 1990), depending on the stiffness of the reinforcement. At this stage of development, whereby subsoil and embankment movement is still going on, and thereby the stresses in the geogrid would still change, it is premature to evaluate the potential failure plane. Nevertheless, within the considered time frame, it was demonstrated that neither of the aforementioned criteria were observed (see Fig. 7.95).

7.12.7 Comparison of Performances of Welded Wire and Polymer Tenax Geogrid Test Embankment

The welded wire reinforced test embankment was previously constructed on the AIT campus (Bergado et al. 1990). Both embankments have similar dimensions and similar backfill soil consisting of weathered Bangkok clay. The main points of comparison are focused on the different behavior of the test embankment such as settlement, lateral movements, vertical base

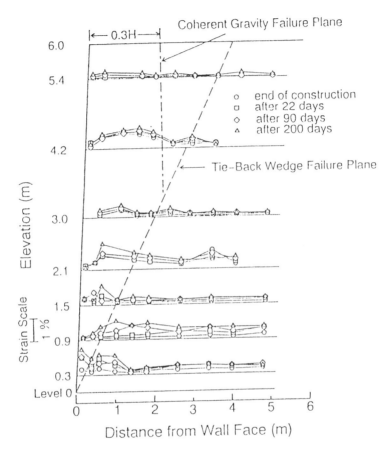

Fig. 7.95 Variation of Strain with Distance from Wall Face at Different Elevations with Surface Settlement Profile

pressures, and tensile forces in the reinforcements. Table 7.7 shows the comparison of embankment behavior at 270 days after the beginning of construction. Similarity of behavior was observed. Differences were observed in the lateral movements and vertical pressures due to slight differences in the construction procedures and techniques.

7.13 PRESENT AND FUTURE OUTLOOK OF MSE SYSTEM

At present, MSE systems are widely used in civil engineering construction due to its competitive costs and flexible construction methods. In the case of conventional wall construction, besides the costly construction materials, the foundation excavation is unavoidable because of the embedment requirement. The excavation is critical when the wall need to be constructed in a crowded city area where any excavation will influence the surrounding structures and transportation system. However, the MSE system can be constructed with minimum excavation and without large equipment. Figures 7.96a,b show a MSE wall during the construction, which is for an overpass bridge approach in Manila in 1991. As can be seen in the figure, during the construction the traffic is not interrupted. It also needs to be noted that comparing with other methods, gravity retaining wall or bridge system, the cost of MSE is as much as one-half of the alternatives.

Table 7.7   Comparative Results on Surface Settlements, Lateral Deformation and Vertical Pressure of Welded Wire Wall and Polymer Geogrid Reinforced Embankments

PARAMETERS	LOCATION	WELDED WIRE	POLYMER GEOGRID
Lateral Deformation (mm)	Top of Embank.	435	360
	3 m below g.s.	135	160
Surface Settlements (mm)	Front	850	650
	Center	700	750
Vertical Pressure (kPa)	Front	1	50
	Center	82	12

Note: Results are taken at the end of about 270 days since beginning of construction

The expansion of urban areas induced the necessity of land reclamation. Sometimes, the reclaimed area is too soft to withstand any loading. In this case, the reinforcements, such as geogrid, geotextiles, and locally available bamboo materials can be efficiently used as base reinforcement. Then, further combination with below ground improvement techniques, such as preloading, vertical drain, etc., can be employed. Figure 7.97 shows a typical example of using bamboo grids to stabilize the slurry deposit under preloading surcharge. The slurry deposit is

Fig. 7.96　Mechanically stabilized earth (MSE) construction in Manila

Fig. 7.97   Bamboo with Geotextile Reinforcement in Laem Chabang Deep Sea Port

part of the hydraulic fill in the reclamation of Laem Chabang Deep Sea Port in Thailand.

Construction activities are intense and vigorous in Southeast Asian countries compared to other regions. Therefore, the geotechnical activities will continue to prosper in the region, and the application of the MSE system will increase in road, airport, and other civil engineering construction projects.

7.14 CONCLUSION

For the last two decades, extensive research and investigations have been carried out to understand the behavior and mechanisms of the MSE system as well as the new technologies of reinforced earth construction. The results lead to the utilization of a variety of reinforcements and backfill materials. Among the numerous schemes, using grid reinforcements with locally available and poor-quality backfill materials has engineering and economic advantages.

Designing an MSE system involves evaluation of required and available resistance force from the reinforcement. The required force is determined from stability analysis of the structure, and the available force is controlled by the soil/reinforcement interaction properties. The direct shear and pullout tests are two commonly used method to investigate the interface behavior between the reinforcement and soil. These two fundamentally different test method simulate two different interface mechanisms, direct shear and pullout. For grid reinforcement, the interface direct shear resistance consists of two components: shear between soil and plane surface of reinforcement and shear between soil to soil at grid opening. The pullout resistance also has two parts, namely: friction resistance from grid longitudinal members and passive bearing resistance from grid transverse members. Several prediction equations are available for estimating the grid pullout resistance. One such prediction has been proposed in this book.

Large amount laboratory and field pullout test results confirmed that the cheaper, locally available, low-quality, cohesive-frictional soils can be effectively used as MSE backfill materials providing compacted at or dry side of optimum water content to over 90% of standard Proctor compaction. The results of the full-scale test MSE wall/embankment at the AIT campus verified the applicability of the MSE system on soft ground, and the information obtained can be used to modify the design guidelines.

Numerical methods, such as the finite element method, are powerful tools for analyzing the behavior of the MSE system, and most effectively for investigating the main influence factors, because the properties of every component in the system and their interaction behavior can be explicitly expressed in the numerical model. Two example analyses presented in this chapter demonstrated the ability of the finite element method to analyze the performance of the MSE system using either extensible Tensar geogrids or inextensible steel grids on soft ground.

The use of high-strength geotextile as base reinforcement can increase considerably the ultimate height of embankment on soft clay (up to 2 m more). The rupture of geotextile occurred at a large deformation of foundation soil. Furthermore, the high-strength geotextile reinforcement can be used effectively on the soft soils that can sustain large deformation during construction.

At a stress level lower than the limit state of unreinforced embankment, the strains in geotextile are controlled mainly by the lateral displacement of the weathered crust beneath the reinforcement. At a low stress level, there were no differences in lateral displacements between the foundations of CE and HGE embankments. Increasing the embankment height, the base reinforcement decreased the lateral movement of the ground. However, it seems that the geotextile has no significant effects on the lateral movements of the soft clay layer below the weathered crust.

The high values of K near the top of the wall are the result of compaction-induced stresses and a design K value less than the envelope should be acceptable in this region. The lateral earth pressure coefficient below a depth of 6.10 m appears to be near the active case (K_a=0.33 for ϕ=30°). The use of the lateral earth pressure relationship recommended for welded wire walls appears to be conservative for RSE walls. The yield condition of the wall must be considered in the choice of the value of K. The whole process was dictated by the rigidity of the wall facing, the flexibility of the reinforcement/wall panel connection, and the pullout resistance of the mats behind the potential failure plane. The vertical pressure distribution at the base of the wall suggests that there were no overturning moments generated.

Performance evaluations were made based on the results of fully instrumented full-scale test embankment. The results were compared to the previous studies of welded wire reinforced test embankments. The location of maximum tensile forces along the geogrid reinforcement does not coincide with either the Coulomb/Rankine Tieback Wedge or the bilinear Coherent Gravity Method due to compression of underlying soft subsoils. The maximum tensile stress in the reinforcement corresponded to only 4.2% of the ultimate strength. The degree of consolidation of 70% was achieved after 200 days from the start of construction. The settlement profiles, both at the surface and at different elevations, are consistently uniform. This fact confirms the capacity of the geogrids to distribute the loads, thus decreasing any differential settlement.

7.15 REFERENCES

Adib, M., Mitchell, J.K., and Christopher, B. (1990), Finite element modeling of reinforced soil walls and embankments, Design and Performance of Earth Retaining Structure, ASCE Geotech. Special Publication No. 25, pp. 409-423.

Ahmed, M.M. (1977), Determination of permeability profile of soft Rangsit clay by field and laboratory tests. M. Eng'g. Thesis No. 1002, Asian Institute of Technology, Bangkok, Thailand.

Anderson, L.R., and Nielsen M.R. (1984), Pullout resistance of wire mats embedded in soil. Report for the Hilfiker Co., from the Civil and Envir. Eng'g. Dept., Utah State Univ., Logan, Utah.

Anderson, L.R., Sharp, K.K., and Harding, O.T. (1987), Performance of a 50-feet high welded wire wall, Soil Improvement- A Ten Year Update, ASCE Geotech. Special Publication 12, pp. 280-308.

Anderson, L.R., Sharp, K.D., Woodward, B.L., and Winward, R.F. (1985), Performance of the Rainier Avenue welded wire retaining wall, Seattle, Washington, Report Submitted to the Hilfiker Co. and Washington State Dept. of Transportation, USA.

Anderson, L.R., and Wong, W.L. (1987), Monitoring field performance of welded wire walls, Research Report to Hilfiker Co., Dept. of Civil and Environmental Eng'g., Utah State Univ., Logan, Utah.

Asakami, H. (1989), The smear effect of vertical band drains, M. Eng'g. Thesis No. GT-88-8, Asian Institute of Technology, Bangkok, Thailand.

Balasubramaniam, A.S., Phien-Wej, N.M., Indraratna, B., and Bergado, D.T. (1989), Predicted behavior of the test embankment on a Malaysian marine clay, Proc. Intl. Symp. on Trial Embankment on Malaysian Marine Clays. Kuala Lumpur, Vol. 2, pp. 1/1-1/8.

Balasubramaniam, A.S., Hwang, Z.M., Uddin, W., Chaudhry, A.R., and Li Y.G. (1978), Critical state parameters and peak stress envelopes for Bangkok clays, Q.J. Eng'g. Geol., Vol. 11, No. 219-232.

Bergado, D.T., Sampaco, C.L., Alfaro, M.C., and Balasubramaniam, A.S. (1988), Welded-wire reinforced earth (mechanically stabilized embankments) with cohesive backfill on soft clay, Second Progress Report Submitted to USAID, Thailand.

Bergado, D.T., Ahmed S., Sampaco, C.L., and Balasubramaniam, A.S. (1990), Settlements of Bangna-Bangpakong Highway on soft Bangkok clay, J. of Geotech. Eng'g. Div., ASCE Vol. 116, No. GT1, pp. 136-155.

Bergado, D.T., Sampaco, C.L., Shivashankar, R., Alfaro, M.C., Anderson, L.R., and Balasubramaniam, A.S. (1991), Performance of a welded wire wall with poor quality backfills on soft clay, Proc. ASCE Geotech. Eng'g. Congress at Boulder, Colorado, pp. 909-922.

Bergado, D.T., Hardiyatimo, H.C., Cisneros, C.B., Chai, J.C., Alfaro, M.C., Balasubramaniam, A.S., and Anderson, L.R. (1992), Pullout resistance of steel geogrids with weathered clay as backfill material. Geotech. Testing J., Vol. 15, No. 1, pp. 33-46.

Bergado, D.T., Long, P.V., Lee, C.H., Loke, K.H., and Werner, G. (1994), Performance of reinforced embankment on soft Bangkok clay with high strength geotextile reinforcement, Geotextiles and Geomembranes, Vol. 13, pp. 403-420.

Binquet, J., and Lee, K. L. (1975), Bearing capacity analysis of reinforced earth slabs, J. Geotech. Eng'g. Div., ASCE, Vol. 101, No. 12, pp. 531-534.

Biot, M.A. (1941). General theory of three-dimensional consolidation, J. of Applied Physics, Vol. 12, pp. 155-164.

Bishop, J.A., and Anderson, L.R. (1979), Performance of a welded wire retaining wall, Research Report Submitted to Hilfiker Co., Dept. of Civil Environmental Eng'g., Utah State Univ., Logan, Utah.

Bonaparte, R., and Christopher, B. R. (1987), Design and construction of reinforced embankments over weak foundation, Transportation Research Record No. 1153, pp. 26-39.

Boulon, M. (1989), Basic features of soil structure interface behavior, Computers and Geotechnics, Vol. 7, pp. 115-131.

Brand, E.W., and Premchitt, J. (1989), Comparison of the predicted and observed performance of Muar test embankment on Malaysian Clays, Proc. Symp. on Trial Embankments on Malaysian Marine clays, Kuala Lumpur, Vol. 2, pp. 1/1-1/8.

Britto, A.M., and Gunn, M.J. (1987), Critical state soil mechanics via finite elements, Ellis Horwood.

Carter, J.P., Booker, J.R., and Davis, E. H. (1977), Finite deformation of an elasto-plastic soil, Intl. J. for Numerical and Analytical Methods in Geomechanics, Vol. 1, No. 1, pp. 25-43.

Carter, J.P., Booker, J.R., and Small, J.C. (1979), The analysis of finite elasto-plastic consolidation, Intl. J. for Numerical and Analytical Method in Geomechanics, Vol. 3, No. 2, pp. 107-130.

Chai, J.C. (1992), Interaction between grid reinforcement and cohesive-frictional soil and performance of reinforced wall/embankment on soft ground, D. Eng'g. Dissertation, Asian Institute of Technology, Bangkok, Thailand.

Chang, J.C., Hannon, J.B., and Forsyth, R.A. (1977), Pull resistance and interaction of earthwork reinforcement and soil, Transportation Research Board Record No. 640, Transportation Research Board, National Research Council, Washington, D.C., pp. 1-7.

Cheney, R.S. (1990), Selection of retaining structures: The owner's perspective, Proc. ASCE Conf. Design and Performance of Earth Retaining Structures, Geotech. Special Publ. 25, Ithaca, New York, pp. 52-65.

Christopher, B.R., Gill, B.S., Giroud, J.P., Juran, I., Schlosser, F., Mitchell, J.K., and Dunnicliff J. (1989), Reinforced soil structures, Vol. 1: Design and Construction Guidelines, Report Prepared for US Federal Highway Administration, 287p.

Clough, G.W., and Duncan, J.M. (1971). Finite Element analysis of retaining wall behavior, J. of Soil Mech. and Found. Eng'g. Div., ASCE, Vol. 97, No. SM12, pp. 1657-1673.

Collios, A., Delmas, P., Gourc, J.P., and Giroud, J.P. (1980), Experiments of soil reinforcement with geotextiles, Proc. Symp. Use of Geotextiles for Soil Improvement, ASCE, pp. 53-73.

Delmas, P., Queyroi, D., Quaresma, M., De Saint Amand and Puech, A. (1992), Failure of an experimental embankment on soft soil reinforced with geotextile: Guiche. Geotextiles, Geomembranes and Related Products, Den-Hoedt (ed), Balkema, Rotterdam.

Desai, C.S., Zaman, M.M., Lightner, J.G., and Siriwardane, H.J. (1984), Thin layer element for interface and joints, Int. J. for Numerical Analytical Method in Geomechanics, Vol. 8, pp. 19-43.

Duncan, J.M., and Chang, C.Y. (1970), Non-linear analysis of stress and strain in soils, J. of Soil Mech. and Found. Eng'g. Div., ASCE, Vol. 96, No. 5, pp. 1629-1653.

Duncan, J.M., and Seed, R.B. (1986), Compaction-induced earth pressure under K_o-condition, J. of Geotech. Eng'g. Div., ASCE, Vol. 112, No. 1, pp. 1-22.

Duncan, J.M., Byrne, P., Wong, K.S., and Mabry, P. (1980), Strength, stress-strain and bulk modulus parameters for finite element analysis of stresses and movements in soil, Geotech. Eng'g. Research Report: UCB/GT/80-01, Dept. of Civil Eng., Univ. of California, Berkeley, August, 1980.

Fowler, J. (1982), Theoretical design considerations for fabric reinforced embankments, Proc. 2nd Intl. Conf. on Geotextiles, Las Vegas, Vol. 3, pp. 665-670.

Gens, A., Carol, I., and Alonso, E.E. (1990), An interface element formulation for the analysis of soil-structure interaction, Computer and Geotechnics, Vol. 9, pp. 3-20.

Goodman, R.E., Taylor, R.L., and Brekke, T.L. (1968), A model of the mechanics of joint rock, J. Soil Mech. and Found. Eng'g. Div., ASCE, Vol. 94, pp. 637-659.

Gray, H. (1936), Stress distribution in elastic solids, Proc. 1st Intl. Conf. Soil Mech. and Found. Eng'g., Vol. 2, p. 157.

Hausmann, M.R. (1976), Strength of reinforced soil, Proc. 8th Aust. Road Research Conf., Vol. 13, pp. 1-8.

Hermann, L.R. (1978), User's manual for REA (General Two Dimensional Soils and Reinforced Earth Analysis Program), Univ. of California, Davis, Calif.

Hird, C.C., and Kwok, C.M. (1986), Prediction for the Stanstead Abbotts trial embankment, based on the finite element method, Proc. of the Prediction Symp. On Reinforced Embankment on Soft Ground, Strand, London.

Hird, C.C., and Pyrah, I.C. (1990), Predictions of the behavior of a reinforced embankment on soft ground, Proc. Symp. Performance of Reinforced Soil Structures, pp. 409-414.

Hird, C.C., and Kwok, C.M. (1989), FE studies of interface behavior in reinforced embankments on soft ground, Computers and Geotechnics, Vol. 8, pp. 111-131.

Huisman, M.G.H. (1987), Design guideline for reinforced embankments on soft soils using Stabilenka reinforcing mats, Enka Technical Report, Arnhem.

Ingold, T.S. (1982), Reinforced earth, Thomas Telford, London.

Ingold, T.S. (1983a), Laboratory testing of grid reinforcement in clay, Geotech. Testing J., Vol. 6, No. 3, pp. 112-119.

Ingold, T.S. (1983b), The design of reinforced soil walls by compaction theory, J. of the Inst. of Struct. Engrs., London, Vol. 61A, No. 7, pp. 205-211.

Ingold, T.S. (1983c), Some factors in design of geotextile reinforced embankments, Proc. 8th Europ. Conf. Soil Mech. and Found. Eng'g., Helsinki, Vol. 2, pp. 503-508.

Janbu, N. (1963), Soil compressibility as determined by oedometer and triaxial tests, Proc. Europ. Conf. Soil Mech. Found. Eng'g., Wiesbaden, Vol. 1, pp. 19-25.

Jewell, R.A. (1980), Some effects of reinforcements on soils, Doctoral Dissertation, Cambridge University, England.

Jewell, R.A. (1990), Reinforcement bond capacity, Geotechnique, Vol. 40, No. 3, pp. 513-518.

Jewell, R.A. (1982), A limit equilibrium design method for reinforced embankments, Proc. 2nd Intl. Conf. on Geotextiles, Las Vegas, Vol. 3. pp. 671-676.

Jewell, R.A., Milligan, G.W.E., Sarsby, R.W., and Dubois (1984), Interaction between soil and geogrids, Proc. Symp. on Polymer Grid Reinforcement in Civil Eng'g., London, pp. 19-29.

Jones, C.J.F.P. (1985), Earth reinforcement and soil structures, Butterworths, London.

Juran, I., and Christopher, B. (1989), Laboratory model study on geosynthetic reinforced soil retaining walls, J. of Geotech. Eng'g. Div., ASCE, Vol. 115, No. 7, pp. 905-926.

Juran, I., Knochennus, G., Acar, Y.B., and Arman, A. (1988), Pullout response of geotextiles and geogrids, Geosynthetics for Soil Improvement, edited by R.D. Holtz, Geotech. Special Publication 18, ASCE, pp. 92-111.

Koerner, R.M. (1986), Designing with geosynthetics, Prentice Hall, Englewood Cliffs, N.J.

Kondner, R.L., and Zelasko, J.S. (1963), A hyperbolic stress-strain formulation of sand, Proc. 2nd Pan American Conf. on Soil Mech. and Found Eng'g.

Lee, K.L., Adams, B.D., and Vagneron, J.J. (1973), Reinforced earth retaining walls, J. Soil Mech. and Found. Div., ASCE, Vol. 99, pp. SM10.

Macatol, K.C. (1990), Interaction of lateritic backfill and steel grid reinforcements at high vertical stress using pullout test, M. Eng'g. Thesis No. GT-89-12, Asian Institute of Technology, Bangkok, Thailand.

Magnan, J.P. (1989), Experience-based prediction of the performance of Muar Flats Trial Embankment to failure, Proc. Int. Symp. on Trial Embankments on Malaysian Marine Clays.

MHA, Malaysian Highway Authority (1989), Proc. of the Intl. Symp. on Trial Embankments on Malaysian Marine Clays, Vol. 1.

Milligan, V., and La Rochelle, P. (1984), Design methods for embankment over weak soils, In Polymer Grid Reinforcement, Thomas Telford, London, pp. 95-102.

Mitchell, J.K., and Villet, W.C.B. (1987), Reinforcement of earth slopes and embankments, National Cooperative Highway Research Program Report 290, Trans. Research Board, National Research Council, Washington, D.C.

Michell, J.K., and Christopher, B.R. (1990), North American practice in reinforced soil systems, Proc. Design and Performance of Earth Retaining Struct., ASCE Geotech. Publ. No. 25, Lambe, P.C. and Hansen, L.A. (eds.).

Murray, R.T., and Farrar, D.M. (1990), Reinforced earth wall on the M25 Motorway at Waltham Cross, Proc. Inst. of Civil Engrs., Part 1, Vol. 88, pp. 261-282.

Ochiai, H., and Sakai, A. (1987), Analytical method for geogrid reinforced soil structures, Proc. Int. Symp. on Geosynthetics, Japanese Chapter of Int. Geotech. Society, pp. 483-686.

Palmeira, E.M., and Milligan, G.W.E. (1989), Scale and other factors affecting the results of pullout tests of grids buried in sand, Geotechnique, Vol. 39, No. 3, pp. 511-524.

Peterson, L.M., and Anderson, L.R. (1980), Pullout resistance of welded wire mats embedded in soil, Research Report Submitted to Hilfiker Co., Civil and Envir. Eng'g. Dept., Utah State Univ., Utah.

Poulos, H.G., Lee, C.Y., and Small, J.C. (1989), Prediction of embankment performance on Malaysian Marine Clays, Kuala Lumpur, Vol. 2, pp. 4/1-4/10.

Poulos, H.G., and Davis, E.H. (1974), Elastic solutions in soil and rock mechanics, John Wiley and Sons, Inc., New York.

Poulos, H.G. (1972), Difficulties in prediction of horizontal deformations in foundations, J. Soil Mech. and Found. Div., ASCE, Vol. 98, No. SM8, pp. 843-848.

Poulos, H.G., Lee, C.Y., and Small, J.C. (1989), Prediction of embankment performance on Malaysian marine clays, Proc. Intl. Symp. on Trial Embankments on Malaysian Marine clays, Kuala Lumpur, Vol. 2, pp. 4/1-4/10.

Prandtl, L. (1921), Eindringungs fesligkeit and fesigkeit von schneiden, Zeitschrift fur Angewandte Mathematik und Mechanik, Vol. 1, No. 15.

Quast, P. (1983), Polyester reinforcing fabric mats for the improvement of the embankment stability, Proc. 8th Europ. Conf. Soil Mech. and Found. Eng'g., Helsinki, Vol. 2, pp. 531-534.

Roscoe, K.H., and Burland, J.B. (1968), On the generalized stress-strain behavior of wet clays, Proc. Eng'g. Plasticity, Cambridge Univ. Press, Cambridge, pp. 535-609.

Rowe, R.K. (1984), Reinforced embankments, analysis and design, J. Geotech. Eng'g. Div., ASCE, Vol. 110, No. 2, pp. 231-246.

Rowe, R.K., and Soderman, K.L. (1985), An approximate method for estimating the stability of geotextile-reinforced embankments, Canadian Geotech. J., Vol. 22, No. 3, pp. 392-398.

Sampaco, C.L., Anderson, L.R., and Robertson, D.G. (1992), Performance of an RSE wall with concrete facing panels, Proc. 28th Symp. on Eng'g. Geol. and Geotech. Eng'g., Boise, Idaho.

Schaefer, U.R., and Duncan, J.M. (1988), Finite analyses of the St. Alban test embankments, ASCE, Geotech. Special Publication No.18, pp. 158-177.

Schlosser, F., and De Buhan, P. (1990), Theory and design related to the performance of reinforced soil, Proc. Symp. Performance of Reinforced Soil Structures, London, pp. 1-14.

Schlosser, F., and Long, N.T. (1972), Comportment de la terre armee dans les ouvrages de soutenement, Proc. 5th ECSMFE Madrid, Vol. 1, No. 111a, pp. 299-306.

Schlosser, F., and Long, N.T. (1973), Recent results in French research in reinforced earth, J. of Const. Div., ASCE, Vol. 100, No. 3, pp. 223-237.

Schmertmann, G.R., Cnew, S.H., and Mitchell, J.K. (1989), Finite element modeling of reinforced soil wall behavior, Geotech. Eng'g. Report, No. UCB/GT/89-01, Dept. of Civil Eng'g. Univ. of California, Berkeley, Calif.

Seed, R.B., and Duncan, J.M. (1986), FE analyses: Compaction-induced stresses and deformations, J. of Geotech. Eng'g. Div., ASCE, Vol. 112, No.1, pp. 23-43.

Shivashankar, R. (1991), Behavior of a mechanically stabilized earth (MSE) embankment with poor quality backfill on soft clay deposits, including a study of the pullout resistances, D. Eng'g. Dissertation, Asian Institute of Technology, Bangkok, Thailand.

Tavenas, F., and Leroueil, S. (1980), The behavior of embankments on clay foundations, Canadian Geotech. J., Vol. 17, pp. 236-260.

Tavenas, F., Jean, P., Leblond, P., and Leroueil, S. (1983), The permeability of natural soft clays, Part II: Permeability characteristics, Canadian Geotech. J., Vol. 20, pp. 645-660.

Taylor, D. W. (1948), Fundamentals of soil mechanics, Wiley and Sons, New York.

Vesic, A.S. (1963), Bearing capacity of deep foundations in sand, Highway Research Record Vol. 39, pp. 112-153.

Vesic, A.S. (1972), Expansion of cavities in infinite soil masses, J. Soil Mech. and Found. Div., ASCE, Vol. 94, No. SM3, pp. 661-688.

Vidal, H. (1969), The principle of reinforced earth, Highway Research Record No. 282, Highway Research Board, National Research Council, Washington, D.C., pp. 1-16.

Yang, Z. (1972), Strength and deformation characteristics of reinforced sand, Ph.D. Dissertation, Univ. of California, Los Angeles, Calif.

Zienkiewicz, O.C., Best, B., Dullage, C., and Stagg, K.G. (1970), Analysis of nonlinear problems with particular reference to jointed rock systems, Proc. 2nd Int. Conf. Society of Rock Mechanics, Belgrade, Vol. 3, pp. 501-509.

SUBJECT INDEX

Active zone 323
AIT test embankment 122, 127, 131, 206, 336, 394
Average strength method 199, 287
Bangkok clay 1, 122, 127, 131, 146, 212, 272, 336, 394, 404
Base pressure 347, 411
Bearing capacity 191, 279, 295; bearing resistance 319; isolated pile 191, 279; pile group 194, 281
Bond coefficient 343
Breakthrough 298
Case-borehole 188
Chao Phraya Plain 5, 208, 218, 306
Cement stabilization 235, 236, 241, 242, 249, 252, 291; cement piles 273
Coefficient of consolidation 108, 109, 170, 174
Compaction 2, 9, 24, 25, 366; surface compaction 9, 10, 11, 12, 24, 26; deep compaction 9, 35
Composite soil 188, 235
Cone penetration test 56, 85; resistance 2, 85
CON2D 127, 212
CONSAX 127, 212
Consolidation (see Soil consolidation)
Construction simulation 363
Creep effect 152
Creep limit 284
CRISP computer program 370, 377
Deep mixing method 1, 9, 270, 273; dry jet mixing 270; wet jet mixing 273
Degree of consolidation 96, 103, 113
Differential settlement, 286, 287
Direct shear resistance 313, 321, 363
Discharge capacity 161, 163
Drain core 93, 114
Drainage 88
Drain equivalent diameter 99
Dry jet mixing method (see Deep mixing method)

Dynamic compaction 62, 66, 69, 76, 80
Earth pressure 325, 384, 392
Effect of compaction 350
Elasto-plastic model 355
Excess pore pressure (see Pore pressure)
External stability 307, 311
Failure mechanism 191, 314, 315, 323
FEM model parameters 367, 377
Field vane shear test 140
Filter criteria 114, 117; jacket 117; permeability 111
Finite element 118, 127, 353, 368, 378
Frictional soil 35, 38
General shear failure 314
Geogrid 404
Geotextile 394; strain 403
Granular pile 186, 205; stone column 186; sand compaction pile 1, 9, 188, 218
Grid reinforcement 312, 314, 320-322, 324, 335, 403
Ground improvement 1, 8, 9, 35, 88, 186, 235; soil improvement 5, 8, 9, 10, 88, 186, 235, 306
Ground subsidence 1, 5-7
Ground vibration 54, 56
Hyperbolic model 354
Internal stability 307, 311
Interaction (see Soil reinforcement interaction)
Jet grouting 272, 275; pressure 276
Lateral displacement 152, 347, 399, 408
Lime stabilization 235, 239, 245, 252; piles 279
Load test 205
Loose sand 1, 36, 44, 48, 76, 83
Lumped parameter method 203
Mandrel 90, 102, 105, 124, 131
Mechanically stabilized earth 306, 353, 384, 388, 411, 413
Mueller resonance compaction 48-50
Model parameters 368, 382
Modified cam clay 127, 367, 377

Nong Ngu Hao 124, 146
Organic contents 249, 250
Permeability (see Soil permeability)
Pile group 194, 280
Pore pressure 73, 90, 345, 408
Prefabricated vertical drains (PVD) 9, 88, 90, 95, 124, 127, 131, 152, 159
Preloading 88
Profile method 199
Pullout displacement 328, 338; resistance 319, 321, 325, 363; test 325, 336, 344
Punching shear failure 316
Quicklime 249
Rate of consolidation 111, 203; settlement 140
Reclamation 5
Reinforced earth 305, 307, 310, 324, 338, 344, 367, 375, 383, 386, 393, 403
Reinforcement stiffness factor 324, 325; tension 351
Resistant zone 323
Salt content 249
Sand drains 9, 90, 122, 124
Sandwick 124
Secondary settlement 204
Settlement 196, 283, 286
SHANSEP 140, 146, 152
Slope stability 3, 140, 146, 387, 403; failures 3, 4, 399
Smear effect 102, 108; zone 102, 105, 108, 166

Soft ground 1, 2, 76, 80, 122, 124, 127, 131, 146, 170, 186, 345, 353, 368, 378, 404; clay 1, 88, 122, 124, 127, 131, 146, 170, 186, 345, 353, 368, 378, 404
Soil consolidation 93, 96, 109, 173, 174, 204, 255; permeability 33, 108, 120, 168, 366; reinforcement interaction 312, 320, 358, 362
Spacing ratio 96
Stability (see Slope stability)
Stage loading 140, 152, 328, 338
Standard penetration test 2, 72, 224
Stiffness ratio 318
Strength increase 140, 146, 152
Terra-Probe 35, 37, 38
Test embankment 122, 124, 127, 146, 176, 205, 273, 368, 376, 394, 404; excavation 212
Triaxial test 259, 268, 301; constant stress ratio test 257
Undrained shear strength 121, 140; stress path 263; behavior 265
Vacuum preloading 124
Vertical drains (see Prefabricated vertical drain or Sand drain)
Vibro compaction 36, 39, 186; compozer 188; replacement 36, 39, 188
Vibroflot 36, 186
Vibroflotation 36-38, 41, 44
Waste containment structures 28, 29
Well resistance 99
Wet jet mixing method (see Deep mixing method)

AUTHOR INDEX

Note: When there is more than one author, only the primary author is listed.

Aboshi, 188, 184, 191, 193, 195, 197, 198, 199, 225
Adib, 324, 353, 362
Ahmed, 367
Akagi, 88, 90, 105, 108, 122, 123
Albakri, 109, 110
Anderson, 323, 343, 387, 391
Ariizumi, 248
Asaoka, 120, 127, 152, 212
Assarson, 238, 239, 240, 245
Balasubramaniam, 88, 124, 125, 248, 251, 367, 377
Barksdale, 188, 191, 192, 194, 196, 198, 199, 200
Barron, 93, 96, 118
Baumann, 36, 39, 186, 187
Belloni, 88
Bergado, 88, 90, 91, 99, 102, 106, 107, 108, 109, 118, 120, 127, 129, 131, 188, 191, 195, 196, 202, 205, 206, 207, 208, 211, 212, 227, 324, 338, 343, 389, 393, 410
Bersabe, 218
Binquet, 402
Biot, 358
Bishop, 387
Bonaparte, 402
Boulon, 362
Bradl, 239, 245
Brand, 308
Brenner, 88, 90
Britto, 357, 358, 365, 370
Broms, 234, 238, 240, 241, 243, 244, 245, 247, 251
Brown, 36, 43
Buttling, 84, 85
Cahulogan, 219
Calladine, 263
Carillo, 111, 203
Carroll, 117
Carter, 358
Casagrande, 90

Cedergreen, 117
Chai, 314, 318, 319, 320, 332, 367
Chambosse, 203
Chang, 305, 307, 328
Chaudry, 247
Chen, 118
Cheney, 305
Cheung, 120
Chida, 234, 269, 272
Choa, 88
Chou, 88
Christopher, 323, 325, 386
Clough, 361
D'Appolonia, 41, 42, 44, 48
D'Orazio, 127, 212
Daniel, 28, 32
Datye, 188
Davidson, 247
Delmas, 402
Dennis, 218
Desai, 359
Diamond, 239, 240
Dimaggio, 203
DJM Research Group, 234, 271
Duncan, 127, 212, 332, 353, 354, 361, 365, 367
Dunham, 223, 377
Easdes, 244, 247
Engelhardt, 36, 186
Enriquez, 205
Fitzhardinge, 79, 81, 82
Fowler, 402
Frelund, 199
Fudo Contruction, 218
Gens, 362
Gibbs and Holtz, 223
Glover, 37, 40
Goodman, 359
Gouw, 44
Gray, 407
Handy, 244
Hansbo, 96, 97, 99, 108, 118, 131

Hausmann, 310
Hermann, 352
Herrin, 239, 244
Hilt, 243, 244, 247
Hird, 352, 358, 362, 364
Holm, 245
Holtz, 12, 108, 111, 112
Honjo, 272
Hughes, 205
Huisman, 402
Ingles, 240
Ingold, 310, 313, 402
Jamiolkowski, 90, 102, 103, 104, 109
Janbu, 90, 353
Jewell, 307, 312, 314, 315, 316, 317, 319, 332, 402
Johnson, 14, 15
Jones, 350
Juran, 320, 322, 323, 344, 410
Kamaluddin, 251
Kamon, 1, 8, 9, 118
Kawamura, 74, 75
Kawasaki, 234
Kellner, 117
Kezdi, 239, 242, 244, 246
Kobayashi, 168
Konder, 353
Kumagai, 72, 73
Ladd, 121, 140, 146, 152
Lambe, 18, 19, 20
Lea, 234
Lee, 88, 247, 307
Leonards, 69, 72, 204
Leong, 219
Long, 127, 208
Lukas, 69, 70, 72
Macatol, 367
Magnan, 377
Massarsch, 48, 49, 50, 52, 53, 54, 55, 56, 57, 58
Mayne, 69, 70, 75, 76
McDowell, 244
Menard, 62, 69
Mesri, 170, 204
Metcalf, 247
Meyerhoff, 223

MHA, 367, 377
Milligan, 402
Mitchell, 28, 31, 32, 69, 80, 248, 307, 312, 386
Miura, 168, 234, 241, 248, 269, 272
Moh, 88, 124, 126
Mori, 63, 64, 65, 77, 78
Mukherjee, 131
Murray, 389
Nakanodo, 163
Narumi, 66, 68
Nicholls, 88
Nutalaya, 218
Ochiai, 332
Okamura, 245
Onoue, 108, 110
Palmeira, 316, 344
Panichayatum, 205
Peterson, 305, 315, 332
Pilot, 90
Poulos, 90, 372, 407
Pradhan, 159, 163
Prandtl, 314
Quast, 402
Rahman, 88
Ranjan, 188
Rao, 188
Rixner, 89, 91, 98, 99, 101, 105, 111, 113
Robertson, 223
Roscoe, 357
Rowe, 358
Ruff, 247
Saitoh, 240, 242, 243
Sampaco, 391
Schaefer, 352
Schlosser, 310, 312, 320, 323, 324, 410
Schmertmann, 335, 353, 366
Schober, 117
Schweiger, 120
Seed, 19, 366
Selvakumar, 218
Sherard, 117
Shinsha, 120
Shivashankar, 367
Sim, 205
Singh, 205

Skempton, 48, 127, 212
Suzuki, 88, 234
Takai, 88
Tanimoto, 88
Tavenas, 365
Taylor, 245, 365
Terashi, 234
Terzaghi, 93, 114
Tjakrawiralaksana, 218
Tominaga, 88
Turnbull, 14
Vesic, 219, 316, 318, 319
Vidal, 305
Volders, 88
Vreeken, 114, 115
Warinsisak, 212
Whitman, 199
Wissa, 243
Woo, 88, 124, 127, 243
Yamanouchi, 269
Yamazaki, 76
Yang, 310
Yuasa, 76
Zienkiewiez, 332, 359